Wine Sensory Faults: Origin, Prevention and Removal

Wine Sensory Faults: Origin, Prevention and Removal

Editors

Fernando M. Nunes
Fernanda Cosme
Luís Filipe-Ribeiro

MDPI • Basel • Beijing • Wuhan • Barcelona • Belgrade • Manchester • Tokyo • Cluj • Tianjin

Editors
Fernando M. Nunes
Chemistry Department
University of Trás-os-Montes
and Alto Douro
Vila Real
Portugal

Fernanda Cosme
Biology and Environment
Department
University of Trás-os-Montes
and Alto Douro
Vila Real
Portugal

Luís Filipe-Ribeiro
Chemistry Research Centre
University of Trás-os-Montes
and Alto Douro
Vila Real
Portugal

Editorial Office
MDPI
St. Alban-Anlage 66
4052 Basel, Switzerland

This is a reprint of articles from the Special Issue published online in the open access journal *Molecules* (ISSN 1420-3049) (available at: www.mdpi.com/journal/molecules/special_issues/wine_sensory_faults).

For citation purposes, cite each article independently as indicated on the article page online and as indicated below:

LastName, A.A.; LastName, B.B.; LastName, C.C. Article Title. *Journal Name* **Year**, *Volume Number*, Page Range.

ISBN 978-3-0365-4998-9 (Hbk)
ISBN 978-3-0365-4997-2 (PDF)

Cover image courtesy of Fernando M. Nunes

© 2022 by the authors. Articles in this book are Open Access and distributed under the Creative Commons Attribution (CC BY) license, which allows users to download, copy and build upon published articles, as long as the author and publisher are properly credited, which ensures maximum dissemination and a wider impact of our publications.

The book as a whole is distributed by MDPI under the terms and conditions of the Creative Commons license CC BY-NC-ND.

Contents

About the Editors vii

Preface to "Wine Sensory Faults: Origin, Prevention and Removal" ix

Štefan Ailer, Silvia Jakabová, Lucia Benešová and Violeta Ivanova-Petropulos
Wine Faults: State of Knowledge in Reductive Aromas, Oxidation and Atypical Aging, Prevention, and Correction Methods
Reprinted from: *Molecules* **2022**, *27*, 3535, doi:10.3390/molecules27113535 1

Francesco Errichiello, Luigi Picariello, Antonio Guerriero, Luigi Moio, Martino Forino and Angelita Gambuti
The Management of Dissolved Oxygen by a Polypropylene Hollow Fiber Membrane Contactor Affects Wine Aging
Reprinted from: *Molecules* **2021**, *26*, 3593, doi:10.3390/molecules26123593 25

Emilio Celotti, Georgios Lazaridis, Jakob Figelj, Yuri Scutaru and Andrea Natolino
Comparison of a Rapid Light-Induced and Forced Test to Study the Oxidative Stability of White Wines
Reprinted from: *Molecules* **2022**, *27*, 326, doi:10.3390/molecules27010326 39

Daniela Fracassetti, Sara Limbo, Natalia Messina, Luisa Pellegrino and Antonio Tirelli
Light-Struck Taste in White Wine: Protective Role of Glutathione, Sulfur Dioxide and Hydrolysable Tannins
Reprinted from: *Molecules* **2021**, *26*, 5297, doi:10.3390/molecules26175297 53

Antonio Morata, Iris Loira, Carmen González and Carlos Escott
Non-*Saccharomyces* as Biotools to Control the Production of Off-Flavors in Wines
Reprinted from: *Molecules* **2021**, *26*, 4571, doi:10.3390/molecules26154571 73

Gary J. Pickering and Andreea Botezatu
A Review of Ladybug Taint in Wine: Origins, Prevention, and Remediation
Reprinted from: *Molecules* **2021**, *26*, 4341, doi:10.3390/molecules26144341 85

Ouli Xiao, Minmin Li, Jieyin Chen, Ruixing Li, Rui Quan and Zezhou Zhang et al.
Influence of Triazole Pesticides on Wine Flavor and Quality Based on Multidimensional Analysis Technology
Reprinted from: *Molecules* **2020**, *25*, 5596, doi:10.3390/molecules25235596 101

Alessandra Rinaldi, Alliette Gonzalez, Luigi Moio and Angelita Gambuti
Commercial Mannoproteins Improve the Mouthfeel and Colour of Wines Obtained by Excessive Tannin Extraction
Reprinted from: *Molecules* **2021**, *26*, 4133, doi:10.3390/molecules26144133 119

Antonio G. Cordente, Damian Espinase Nandorfy, Mark Solomon, Alex Schulkin, Radka Kolouchova and Ian Leigh Francis et al.
Aromatic Higher Alcohols in Wine: Implication on Aroma and Palate Attributes during Chardonnay Aging
Reprinted from: *Molecules* **2021**, *26*, 4979, doi:10.3390/molecules26164979 133

Fernanda Cosme, Sara Gomes, Alice Vilela, Luís Filipe-Ribeiro and Fernando M. Nunes
Air-Depleted and Solvent-Impregnated Cork Powder as a New Natural and Sustainable Fining Agent for Removal of 2,4,6-Trichloroanisole (TCA) from Red Wines
Reprinted from: *Molecules* **2022**, *27*, 4614, doi:10.3390/molecules27144614 155

About the Editors

Fernando M. Nunes

Fernando M. Nunes is an Associate Professor with Habilitation of Food and Analytical Chemistry at the Chemistry Department of the University of Trás-os-Montes e Alto Douro and an integrated researcher of the Chemistry Research Centre-Vila Real (CQ-VR). His research is focused on food chemistry, namely coffee and wine, melanoidins, polysaccharides and phenolic compounds and food analysis. He is an author/co-author of more than 150 articles and 15 patents and supervisor of post-doc and doctoral dissertations. As a principal investigator or researcher, he participated in several research projects in Food Science.

Fernanda Cosme

Fernanda Cosme is an Assistant Professor of Food Science/Oenology at the University of Trás-os-Montes and Alto Douro and an integrated researcher of the Chemistry Research Centre-Vila Real (CQ-VR). She graduated in Agro-Industrial Engineering from the Technical University of Lisbon and has a Ph.D. in Food Science at UTAD. The main research interests are wine stabilization, quality, safety, and ageing concerning the wine technology and stabilization process. She is an author/co-author of several articles in peer-reviewed scientific journals and book chapters and participated in scientific congresses. Supervisor of Masters and Ph.D. Thesis and participated in research projects in Food Science/Oenology.

Luís Filipe-Ribeiro

Luís Filipe-Ribeiro has a degree in oenology, a Master's in viticulture and oenology, a Doctorate in chemical and biological Sciences and a Post-Doc in oenological sciences. He is an invited oenology Professor at the University of Trás-os-Montes and Alto Douro since 1997–1998, where he participated in many subjects (vinification, wine stabilization, analytical control, sensorial analysis, winery building and equipment). He is a researcher in the Chemistry Research Centre of Vila Real, where he participates in many projects, research works, and student orientations. He is the Technical Director of SAI Enology and is responsible for the Laboratory and R&D+i. He is the author of several scientific papers (SCI), technical publications, and patents, and he participates regularly in international and national congresses as a speaker (orcid.org/0000-0002-6744-4082).

Preface to "Wine Sensory Faults: Origin, Prevention and Removal"

Wine is highly appreciated for its distinctive sensory characteristics, including its colour, aroma, and taste. However, unwanted microbiological activity, unbalanced concentrations of certain compounds resulting from unbalanced grape chemical compositions, and inadequate winemaking practices and storage conditions can result in sensory defects that significantly decrease wine quality. Although preventing wine defects is the best strategy, they are sometimes difficult to avoid. Therefore, when present, several fining agents or additives and technologies are available or being developed with different performances regarding their impact on wine quality. Wine stabilisation refers to removal and prevention strategies and treatments that limit visual, olfactory, gustatory, or tactile wine defects, as well as increase wine safety and stability through fining and the application of different operations carried out in wineries (filtration, pasteurisation, electrodialysis, and cold stabilisation) and the use of emerging technologies (electron-beam irradiation, high hydrostatic pressure, pulsed electric fields, ultrasound, pulsed light). Future trends in this field involve using more sustainable and environmentally friendly fining agents and technologies and developing treatments with better performance and specificity.

This Book of the Special Issue focuses on different aspects of wine sensory faults, their origin, prevention, and removal, and their impact on wine sensory quality. This Special Issue, composed of ten valuable accepted and published articles, divided into seven original articles and three reviews, provides an overview of the main wine sensory faults describing their origin, prevention, and removal strategies. Topics include origins of wine sensory faults, the impact of the wine faults on wine quality and safety, prevention of wine faults either by viticultural or oenological practices, the performance of available fining agents and new potential fining agents for the removal of wine defects, their selectivity, and impact on wine quality, methods to estimate wine stability. The Special Issue collected contributions from researchers from Universities and Research Centres from different parts of the world, namely Italy, Spain, Portugal, Moldova, North Macedonia, Slovakia, Canada, the USA, Australia, and China, establishing the interest of the international scientific community toward the aims mentioned above and scopes.

Briefly, Ailer, Jakabová, Benesova, and Ivanova-Petropulos, reviewed the latest scientific findings and recommendations for the prevention of the reductive aromas, mainly caused by excessive H_2S and other volatile sulphur compounds, of browning associated with the enzymatic and non-enzymatic catalysed oxidation of polyphenols, and atypical ageing, associated with the stress and lack of nutrients and moisture in green land cover in the vineyard. In the original contribution from Errichiello, Picariello, Guerriero, Moio, Forina, and Gambuti, the management of the dissolved oxygen content in wines by a polypropylene hollow fibre membrane contactor apparatus was performed. After ageing (11 months), red wines with high oxygen content resulted in the massive formation of polymeric pigments and BSA reactive tannins, as opposed to wines with low oxygen levels, demonstrating that the membrane contactor can be a successful tool to manage dissolved oxygen in wines as to prevent their oxidative spoilage. Celotti, Lazaridis, Figelj, Scutaru, and Natolino, evaluated the use of a portable prototype instrument for light irradiations at different wavelengths and times to evaluate the oxidative stability of white wines and the effect of some oenological adjuvants on wine stability. The sensorial analysis revealed that white and light blue were the most significant after only 1 h of irradiation. The experimental results showed that hydrogen peroxide could enhance the effect of light treatment. On the other hand, light exposure of white wine can cause a light-struck taste, a fault induced by riboflavin and methionine activation leading to the formation of volatile sulphur compounds (VSCs), including methanethiol (MeSH) and dimethyl disulphide (DMDS). Fracasseti, Limbo, Messina, Pellegrino, and Tirelli, studied the impact of different antioxidants, i.e., sulphur dioxide (SO_2), glutathione (GSH), and chestnut tannins (CT), either individually or in various combinations, on preventing light-struck taste (LST). The presence of antioxidants limited the formation of light-struck taste as lower concentrations of volatile sulphur compounds. The order of their effectiveness was CT ⩾ GSH >SO_2. The results indicate tannins as an effective oenological tool for preventing LST in white wine.

Morata, Loira, González, and Escott reviewed the use of non-Saccharomyces yeasts, such as *Lachancea thermotolerans*, which results in effective acidification through the production of lactic acid from sugars and yeasts with hydroxycinnamate decarboxylase (HCDC) activity can be helpful to promote the fermentative formation of stable vinylphenolic pyranoanthocyanins, reducing the amount of ethylphenol precursors that can be used as natural solutions for preventing the formation of undesirable off-flavours in wines.

Pickering and Botezatu reviewed a range of vineyard practices that seek to reduce *Coccinellidae* densities, as well as both "standard" and novel wine treatments aimed at reducing alkyl-methoxypyrazine load responsible for the recently recognised faults known as Ladybug taint possessing excessively green, bell pepper-, and peanut-like aroma and flavour.

In their original contribution, Xiao, Li, Chem, Li, Quan, Zhang, King, and Dai studied the effect of triazole pesticides that are widely used to control grapevine diseases. However, they can significantly affect the ester and acid aroma components and change the wines' flower and fruit flavour. This change was attributed to changes in the yeast fermentation activity caused by the pesticide residues. underlining the desirability of stricter control by the food industry over pesticide residues in winemaking.

Rinaldi, Gonzalez, Moio, and Gambuti investigate using three different commercial mannoprotein-rich yeast extracts (MP, MS, and MF) to reduce high bitterness and astringency in finished wines resulting from pressing marcs and extended maceration techniques that increase the extraction of phenolic compounds. Mannoproteins had a different effect depending on the wine's anthocyanin/tannin (A/T) ratio. When tannins are strongly present (extended maceration wines with A/T = 0.2), the MP conferred mouthcoating and soft and velvety sensations and colour stability to the wine. At A/T = 0.3, as in marc-pressed wines, both MF and MP improved the mouthfeel and colour. However, in free-run wine, where the A/T ratio is 0.5, the formation of polymeric pigments was allowed by all treatments and correlated with silk, velvet, and mouthcoating subqualities.

The selection of variants of AWRI796 yeast strains can be used to modulate the formation of higher alcohols 2-phenylethanol, tryptophol, and tyrosol and methionol, as well as other volatile sulphur compounds derived from methionine, highlighting the connections between yeast nitrogen and sulphur metabolism during fermentation, and modulate the dynamic changes in wine flavour over ageing (Cordente, Nandorfy, Solomon, Schulkin, Kolouchova, Francis, and Schmidt).

Cork powder after extractives removal and air removal by ethanol impregnation was studied in its efficiency in removing 2,4,6-Trichloroanisole (TCA) from contaminated red wines. This potential fining agent removed 91% of TCA from wines containing 6 ng/L of TCA at an optimised cork powder application of 0,25g/L. The impact on wine colour, phenolic composition and volatile compounds was low, making this modified cork powder a potential and sustainable fining agent to cope with this fault (Cosme, Gomes, Vilela, Filipe-Ribeiro, and Nunes).

As a final note, the Guest Editors would like sincerely to thank all the authors for selecting this Special Issue to publish the results of their hard research work or reviews, as well as to the reviewers and the assistant editors for their precious support, that will contribute to the success of these Special Issue and Book.

Fernando M. Nunes, Fernanda Cosme, and Luís Filipe-Ribeiro
Editors

Review

Wine Faults: State of Knowledge in Reductive Aromas, Oxidation and Atypical Aging, Prevention, and Correction Methods

Štefan Ailer [1], Silvia Jakabová [2,*], Lucia Benešová [2] and Violeta Ivanova-Petropulos [3]

1. Institute of Horticulture, Faculty of Horticulture and Landscape Engineering, Slovak University of Agriculture in Nitra, 94976 Nitra, Slovakia; stefan.ailer@uniag.sk
2. Institute of Food Sciences, Faculty of Biotechnology and Food Sciences, Slovak University of Agriculture in Nitra, Tr. A. Hlinku 2, 94976 Nitra, Slovakia; xbenesova@uniag.sk
3. Faculty of Agriculture, University "Goce Delčev"-Štip, Krste Misirkov 10-A, 2000 Štip, North Macedonia; violeta.ivanova@ugd.edu.mk
* Correspondence: jakabova@is.uniag.sk

Abstract: The review summarizes the latest scientific findings and recommendations for the prevention of three very common wine faults of non-microbial origin. The first group, presented by the reductive aromas, is caused mainly by excessive H_2S and other volatile sulfur compounds with a negative impact on wine quality. The most efficient prevention of undesirable reductive aromas in wine lies in creating optimal conditions for yeast and controlling the chemistry of sulfur compounds, and the pros and cons of correction methods are discussed. The second is browning which is associated especially with the enzymatic and non-enzymatic reaction of polyphenols and the prevention of this fault is connected with decreasing the polyphenol content in must, lowering oxygen access during handling, the use of antioxidants, and correction stands for the use of fining agents. The third fault, atypical aging, mostly occurs in the agrotechnics of the entire green land cover in the vineyard and the associated stress from lack of nutrients and moisture. Typical fox tones, naphthalene, or wet towel off-odors, especially in white wines are possible to prevent by proper moisture and grassland cover and alternating greenery combined with harmonious nutrition, while the correction is possible only partially with an application of fresh yeast. With the current knowledge, the mistakes in wines of non-microbial origin can be reliably prevented. Prevention is essential because corrective solutions for the faults are difficult and never perfect.

Keywords: wine faults; reductive aromas; browning; atypical aging; preventive measures; corrective solutions

1. Introduction

We are currently witnessing a sensory wine revolution. There are liberal styles and fractions, where oxidation, turbidity, or excessive content of phenolic substances are accepted even in white wine. Protein-dependent turbidity and crystalline sediments do not need to be classified as wine faults in the current liberal conditions [1,2]. If there is a market for such wines, and they are produced according to clear applicable rules, it is necessary to respect them. Despite this liberal era, the two-thousand-year history has set and shaped certain rules. It is documented and published in professional and scientific literature, what can be considered a faulty wine with the impacts on the wine sensory profile [3–7]. Consuming faulty wine does not bring pleasure to the consumer, it causes unpleasant feelings, and the consumer does not ask for another sip [4,8].

Wine faults in traditional wine-growing countries are divided into those that are not caused by the activity of microorganisms, which can be relatively reliably corrected (oxidation, atypical aging, reductive aromas, various odor disorders, non-harmonic ratio of components), and faults caused by microbial activity with consequences that cannot be completely remedied (vitrification, brett, refermentation, mouse taint, mannitic fermentation, undesired decomposition of acids) [4,9–11].

Wine faults of non-microbial origin are caused either by various physical and chemical processes in the vineyard, the wine, or by contamination from the environment. Under these influences, wine changes its appearance, aroma, and taste. The most effective way to prevent the faulty wine in stock is good agricultural technology and the precision and consistency of professional staff [4,11].

The precursors for wine faults are physical, chemical, or microbiological, arising from incorrect agricultural technology, low ethanol content, high acid content, high content of phenolic substances (in white and rosé wines), and incorrect sulfurization regime, improper storage, and oxygen regime [4,10,12,13].

The objective of this review was to contribute to the current knowledge of common wine faults related to reductive aromas, browning, and atypical aging to summarize the latest knowledge on these topics, particularly in terms of sensory attributes, chemical background, and the preventive and corrective measures for wines during fermentation and after bottling. The review contributes to the knowledge of the potential for managing the faults during winemaking and wine storage.

2. Reductive Faults

Sulfur, as an important element in biological systems, is metabolized in various compounds and is responsible for the desired but also sometimes unpleasant aromas [14–16]. The complex chemical changes during wine fermentation include sulfur-related chemical pathways connected with wine-yeast metabolism. Along with the desirable volatile sulfur compounds that are perceived positively, such as 3-mercaptohexan-1-ol, 3-mercaptohexyl acetate, 4-mercapto-4-methylpentan-2-one, and 4-mercapto-4-methylpentan-2-ol with citrus, grapefruit and passionfruit tones in wine aromas, other volatile sulfur compounds contributing to so-called reductive aromas are likely generated through the same winemaking techniques [15,17]. Most of these undesirable volatile sulfur compounds are produced by yeast in sulfate assimilatory and dissimilatory reduction pathways [15,18,19] but H_2S and mercaptans can also be formed via an alternative biochemical route from the other sources such as glutathione or sulfane sulfur compounds which are not well elucidated [14,16,20].

Despite the implementation of good winemaking practices, it is not uncommon for reductive aromas to appear in wine during vinification [21]. Goode and Harrop [22] reported that reductive faults are responsible for 30% of all faults in commercial wines and should be considered seriously due to the significant economic impact on winemakers. The discernment of undesirable sulfur off-aromas due to higher concentrations of the above-mentioned compounds in finished wine has a negative effect and devalues a wine's quality, and is a possible reason for the rejection of a faulty wine by consumers. Volatile sulfur compounds thus present a challenge for modern-day winemaking. It is desired to limit (or eliminate) the production of undesirable H_2S and thiols but at the same time, maintain and enhance the production of the favorable volatile thiols [18].

2.1. Sensory Attributes of Reductive Faults

Besides the desired S-containing compounds that are important for wine quality, some substances, even in very low concentrations, cause off-odors and strong undesired aromas [15,23]. Sensory attributes associated with reductive aromas are rotten egg, putrefaction, sewage-like, rotten cabbage, onion, and burnt rubber. The most important volatile sulfur compounds (VSCs) linked with these descriptors are hydrogen sulfide (H_2S), methanethiol (MeSH), and ethanethiol (EtSH) [24]. Other S-containing compounds with a negative effect on the sensory properties of wine are dimethylsulfide and benzenemethanethiol. Their impact on wine aroma and flavor is due to their high volatility, reactivity, and low threshold concentrations (Table 1). These compounds are well known and belong to the commonly occurring problems in winemaking [15,16,25] and present an interesting topic.

Table 1. Selected sulfur volatile compounds and their detection thresholds.

Compound	Aroma Description	Odor Detection Threshold (µg L^{-1})	References
Hydrogen sulfide	Rotten egg, sewage-like, vegetal	1.1–1.6	[24,26]
Methanethiol	Cooked cabbage, onion, putrefaction, rubber	1.8–3.1	[24,27]
Ethanethiol	Onion, rubber, natural gas, faecal, earthy	1.1	[27,28]
Dimethylsulfide	Asparagus, corn, molasses, boiled cabbage, canned corn, blackcurrant, truffle	25	[28]
Diethylsulfide	Cooked vegetables, onion, garlic, rubber	0.90	[27,28]
Dimethyl disulfide	Cooked cabbage, intense onion	29	[27,28]
Diethyl disulfide	Onion, garlic, burnt rubber	4.3	[27,28]
3-(methylthio)-1-propanol (methionol)	Cauliflower, cabbage, potato	500	[29]

The contribution of organic and inorganic sulfur-containing compounds to aroma characteristics and their impact on wine aroma perception are associated with problems or wine faults if they are present in concentrations higher than their odor threshold [15,24] (Table 1).

2.2. Biochemical Background of Volatile Sulfur Formation

Sulfur can be present in eight oxidation states from S (−II) up to S (+VI) [16]. Due to this, sulfur can participate in an array of oxidation, reduction, and disproportionation reactions that are inorganic, and, in addition, these reactions are connected with microbial metabolism. In the process of fermentation, pesticide residues on grapes containing elemental sulfur are not metabolized only to hydrogen sulfide, but it is possible to form precursors that generate H$_2$S in the post-fermentation stage after bottling. The known precursors are glutathione tri- and polysulfanes (Glu-S-Sn-S-Glu) and recently also tetrathionate (S$_4$O$_6{}^{2-}$) was identified [16].

The wine fermentation process is supported by wine yeast and involves changes in the composition of sulfur compounds that are connected with the biosynthesis of S-containing amino acids cysteine and methionine through the sulfate assimilatory reduction pathway [16,18,30–33].

The chemical and microbiological transformations of sulfur-containing substances are accompanied by sensorial changes and not all of them are desirable in the final product [16,24]. Wine, accompanied by the smell of rotten eggs, indicates the presence of hydrogen sulfide (H$_2$S), and its derivatives (methanethiol, ethanethiol, etc.) which are products of yeast metabolism during fermentation and other transformations during the post-bottling stage [16,24].

2.2.1. Formation of Hydrogen Sulfide, Methanethiol, and Ethanethiol

Explaining the formation of H$_2$S in the fermentation process of grapes has been a research topic for many authors, e.g., Rauhut [34], Swiegers and Pretorius [18], Ugliano and Henschke [35], Ugliano et al. [36], Cordente et al. [37], etc. The presence and concentration levels of S-containing organic compounds such as aminoacids and peptides (cysteine, methionine, S-adenosylmethionine, and glutathione) are essential for the wine yeast metabolism and growth. An insufficient concentration of these compounds in the yeast diet leads to their synthesis from inorganic sulfur sources by the yeast cells. The first step of this synthesis is the reduction of inorganic ions of sulfites and sulfates to hydrogen sulfide, which is a precursor of sulfur-containing amino acids. The formation of hydrogen sulfide is present in every fermentation process, but its further use in metabolic pathways of yeast is closely dependent on the levels of other substrates—nitrogenous substances that are essential in the formation of S-containing aminoacids and their peptides and proteins. Excessive H$_2$S production by yeast is affected by the three main factors during wine fermentation: assimilable nitrogen, sulfur dioxide, and yeast strain [38].

This process of H$_2$S formation is present in every wine, however, its levels differ based on the above-mentioned conditions. The lack of nitrogenous substances in the yeast diet is a trigger for excessive production. This explanation was supported by the studies of Ugliano et al. [36] and Müller et al. [16] who reported that sulfur sources, such as fungicide residues containing elemental sulfur, hydrogen sulfide, and methanethiol, etc., that are formed by yeasts are considered to be the main substrates for the generation of latent precursors of off-odor compounds during the vinification. Ferrer-Gallego et al. [39] mentioned that sulfur dioxide is connected with the formation of hydrogen sulfide via yeast metabolism.

According to several studies [32,36], the lack of nitrogenous substances in the must is considered the main reason for hydrogen sulfide formation. On the other hand, some experiments proved that H$_2$S is also formed in conditions where assimilable nitrogen is present or supplemented [12,32,36]. Ugliano et al. [12] stated that nitrogen supplementation in wine to influence the formation of H$_2$S stem is from the initial yeast assimilable nitrogen and yeast properties that produce H$_2$S. The explanation of the excessive formation of H$_2$S is based on the permanent exposition of vineyards to stress conditions (drought, malnutrition, overload).

The biochemical pathway for the formation of hydrogen sulfide can be described through the wine yeast metabolism (Figure 1). Inorganic (sulfate, sulfite, sulfur dioxide), as well as organic sulfur compounds (cysteine and glutathionine), are reactants for the formation of H$_2$S [18,38]. The biochemical pathway in *S. cerevisiae* was elucidated by Yamagata [40] and Rauhut [34] as a sulfate reduction sequence (SRS) pathway. Current knowledge of the regulatory mechanisms are described in the work of Guidi et al. [38] and Müller et al. [16].

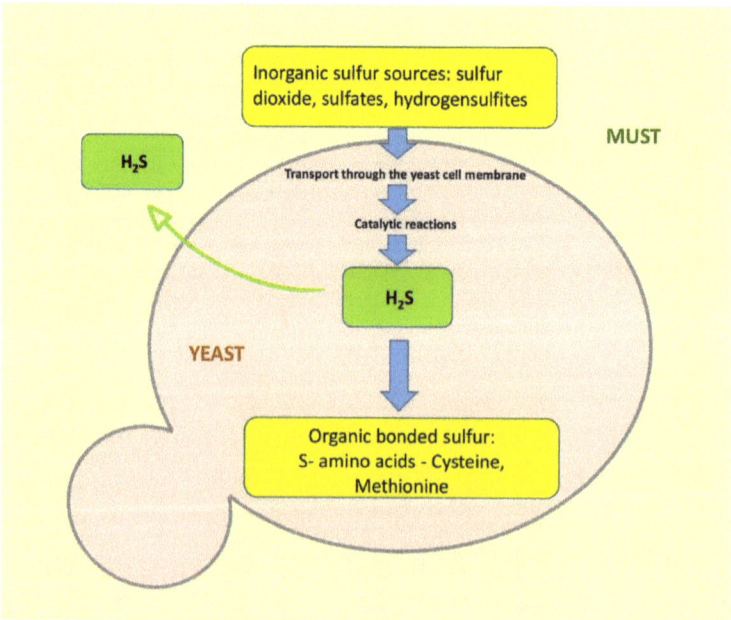

Figure 1. Schematic sulfur reduction pathway and formation of amino acid biosynthesis in *S. cerevisiae*, based on [34,38,40].

Inorganic sulfur sources are normally present in the grape must, but the must does not contain organic-binding sulfur which leads to the biosynthesis of sulfur compounds that are important for wine yeast [41–43]. Basically, H_2S is formed either from the HS^- ions which present a metabolic intermediate from sulfate and sulfite reduction that are important for the synthesis of organic sulfur compounds. In this step of metabolism, nitrogen supply is essential for further reactions in which HS^- ions are transformed especially to the S-aminoacids methionine and cysteine through the O-acetylserine and O-acetylhomoserine, which are formed during nitrogen metabolism [41–43]. A lack of nitrogen sources or its unsuitability causes an end to the biochemical reaction with the formation of hydrogen sulfide, which first accumulates in the yeast cell and then diffuses through the membrane into fermenting must [19,30,44–46].

The sulfur reduction pathway, especially in longer maturation, allows the conversion of cysteine, methionine, and glutathione in H_2S [47]. However, in the growth phase of yeast-free S-amino acids, they are bound to the proteins and as the fermentation proceeds, these free molecules can be released from yeast into the finished wine. Degradation of glutathione into amino acids happens in the conditions of cellular nitrogen deficiency [48], and the deficiency of nitrogen can also lead to the release of H_2S from cysteine [18].

The role of genetic-based ability in the formation of H_2S and other volatile sulfur compounds was proved in different yeast strains [18,39,49,50].

Besides the hydrogen sulfide, yeasts also assimilate through the sulfur metabolism of MeSH, EtSH, etc., [18,48]. Sulfhydryls, methanethiol, and ethanethiol are also responsible for sulfurous unpleasant off-odors due to their low odor thresholds and suppression of fruity and floral wine aromas [24,51].

The formation of methanethiol is possible from methionine via transamination and activity of demethiolase [18,31,51]. Methanethiol is able to be esterified to metylthioacetate [23,52]. Ethanethiol was explained to be formed in vitro in the reaction of hydrogen sulfide with ethanol or acetaldehyde [50].

In the post-bottling stage, hydrogen sulfide levels in wine are influenced by several determinants [17]. Their formation in the post-bottling stage is associated with the possible decomposition of cysteine followed by the accumulation of hydrogen sulfide in wine [53]. An important role is probably played by the concentration of oxygen after fermentation and during storage. A lower concentration affects the higher generation of hydrogen sulfide during post-bottling [17,54,55]. Nguyen et al. [56] investigated the influence of micro oxidation on Cabernet Sauvignon wine in the oxygen doses of 5–20 mg L^{-1} for one month in combination with malolactic fermentation. They observed some decreases in the levels of sulfur off-odors.

Sulfhydryl compounds are able to react with wine components, and the reductive components can be captured in the form of precursors. During the wine storage, decomposition of these precursors may occur resulting in the release of malodorous compounds. The precursors can be additionally induced by wine treatments (e.g., aeration, copper fining), which are performed to avoid or decrease reductive aromas in the wine before bottling [16].

Factors that impact the formation of H_2S are as follows: the presence of a higher concentration of elemental sulfur, sulfurization with sulfur dioxide, the presence of compounds with organically bonded sulfur, a deficiency of pantothenic acid, and the existence of amino acids [18,31,57].

2.2.2. Post-Fermentation Reductive Aromas—Thiols and Disulfides

The odor of hydrogen sulfide spreads from yeast from bottom to top. Post-fermentation generation of methanethiol and hydrogen sulfide has been explained based on several hypotheses including non-enzymatic reactions. The formation of methanethiol and ethanethiol has been proposed by the non-enzymatic reduction of symmetrical disulfides, thioacetate, and thioether hydrolysis, and decomposition of S-amino acids [58].

Therefore, the fermenting wine from the lower layers of the container and the yeast sludge must be checked regularly, at least once every three days. If hydrogen sulfide is not removed from the wine in time, more complex sulfur compounds ethanethiol and disulfides are formed that create a strong odor of rotten onions and feces of various intensities (mercaptans, post-fermentation sulfur taint). In the study by Kreitman et al. [59], in anoxic conditions during storage, the concentration of both hydrogen sulfide and methanethiol increased. Application of Cu(II) to the wine in amounts over 2-fold molar higher than the concentration of volatile sulfur compounds (H_2S, MeSH, EtSH), resulted in the complete removal of all sulfhydryls [58]. In the same study, the addition of tris(2-carboxyethyl) phosphine was poorly efficient in releasing hydrogen sulfide from its copper complex. The efficiency of other copper chelators was studied in order to release sulfhydryls from their complexes, e.g., bathocuproinedisulfonic acid [58].

Ethanethiol can be removed from wine relatively reliably by applying copper salts. In the post-bottling study of Bekker et al. [55], the impact of Cu^{2+} in combination with sulfur dioxide (SO_2) was followed in relation to hydrogen sulfide (H_2S) formation in Shiraz and Verdelho wines. Treatment with copper sulfate in combination with oxygen exposure and glutathione in the post-bottling study on hydrogen sulfide and methanethiol was studied by Ugliano et al. [54]. Copper addition resulted in H_2S accumulation during the second 3 months of storage and its highest concentrations were observed, especially in variants with glutathione and copper treatment in low-oxygen conditions. However, copper fining is generally considered as a manner of sulfhydryl removal [60,61], in conditions with a low-oxygen exposure which seems to promote the increase of H_2S concentration in wine in the post-bottling stage [54,59].

According to Franco-Luesma and Ferreira [51], both de novo formation of H_2S from precursor compounds as well as the release of H_2S from metal complexes contribute to the final concentration of H_2S formation in wines post-bottling, with the release from metal complexes responsible for the majority of H_2S produced in red wines, and de novo formation responsible for the majority of H_2S produced in white wines and rosé wines. Additionally, Franco-Luesma [62] suggests that the release of free H_2S and MeSH from bound sources is a function of a decrease in the redox potential of wines.

Disulfides do not react or react only very weakly with copper salts, moreover, they are precursors of further ethyl mercaptan formation. Disulfides can be partially removed from wine by adsorbents, the best of which have the ability to bind odors. Activated carbon, as the main component of adsorbent was studied by Huang et al. [63] and was proved as efficient in the elimination of three types of disulfides (diethyl disulfide, dimethyl disulfide, and dimethyl trisulfide). It is a drastic intervention when besides the off-odors bouquet substances, other valuable substances are removed from the wine. Therefore, analysis of sulfur-containing compounds is important for wine quality control and research [16].

2.3. Preventive Measures of Formation Reductive Aromas

The most probable reason for the formation of sulfur-related flaws is a combination of circumstances: strong sulfurization of must or mash before fermentation, deficiency of non-essential aminoacids [36], and incorrect application of nitrogenous substances (nutrition) during fermentation. The best prevention of its occurrence is, therefore, the prevention and elimination of these factors [64].

The treatment of containers with sulfur dioxide vapors has to be ensured with slow-burning sulfur slices. The slices need to be ignited from above. If we ignite the slice from the bottom part, due to the generation of a large amount of heat during rapid combustion, a part of the elemental sulfur does not burn, rather melts and drips to the bottom of the vessel. Such unburned sulfur can later become the source of sulfur for the formation of H_2S by yeast. Sulfur slices must therefore burn slowly and elemental sulfur must be completely oxidized into sulfur dioxide [64].

Several approaches in the elimination of undesirable off-odors from hydrogen sulfide were performed and proposed: (1) macro-oxygenation and aeration [21,55,65–67], (2) yeast assimilable nitrogen (monitoring, supplementation in wine) [15,21,36,68,69], (3) application of fining agents (especially copper fining) [21,58,70,71], and (4) application of lees [21,72].

The classical approach to decreasing the content of H_2S in wines is aeration [65,66]. This method allows hydrogen sulfide removal with the immediate bottling of wine from lees with air and its continuous sulfurization (in a dose of 20 mg SO_2 L^{-1}). This procedure involves three processes: oxidation of hydrogen sulfide with atmospheric oxygen, venting of the hydrogen sulfide in the gas phase, and its partial reaction with sulfur dioxide, to form elemental sulfur. Aeration is successful in removing sulfhydryls as well as thioacetates [55]. However, simple aeration is not so effective for organic sulfides such as methanethiol and ethanethiol or heavier mercaptans. Anoxic conditions during the wine storage contribute to the permanent or increasing presence of these compounds resulting in the reductive character of the finished wine [67]. Targeted oxygenation also has beneficial effects on a healthy, smooth fermentation process, thus limiting the formation of hydrogen sulfide. The oxygenation must be done at temperatures below 16 °C. Enrichment of must with oxygen during oxygenation plays an important role in the formation of sterols and unsaturated fatty acids, which are important factors for the harmonious multiplication of yeasts and the integrity of their cell walls. In white wines made from oxidized must, hydrogen sulfide is formed in excessive quantities only rarely [73].

Remediation strategies, such as the addition of diammonium phosphate during fermentation, copper fining, the addition of fresh lees or lees products to wine, and aeration of the must during and after fermentation, are commonly employed in an effort to prevent the formation or to remove undesirable volatile sulfur compounds [21].

2.4. Corrective Solutions in Reductive Aromas

If the hydrogen sulfide odor in the wine is very strong (which means it was detected late), simple aeration and clarification may not be sufficient. Remediation is usually based on the selectivity of certain methods directed toward the sulfur compounds responsible for the reductive faults.

Certain treatments are selective in their ability to remove different types of sulfur species (i.e., sulfhydryls, disulfides, thioacetates, dialkyl sulfides), while other treatments may have associated risks [55]. Removal of hydrogen sulfide and methanethiol was attempted by the application of reducing agents, metal chelators, and maintaining lower levels of oxygen in red and white wines. Application of copper salts (sulfate or citrate) is a common way of this treatment [16] and was studied by many researchers (e.g., Smith et al. [15]; Bekker et al. [40]; Kreitman et al. [58]; Kreitman et al. [59]; Allison [66], etc.) Th addition of copper salts after fermentation results in the formation of non-volatile complexes and anoxic conditions in combination with other factors that can release sulfhydryls with an impact on promoting the off-odors. Treatment of finished wines with copper sulfate is due to the formation of stable complexes with sulfur compounds in order to eliminate the generation of H_2S and methanethiol [15]. The precipitation of H_2S with copper ions produces insoluble copper sulfide, which is an odorless compound. This will remove the hydrogen sulfide from the wine without aeration.

The use of copper sulfate poses an interesting dilemma to the winemaker as it is used to treat wines tainted with H_2S and mercaptans, but at the same time, it reduces the concentration of the desirable volatile thiols as the Cu^{2+} ion does not discriminate between the two classes of sulfur compounds [18]. Copper fining is also not efficient for sulfides, disulfides, and thioacetates, and is limited to H_2S and thiol removal [24]. The formation of disulfides and trisulfides contributing to additional off-odors due to copper fining, as well as problems associated with wine instability and additional wine-processing logistics, should also be taken into account as an unwanted impact of copper fining [24].

A comparative study was performed by Bekker et al. [21] in order to propose the most efficient remediation technique for maintaining fruit attributes and eliminating the reductive tones. In the study on the Shiraz wine variety, the fermentation process was combined with the use of different chemical and microbiological approaches. Techniques based on the application of lees, diammonium phosphate, and copper resulted in a fall in the fruit character of wines, and moreover, enhanced reductive aromas. A commonly described adverse effect of copper fining lies in the risk of copper reactivity with thiols in wine resulting in loss of the varietal character of wine [12]. Moreover, the elimination of copper sulfide from treated wine is sometimes needed. Its removal has been studied with the use of bentonite [71].

However, even with the chemical removal of the sulfur taint, it is necessary for the treated wine to be separated from the yeast sludge by bottling. The dose of the copper preparation to the wine must be determined in the laboratory conditions so that we do not unnecessarily burden the wine. Copper is a heavy metal and therefore it is always necessary to consider whether its use is really necessary. Especially winemakers without experience and laboratory equipment may have copper residues in the wine and then it is questionable whether this wine serves the health of consumers. When using copper preparations, it must be kept in mind that the statutory maximum permissible copper value in wine of 1 mg Cu^{2+} L^{-1} must be complied with for health safety reasons (EC Commission Regulation 606/2009) [74]). In practice, a maximum value of 0.6 mg L^{-1} is calculated. When calculating the dose of copper preparations, it should be kept in mind that the wine also contains a certain amount of copper, which comes from preparations for the treatment of grapes against fungal diseases. In a three-year survey, Ailer [64] found values of 0.06 to 0.61 mg Cu^{2+} L^{-1} in Pinot Blanc and St. Laurent varieties.

3. Wine Browning (Oxidation)

Wine oxidation is one of the major problems encountered in winemaking, and occurs in red, rose, and is especially evident in white wines. Browning is a quality defect in wine that is the result of a complex series of oxidation reactions that take place during processing, aging, and storage, which give rise to a brown color that increases color intensity, decreases brightness, and raises the browning index [75,76].

3.1. Sensory Attributes of Wine Browning

Wine browning is a result of complex non-enzymatic, oxidative, and also enzymatic changes, especially in white wines. Polyphenols in the white and red wines contribute to wine quality and influence further reactions leading to the development of not only desirable sensorial properties.

They play a key role in browning which has a significant effect on wine's organoleptic characters and antioxidant properties. The result of the series of reactions increases the color intensity, decreases the brightness index, and raises the browning index. The change of wine color is characterized as a gradual replacement of the initial pale-yellow color to a brown-yellow color developed from access to oxygen [77]. Besides the color alteration, non-enzymatic oxidation may result in the appearance of a smell of rotten fruit, wood, or cooked vegetables [76,78]. As a result of the browning process, there is also a loss of varietal aroma and flavor, and a development of bitterness and astringency that can cause serious changes in the physical appearance and aroma properties of wine [76,79–81].

3.2. Biochemical Background of Wine Browning

From a sensory perspective, controlled oxidation could be beneficial for red wine by enhancing and stabilizing color and reducing astringency, however, the quality of white wine is generally damaged by excessive exposure to air [73,82,83]. Wine phenolics belong to two main groups: flavonoids (anthocyanins, flavan-3-ols, flavonols, and dihydroflavonols) and nonflavonoids (hydroxybenzoic and hydroxycinnamic acids and derivatives, stilbenes, and volatile phenols). The most common flavonoids are flavan-3-ols and flavonols (and anthocyanins in red wines) (Figure 2). Flavan-3-ols, mainly catechins and catechin–gallate polymers, produce a number of oxidation products that could be regarded as browning agents in white wines [84]. Regarding the group of flavonols, quercetin derivatives are the main components present in white wines. The major nonflavonoid phenolic compounds in white wines are hydroxycinnamic acid derivatives (Figure 3) which are easily oxidized components contributing to the browning of wines during aging.

Flavonoids
Flavan-3-ols

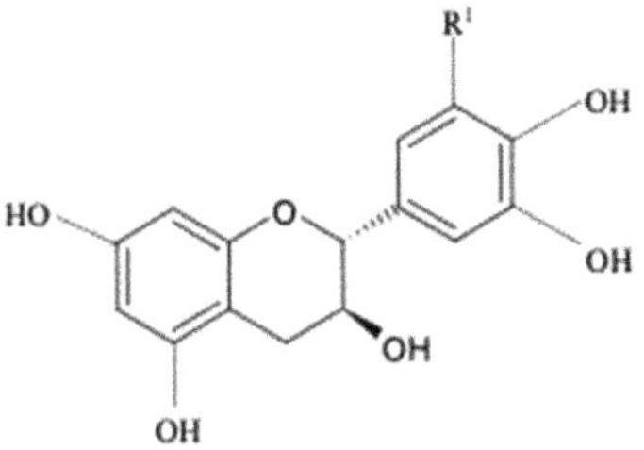

(+)-Catechin (R¹ = H)
(+)-Gallocatechin (R¹ = OH)

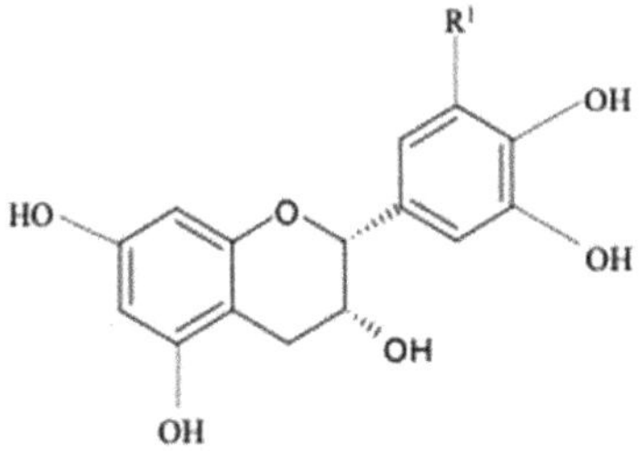

(−)-Epicatechin (R¹ = H)
(−)-Epigallocatechin (R¹ = OH)

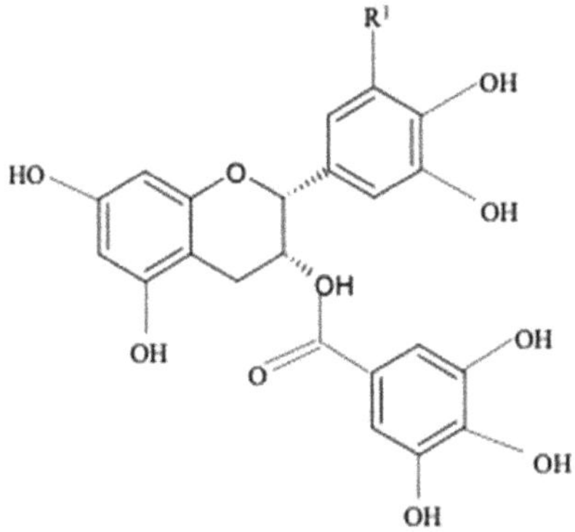

(−)-Epicatechin gallate (R¹ = H)
(−)-Epigallocatechin gallate (R¹ = OH)

Flavonols

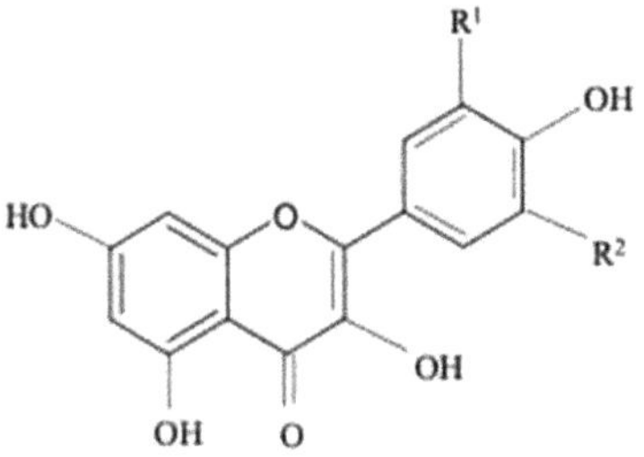

Kaempferol (R¹ = R² = H)
Quercetin (R¹ = OH, R² = H)
Myricetin (R¹ = R² = OH)

Anthocyanins

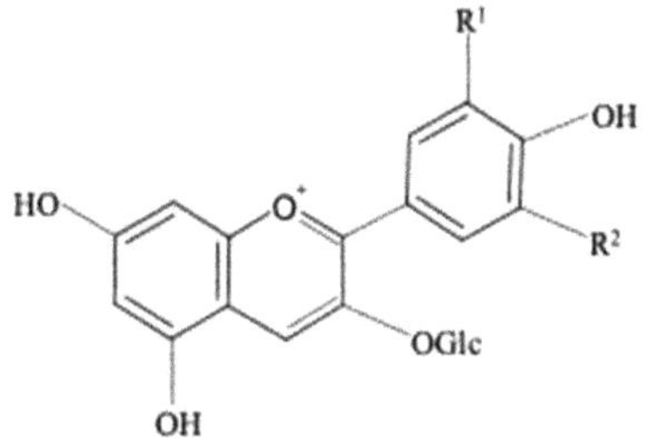

Cyanidin 3-glucoside (R¹ = OH, R² = H)
Peonidin 3-glucoside (R¹ = OCH₃, R² = H)
Delphinidin 3-glucoside (R¹ = R² = OH)
Petunidin 3-glucoside (R¹ = OH, R² = OCH₃)
Malvidin 3-glucoside (R¹ = R² = OCH₃)

Figure 2. Chemical structure of the most important flavonoids in wine.

Nonflavonoids
Derivatives of benzoic and cinnamic acid

	R¹	R²	R³	
Benzoic acid	H	H	H	Cinnamic acid
p-Hydroxybenzoic acid	H	OH	H	p-Coumaric acid
Protocatechuic acid	OH	OH	H	Caffeic acid
Vanillic acid	OCH₃	OH	H	Ferulic acid
Gallic acid	OH	OH	OH	
Syringic acid	OCH₃	OH	OCH₃	Sinapic acid

Stilbenes

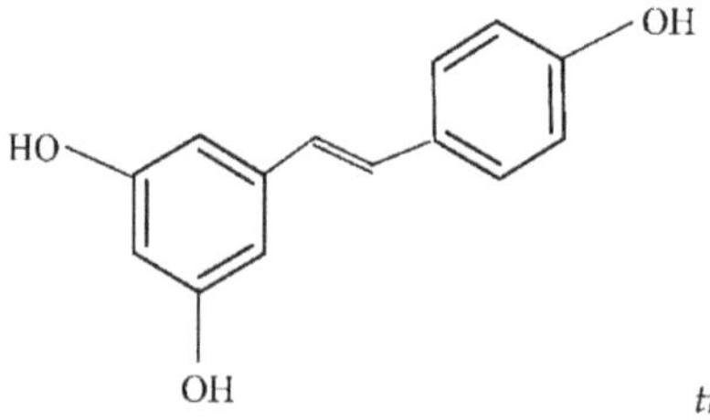

trans-resveratrol

Figure 3. Chemical structures of the most important nonflavonoids participating in enzymatic oxidation in wine.

The grape phenolic composition is affected by several factors such as grape variety, ripening stage, climate, soil, place of growing, and vine cultivation. Winemaking technologies, such as maceration time, temperature, the intensity of pressing, yeast, and SO_2-doses, together with enological practices and aging, also modify the phenolic content in wine [85,86]. Usually, white wines are made without aeration in order to avoid extensive contact with oxygen, preventing browning of the wine and deterioration of the overall quality. Controlled low temperatures (14–18 °C) are usually applied for white wine production.

Then, the phenolic substances present, although not volatile in themselves, catalyze the decomposition of the aromatic profile of the wine and the oxidation processes. In the oxidation processes in wine, quinones and peroxides are formed in the presence of phenolic substances. The result is wine browning and an oxidized flavor. In the production of red wines, the mash is macerated throughout the fermentation. For this reason, the content of polyphenols (mainly anthocyanins) in red wines is higher, 1–5 g per liter, compared to white wines, which contain 0.2–0.5 g of polyphenols per liter [87].

Enzymatic and Nonenzymatic Oxidation

During the winemaking process, oxygen is an important factor in creating quality wines as it takes part in enzymatic and nonenzymatic oxidation reactions. Enzymatic oxidation almost entirely occurs in the grape must, while non-enzymatic reactions can happen both in grape must and wine [82,88,89].

Enzymatic oxidation is mediated by two distinct polyphenol oxidases which give rise to the formation of quinones (brown compounds): (i) A catechol oxidase, derived from healthy grapes, has both cresolase activity (hydroxylation of monophenols to ortho-diphenols) and catecholase activity (oxidation of ortho-diphenols to ortho-quinones), and; (ii) A laccase of fungal origin that does not have cresolase activity but does catalyze the oxidation of several types of phenols, in particular para-diphenols [82]. Enzymatic oxidation of selected phenolic acids in wine is shown in Figure 4.

During the non-enzymatic oxidation process, also called chemical oxidation of wine, the oxidative process is favored by the oxidation of polyphenols containing an ortho-dihydroxybenzene moiety (a catechol ring) or a 1,2,3-trihydroxybenzene moiety (a galloyl group), such as (+)-catechin/(−)-epicatechin, gallocatechin, gallic acid, and its esters, and caffeic acid, which are the most readily oxidized wine constituents [82]. These substrates are sequentially oxidized to semiquinone radicals and benzoquinones while oxygen is reduced to hydrogen peroxide. The whole process is mediated by the redox cycle of Fe^{3+}/Fe^{2+} and Cu^{2+}/Cu^{+} [90]. Other compounds with more isolated phenolic groups such as malvidin, p-coumaric acid, and resveratrol are oxidized at higher potentials.

Caftaric acid (caffeoyl tartaric acid) is the most abundant phenol in must. The polyphenol oxidases have a high affinity for this acid, so the corresponding ortho-quinone is the main enzymatic oxidation product of must. In addition to the high concentration of the quinone, the caftaric acid quinone/caftaric acid redox couple has a high redox potential. This molecule is therefore highly reactive, meaning that it participates in other redox reactions such as the oxidation of ascorbic acid and sulfites, and also of orthodiphenols that are not substrates of polyphenol oxidases. Another less abundant but not less important quinone is p-coumaryl tartaric acid, which is derived from the oxidation of coumaryl tartaric acid (coutaric acid) [91].

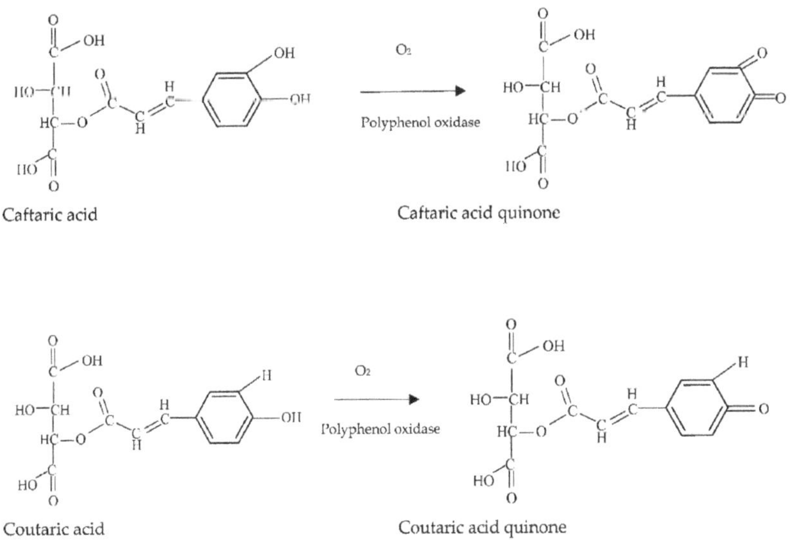

Figure 4. Enzymatic oxidation reactions of caftaric and coutaric acids in must and wine, based on Moreno and Peinado [91].

The enzyme polyphenol oxidase (PPO) reacts with naturally present phenolic compounds and catalyzes their oxidation into highly reactive primary quinones. The phenol/quinone redox couple has a high redox potential, meaning that quinone tends to be reduced, leading to the oxidation of other phenolic compounds, notably oligomeric flavanols and anthocyanins. This phenomenon is known as coupled oxidation. Quinones formed in this manner are called secondary quinones. They are highly unstable and give rise to condensation products (with their reduced form or with caftaric acid) [92].

Several compounds in must and wine have natural antioxidant activity and react with primary quinones, reducing their oxidizing action. One of these compounds is glutathione which is an important tripeptide in must and is highly abundant in certain grape varieties (Figure 5). Glutathione reacts with the quinone of caftaric acid and forms the acid 2-S-glutathionyl caftaric acid, also known as grape reaction product (GRP). The formation of GRP, which is colorless and does not provide a substrate for polyphenol oxidase, paralyzes the formation of brown products, as it inactivates the quinone of caftaric acid and therefore does not give rise to coupled oxidation reactions. GRP, however, can be oxidized in the presence of an excessively high concentration of caftaric quinones once glutathione is depleted, giving rise to intense browning. The browning of grapes must therefore depend on the relative proportions of glutathione and hydroxycinnamic acids [93–95].

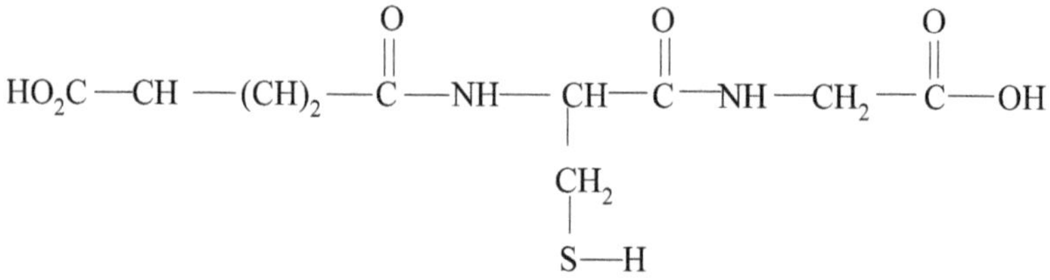

$$HO_2C-CH-(CH_2)_2-\overset{O}{\overset{\|}{C}}-NH-\underset{\underset{S-H}{\overset{|}{CH_2}}}{\overset{|}{CH}}-\overset{O}{\overset{\|}{C}}-NH-CH_2-\overset{O}{\overset{\|}{C}}-OH$$

Figure 5. Chemical structure of glutathione.

3.3. Preventive Measures of Wine Browning

Wine is a highly complex matrix containing many potentially oxidizable compounds, including phenolic compounds, certain metals, tyrosine, and aldehyde, of which the flavonoids (e.g., dihydroxyphenolic compounds) and nonflavonoids (e.g., caffeic acid) can lead to brown products [92,96,97]. Therefore, during storage and treatment during cellar handling, it is necessary to prevent the access of atmospheric oxygen to the wine. With the reducing environment and convenient sulfurization regime, it is possible to prevent wine browning and thus reduce the use of antioxidants, especially sulfur dioxide. Since the main compounds that cause oxidation are phenolics, the intensity of browning (oxidation) depends on the content of phenolic compounds, the degree of wine exposure to atmospheric oxygen, and the degree of neglect of sulfurization [73,98].

Prevention of browning is possible by proper management of phenolic substances during the grape harvesting, pressing of grapes, the subsequent processing of must, regular filling up of the containers, application of a shielding gas atmosphere, and sulfurization. Susceptibility of wine to browning is possible to determine by the analysis of the polyphenolic content, or by using an accelerated test for browning capacity [75].

In anaerobic conditions, redox systems are in equilibrium at certain time intervals. Any supply of oxygen to the wine immediately disturbs the equilibrium state. Oxygen diffuses into musts or wines and reacts with easily oxidizable compounds and oxidizes them. In these reactions, peroxides can be formed that affect other oxidation and radical processes in the wine. There are always iron and copper cations in the wine, which break down the peroxides into the water and active oxygen. The released active oxygen also oxidizes those compounds which are not oxidizable by the molecular oxygen O_2. Compounds such as L-ascorbic acid, oxoacids, amino acids, and phenolic substances, especially anthocyanins, catechins, and yellow and green dyes, are oxidized in the wine in contact with atmospheric oxygen. These processes are partially desirable in the production of some wines (e.g., Tokaj, Madeira, Sherry, Port, or Xerez wines), where a higher level of oxidation compounds is expected and positively evaluated [6], but these oxidations are undesirable in the production of varietal and sparkling wines.

SO_2 is one of the most important agents that is added to wines to prevent these processes. The use of SO_2 in winemaking is due to its ability to be an effective antioxidant, preventing the activity of the oxidases, as well as its antimicrobial property [85]. However, the addition of SO_2 to wines can give the wine undesired flavors and aromas and can raise health-related objections due to serious allergic reactions incurred by sulfite-sensitive individuals, such as headaches, abdominal pain, and dizziness [79,92]. Ascorbic acid (vitamin C) and its optical isomer, have been widely used as an antioxidant in winemaking and are considered an alternative to sulfites, especially for white wine production, primarily because of its ability to scavenge molecular oxygen [96] and minimize the oxidative spoilage (browning) of white wine. Baroň [99–101] states that ascorbic acid has an immediate effect due to its high antioxidant activity. However, it protects the must or wine only against short intensive aeration, but not against long-term oxidation. Therefore, it is not effective in the long-term storage of wine. To date, a single replacement product performing the same roles as SO_2 has not been found.

Management of oxygen is important for the browning potential of grape juice or wine. During the pressing process, exposure to oxygen was studied by Day et al. [102]. The authors stated that pressing management affects the wine composition more than handling management and it is possible to influence the concentration of particular classes of chemicals. Controlled oxygenation in the press can be used in decreasing the concentration of polyphenols. Targeted oxygenation of must or mash as a way of decreasing polyphenols in white wines without impact on sensorial changes of wines was published by Ailer et al. [98] and Pokrývková et al. [73]. If we oxidize mash or must by its exposure to atmospheric oxygen for a while without the use of sulfur dioxide or any other antioxidants, we remove excess phenolic substances from it. This means that antioxidants are not used in the grape processing technology until the must has been clarified. Phenolic substances in must or mash are oxidized with atmospheric oxygen and sedimented during sludge removal. The sulfur dioxide is used for the first time after juice clarification. In the later stages of maturation and in the final product, wine with a minimum content of phenolic substances is less prone to oxidation [73].

A protective role against browning has been attributed also to ellagitannins which may be included in the oxidation reactions of both red and white wines. The compounds ellagitannins and ellagic acid are naturally present in wood barrels and wood chips or may be used as oenological tannins. Their role in the wine oxidation process is due to their ability to quickly absorb dissolved oxygen and support the hydroperoxidation of wine components. Thus, these compounds positively influence wine color attributes and protect it from the browning process [103,104].

Oxygen harms young wine directly (oxidizes it) as well as indirectly. Aerobic conditions may help to develop unwanted microflora in the wine and irreversibly reduce its quality. There will be some oxygenation during each wine handling, but it is also the duty of the oenologist to keep it to a minimum. Polyphenols are characterized by their high antioxidant activity. This positive activity consists in scavenging free radicals and contributed to the phenomenon of the "French paradox" [105,106]. In the technology of wine production, it is necessary to take into account the beneficial aspects of phenolic substances, especially in red wine, but also their relative harmfulness in white and rosé wines. Young wine is not prone to oxidation as it still contains enough free SO_2 from sulfurization and also residual CO_2. However, at some point after the wine has been bottled, the sulfur dioxide is transformed into a bound form and ceases to act.

3.4. Corrective Solutions in Browning

Oxygen reactions with phenolics are catalyzed by oxidative enzymes to form peroxides and quinones, which further chemically oxidize ethanol to sensory negative acetaldehyde.

Highly oxidized browned wine is no longer enough to treat only with sulfur dioxide. In the winemaking, several fining agents were used to decrease the level of brown compounds, mainly active charcoal and polyvinylpolypyrrolidone (PVPP). However, even though the efficiency of these agents is high, the side effect of their use is a change in the wine's sensory characteristics, especially if they are applied at high levels [107].

The wine can be treated with activated carbon in a dose of 15–30 g per hectoliter of wine. It is a drastic intervention when, besides the polyphenols, we also remove bouquet substances and other valuable components from the wine. A gentler way to remedy oxidation is to use polyvinylpolypyrrolidone (PVPP). Its impact on changes in chemical composition and wine properties has been described by several authors [108,109].

Another possible way to remove oxidation is to re-ferment it with yeast [110], (so-called fermentation through "4"). Yeast can do a great service to an oxidized wine—they use part of the oxygen to produce sterols and long-chain fatty acids important for their growth. This will restore the wine's reductive color and freshness.

4. Atypical Aging

Even maintaining all the conditions of correct vinification and using a seemingly faultless grape, sometimes the required result is not achieved. For instance, in the form of negative changes in the sensory wine profile. When this occurs, it is called "atypical wine aging". Atypical aging (ATA) is not considered a common fault but based on its presence and contribution to the olfactory perception of wine from the view of its chemical and sensory characteristics, it is still a hot topic [111]. The causes of its occurrence can be found in the vineyard, from where the basic ingredient for wine-producing comes [13].

Atypical aging (ATA) began to emerge in the 1990s when integrated grape production expanded significantly coupled with the extensive grassland of vineyards. Its origin is attributed to the competition of green land cover for the vineyard and the associated stress [112].

Affected wines are characterized by a faulty aroma, referred to as a "Fox ton", "wet rag", "naphthalene", and "bean sprouts" [113]. At the same time, these faulty aromas completely suppress the desired, fruit-floral aroma. Disharmony in the uptake of macro- and micronutrients from the soil also causes taste faults, with wine having an unpleasant and short persistence [13].

4.1. Sensory Attributes of ATA

4.1.1. Threshold Levels

Recognition of ATA off-flavors is due to the wine complexity not being set to the threshold value of AAP concentration, however, several studies into the possible connection between the concentration of AAP and the appearance of off-odors have not confirmed a clear positive correlation [111,112,114,115]. Although in general, AAP is perceived as the main factor responsible for ATA development. Sensory recognition of ATA based on the AAP concentrations was investigated in several studies, however, a very important role is played by the wine type and its typical aroma. In general, threshold concentrations of AAP vary from 0.5 to 1.0 µg L^{-1}, and in red wines, aroma alteration is observed in concentrations over 1.5 µg L^{-1} [116]. Typical off-odors can be masked by the intensive fresh and fruity wine aroma [111]. Even lower threshold concentrations (less than 0.5 µg L^{-1}) have been observed in meager and light wines [112,114].

This fault is mentioned directly in the scoring system of some sensory wine competitions among the fundamental faults of wine. The organizers thus warn the assessors to take the ATA phenomenon seriously [111,117].

Olfactory perception of ATA and its connection with the AAP levels was investigated in Croatian wines by Alpeza et al. [111]. Wines with the occurrence of ATA were evaluated by the sensory analysis by the methods of "100 points" and "Yes/No" and AAP analyses were done by GC-MS. The results of AAP analyses corroborated the presence of ATA in all samples; the concentrations of AAP were 0.3–4.4 µg L^{-1}. The authors stated that the occurrence of ATA may be associated with the regional, climatic conditions in a particular vintage. According to this study, the intensity of perception of off-flavors corresponding to ATA did not highly correlate with the concentration of AAP as the main chemical descriptor [111].

4.1.2. Classification of the ATA Sensory Attributes

The sensory profile of the wine aroma connected with ATA perception can be divided into two groups [118,119].

The off-odors of the first group are described as chemical tones of naphthalene beads, naphthalene, soap scents, laundry detergents, furniture polishes, shoe pastes, old wax, jasmine scent, acacia blossom, lemon blossom, and dry laundry. The aroma is even more intense with the increasing content of SO_2 [111,118,120].

The smells of wet towels, wet wool, dirty dishes, dishwashers, and dried urine are classified in the second group. These odors imply a transition to the reductive stage in sensory profiling and can make it difficult to identify the fault. In both cases, the fruit, flower, or mineral variety character disappears, partially due to acid-catalyzed ester hydrolysis and oxidation of monoterpenes and the ATA symptoms dominate [118]. Wine loses color over time and becomes lighter, its taste is thin and empty, with a typical metallic bitterness [111]. The aromas from the first group are linked with the levels of AAP in wine and it is possible to simulate them with the addition of the compound [111].

On the other hand, it is difficult to imitate the scents of the second group in wine as they are assumed to come from IAA metabolites (e.g., indole and skatole) and contribute, however; to concentrations below the threshold levels and unpleasant tones. Especially skatoles should be given higher attention due to their fecal odor [112,118,121].

4.2. Biochemical Background of ATA Development

The compound that has been proved to have the main responsibility for atypical aging is 2-aminoacetophenone (AAP), even at a concentration level of 1 µg L^{-1} of wine. On the other hand, even though AAP, appears as the main chemical descriptor, it is not a rule that an increase in AAP concentration results in the enhanced development of ATA [111,118].

AAP is formed from indolyl-3-acetic acid (IAA), a taste- and sensory-neutral compound, after sulfurization of young wine and L-tryptophan [111,122–125]. Biosynthesis of IAA is proposed by two main pathways in vines: the tryptophan-dependent and tryptophan-independent pathways. In the tryptophan independent pathway of IAA biosynthesis are indole or indole-3-glycerol phosphate as the main precursor [126,127].

Indole-3-acetic acid (IAA) is an essential plant auxin, occurring in all stages of plant growth and development [118].

In the tryptophan-dependent pathway, the biosynthesis is based on the conversion of tryptophan (TRP). Zhao [128] described it in a two-step conversion as shown in Figure 6.

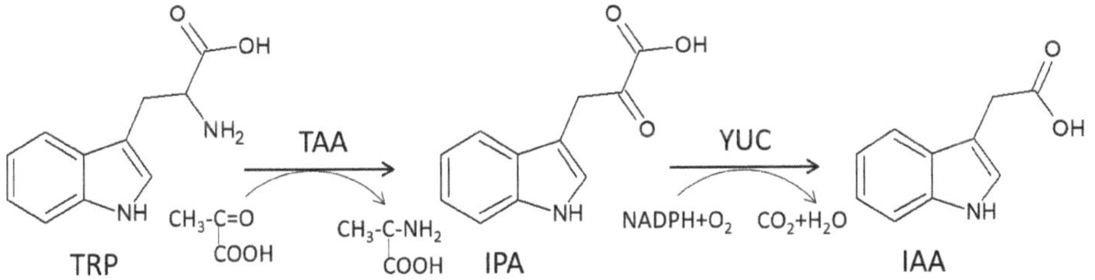

Figure 6. Transformation of Tryptophan to Indole-3-Acetic Acid in Plants, based on [128].

In the first step, TRP is transformed to indole-3-pyruvate (IPA) by replacing the amino group from tryptophan and the creation of IPA with the alfa-keto-carboxylic group. The process is accompanied by the TAA family of amino transferase. In the second step, IAA is formed from IPA in the presence of the YUCCA (YUC) family of flavin-containing monooxygenases. In this step, NADPH reacts with oxygen and is catalyzed by the YUC flavin-containing monooxygenases [128].

IAA is a plant hormone that is stored in grape berries in higher concentrations as a result of plant stress. The presence of superoxide and hydroxyl radicals, which are formed just after the addition of sulfur dioxide to young wine, is usually necessary for the formation of AAP from IAA, usually during the first cellar handling.

Christoph et al. [122] and Nardin et al. [126] described the degradation of IAA (Figure 7) as follows: IAA is degraded in the fermentation process and during wine maturing. The first step is a cleavage of the pyrrole ring and formation of 3-(2-formylaminophenyl)-3-oxopropanoic acid, which is triggered by superoxide radical that originated mostly from the reaction of sulfite to sulfate [122]. A further step is the decarboxylation of 3-(2-formylaminophenyl)-3-oxopropanoic acid to produce formyl-2-aminoacetophenone and finally AAP. Alternatively, oxidized indole-acetic acid [129] can be formed. Conditions for metabolic conversion of IAA to AAP are associated with natural factors such as cooler climate, the activity of yeast, the availability of nutrients and free SO_2, and other factors such as ripeness, hydric stress, and irrigation [111,116,118,126]. Thus, the matrix effect seems to be more important than the initial concentration of IAA. The example provides red wines, that usually do not suffer from ATA, however, they contain approx. 10-times higher IAA levels than white wines [118]. The explanation could be the high antioxidant capacity coming from the higher concentration of phenols in red wines [126].

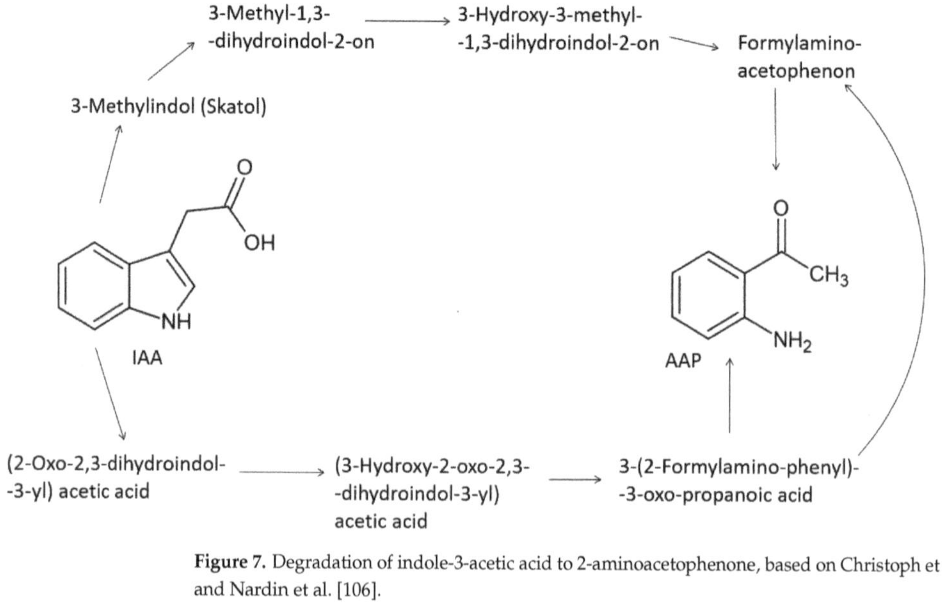

Figure 7. Degradation of indole-3-acetic acid to 2-aminoacetophenone, based on Christoph et al. [122] and Nardin et al. [106].

It is an unfortunate fact that until the young wine is sulfurized with a normal, necessary dose of sulfur dioxide, the potential for atypical aging cannot be determined with certainty. ATA does not occur in red wine because, after the first sulfurization of young wine, superoxide and hydroxyl radicals react primarily with tannins and not with indolyl-3-acetic acid.

4.3. Preventive Measures of ATA

The contribution of sulfites to the conversion of IAA to AAP was studied by Christoph et al. [122] and Hoenicke et al. [129]. Spiking of the model winelike solution with sulfites resulted in the disappearance of 50% of the IAA and the formation of AAP in the range from 0.5 to 20 mol %. The conversion rate was lower in the case when the wines were spiked with IAA. The authors assume the availability of free SO_2 to be important to induce the conversion [122,129].

According to Schneider [118], degradation of IAA occurs even after a brief exposure of wine to trace oxygen concentration, thus even if in the cellar operation where the oxygen absorption is minimized, the reaction cannot be excluded. As a strong radical scavenger to prevent the formation of ATA, the use of ascorbic acid additions in white wines (similarly to the use of tannins in red wines) was confirmed to be efficient. Viticulture stress factors including drought, UV-B radiation, nutrient deficiencies, over-cropping, and premature harvest are at the very origin of a wine matrix prone to producing ATA. Enological factors play a minor role, although skin contact and yeast strain have some impact as far as they affect the presence of oxygen radical scavengers like polyphenols and yeast metabolic products [11,118].

The most important preventive measures include the optimization of the moisture regime (regulation of grass cover in vineyards, irrigation of vineyards) and the optimization of vineyard fertilization, especially with nitrogen [115,130].

No reliable analytical techniques are yet available to predict the susceptibility of wine to ATA and viticultural measures cannot confidently prevent its occurrence [123,130]. Therefore, an accelerated aging test has been proposed and put into practice [131]. This is a test where ascorbic acid is added to one of two flasks of pure wine. Both flasks are stored at 37 to 45 °C for three to four days. After cooling, both samples are evaluated for odor. If the sample without the addition of ascorbic acid shows ATA, then the wine is susceptible to its formation.

4.4. Corrective Solutions of ATA

Defects in nutrition can be operatively solved by applying immediately acceptable macro- and microelements by foliar application of nutrition. The effectiveness of several antioxidant adjuvants against the possible development of ATA was studied by Nardin et al. [126]. The use of ascorbic acid and grape tannin resulted in the reduced production of IAA precursors. Th eapplication of Galla tannin had a protective role especially during the storage period, as despite the IAA content, the formation of AAP was eliminated. A promising capability in the prediction of the possibility of AAP formation in wine in the process of fining was shown in ANCOVA linear modeling, using the grape variety, the IAA content before aging, and the antioxidant treatment of the must.

However, even though the literature offers partial solutions for ATA prevention, a clear method for removing AAP from wine is not confirmed. We believe that this is a clear argument that it cannot be removed without the associated costs and damages.

Clarification with fresh healthy yeast or by repeated fermentation (via "4") can alleviate the presence of this defect due to the masking effect on the ATA off-aromas perception [118].

5. Conclusions

The wine faults discussed in the review are still hot topics in the winemaking industry. Consumption of faulty or sick wine does not cause a pleasant experience, it causes unpleasant feelings, and we do not ask for another sip. This results in a direct negative economic impact on the producer, but the brand image also suffers significantly. Yeast products, incorrect sulfurization regime, high content of undesirable phenolic substances, oxygen regime, and growing stress are precursors for the occurrence of defects of non-microbial origin that were the interest of this review. The latest scientific background provides findings and recommendations, particularly for the prevention of these three wine faults. The primary approach is to avoid growing stress and the use of flawless grapes. Optimal conditions for yeast and control mechanisms in winemaking are considered the best solutions to prevent the sulfur off-odors from occurring. Elimination of polyphenols as the precursors responsible for browning together with the use of antioxidants and lowering the oxygen access during winemaking create an efficient approach for browning prevention. Atypical aging prevention lies in the optimization of moisture and nutrition regime, and avoiding growth stress from excessive use of grassland cover and a fresh yeast can be used as a partial solution for ATA masking. With current knowledge, the wine faults can be reliably prevented. In the winemaking process, an essential role is played by the current analytical methods which provide important information for the early identification of potential problems. The review also provides valuable information, e.g., for producers of specific noble yeasts to eliminate the development of the faults. A further precondition for not having faulty wine in stock is the precision and consistency of the staff involved in the oenological procedures. Prevention is essential because troubleshooting is costly, difficult, and never perfect.

Author Contributions: Conceptualization, Š.A.; writing—original draft, S.J., Š.A. and V.I.-P.; visualization, L.B.; writing—review and editing, Š.A., S.J., V.I.-P. and L.B. All authors have read and agreed to the published version of the manuscript.

Funding: This work was funded by Vedecká Grantová Agentúra MŠVVaŠ SR a SAV (1/0239/21) "Modern analytical approaches to the identification of health safety risks and dual quality of selected foods".

Institutional Review Board Statement: Not applicable.

Informed Consent Statement: Not applicable.

Data Availability Statement: Not applicable.

Acknowledgments: We would like to thank to the Operational Program Integrated Infrastructure: Demand-driven research for the sustainable and innovative food, Drive4SIFood 313011V336, co-funded by the European Regional Development Fund for administrative and technical support.

Conflicts of Interest: The authors declare no conflict of interest. The funders had no role in the preparation of the review.

Sample Availability: Not applicable.

References

1. Ribeiro, T.; Fernandes, C.; Nunes, F.M.; Filipe-Ribeiro, L.; Cosme, F. Influence of the structural features of commercial mannoproteins in white wine protein stabilization and chemical and sensory properties. *Food Chem.* **2014**, *159*, 47–54. [CrossRef] [PubMed]
2. Batista, L.; Monteiro, S.; Loureiro, V.B.; Teixeira, A.R.; Ferreira, R.B. The complexity of protein haze formation in wines. *Food Chem.* **2009**, *112*, 169–177. [CrossRef]
3. Jackson, R.S. *Wine Tasting: A Professional Handbook*, 3rd ed.; Academic Press: Cambridge, MA, USA, 2016; p. 415. ISBN 978-0-12-801813-2.
4. Ailer, Š. *Vinárstvo & Somelierstvo (Winery & Sommelier Proficiency)*; Agriprint: Olomouc, Czech Republic, 2016; p. 197. ISBN 978-80-87091-63-0.
5. Grainger, K. Faulty & undrinkable. *IWFS* **2021**, *136*, 6–9.
6. Franco-Luesma, E.; Honoré-Chedozeau, C.; Ballester, J.; Valentin, D. Oxidation in wine: Does expertise influence the perception? *LWT* **2019**, *116*, 108511. [CrossRef]
7. MacNeil, K. *The Wine Bible*, 2nd ed.; Workman Publishing: New York, NY, USA, 2015; p. 1009. ISBN 978-0-7611-8715-8.
8. Steel, C.C.; Blackman, J.W.; Schmidtke, L.M. Grapevine bunch rots: Impacts on wine composition, quality, and potential procedures for the removal of wine faults. *J. Agric. Food Chem.* **2013**, *61*, 5189–5206. [CrossRef] [PubMed]
9. Poláček, Š.; Tomáš, J.; Vietoris, V.; Lumnitzerová, M. *Vinárstvo, Someliérstvo a Enogastronómia*; Slovak University of Agriculture in Nitra: Nitra, Slovakia, 2018; p. 319. ISBN 978-80-552-1910-3.
10. Grainger, K. *Wine Faults and Flaws: A Practical Guide*; John Wiley & Sons: Chichester, UK, 2021; p. 531. ISBN 978-1-118-97906-8.
11. Hudelson, J. *Wine Faults: Causes, Effects, Cures*, 1st ed.; Board and Bench Publishing: San Francisco, CA, USA, 2010; p. 87. ISBN 978-1-934259-63-4.
12. Ugliano, M.; Kolouchova, R.; Henschke, P.A. Occurrence of hydrogen sulfide in wine and in fermentation: Influence of yeast strain and supplementation of yeast available nitrogen. *J. Ind. Microbiol. Biotechnol.* **2011**, *38*, 423–429. [CrossRef] [PubMed]
13. Ailer, Š.; Jedlička, J.; Paulen, O. Study of the extreme agro-technical and agro-climatic factors influence in vineyard for a qualitative wine potential. *Eur. Int. J. Sci. Technol* **2013**, *2*, 126–132.
14. Winter, G.; Henschke, P.A.; Higgins, V.J.; Ugliano, M.; Curtin, C.D. Effects of rehydration nutrients on H2S metabolism and formation of volatile sulfur compounds by the wine yeast VL3. *AMB Expr.* **2011**, *1*, 36. [CrossRef]
15. Smith, M.E.; Bekker, M.Z.; Smith, P.A.; Wilkes, E.N. Sources of volatile sulfur compounds in wine. *Aust. J. Grape Wine Res.* **2015**, *21*, 705–712. [CrossRef]

16. Müller, N.; Rauhut, D.; Tarasov, A. Sulfane sulfur compounds as source of reappearance of reductive off-odors in wine. *Fermentation* **2022**, *8*, 53. [CrossRef]
17. Bekker, M.Z.; Smith, M.E.; Smith, P.A.; Wilkes, E.N. Formation of hydrogen sulfide in wine: Interactions between copper and sulfur dioxide. *Molecules* **2016**, *21*, 1214. [CrossRef]
18. Swiegers, J.H.; Pretorius, I.S. Modulation of volatile sulfur compounds by wine yeast. *Appl. Microbiol. Biotechnol.* **2007**, *74*, 954–960. [CrossRef]
19. Giudici, P.; Kunkee, R.E. The effect of nitrogen deficiency and sulfur-containing amino acids on the reduction of sulfate to hydrogen sulfide by wine yeasts. *Am. J. Enol. Vitic.* **1994**, *45*, 107–112.
20. Jordão, A.M.; Cosme, F. *Grapes and Wines: Advances in Production, Processing, Analysis and Valorization*, 1st ed.; InTech: Rijeka, Croatia, 2018; p. 386. ISBN 978-953-51-3833-4.
21. Bekker, M.Z.; Espinase Nandorfy, D.; Kulcsar, A.C.; Faucon, A.; Bindon, K.; Smith, P.A. Comparison of remediation strategies for decreasing 'Reductive'Characters in shiraz wines. *Aust. J. Grape Wine Res.* **2021**, *27*, 52–65. [CrossRef]
22. Goode, J.; Harrop, S. Wine faults and their prevalence: Data from the world's largest blind tasting. In Proceedings of the 20th Entretiens Scientifiques Lallemand, Horsens, Denmark, 15 May 2008.
23. Rauhut, D. Volatile sulfur compounds: Impact on "Reduced Sulfur" flavor defects and "Atypical Aging" in wine. In Proceedings of the 31st New York Wine Industry Workshop, Geneva, NY, USA, 3–5 April 2002; New York State Agricultural Experiment Station, Cornell University: Geneva, NY, USA, 2002.
24. Siebert, T.E.; Solomon, M.R.; Pollnitz, A.P.; Jeffery, D.W. Selective determination of volatile sulfur compounds in wine by gas chromatography with sulfur chemiluminescence detection. *J. Agric. Food Chem.* **2010**, *58*, 9454–9462. [CrossRef]
25. Müller, N.; Rauhut, D. Recent developments on the origin and nature of reductive sulfurous off-odours in wine. *Fermentation* **2018**, *4*, 62. [CrossRef]
26. Siebert, T.E.; Bramley, B.; Solomon, M.R. Hydrogen sulfide: Aroma detection threshold study in red and white wine. *AWRI Tech.* **2009**, *183*, 14–16.
27. Lu, Y.; Fong, A.S.Y.L.; Chua, J.-Y.; Huang, D.; Lee, P.-R.; Liu, S.-Q. The possible reduction mechanism of volatile sulfur compounds during durian wine fermentation verified in modified buffers. *Molecules* **2018**, *23*, 1456. [CrossRef]
28. Goniak, O.J.; Noble, A.C. Sensory study of selected volatile sulfur compounds in white wine. *Am. J. Enol. Vitic.* **1987**, *38*, 223–227.
29. Mestres, M.; Busto, O.; Guasch, J. Analysis of organic sulfur compounds in wine aroma. *J. Chromatogr.A* **2000**, *881*, 569–581. [CrossRef]
30. Jiranek, V.; Langridge, P.; Henschke, P.A. Validation of bismuth-containing indicator media for predicting H2S-producing potential of saccharomyces cerevisiae wine yeasts under enological conditions. *Am. J. Enol. Vitic.* **1995**, *46*, 269–273.
31. Spiropoulos, A.; Tanaka, J.; Flerianos, I.; Bisson, L.F. Characterization of hydrogen sulfide formation in commercial and natural wine isolates of saccharomyces. *Am. J. Enol. Vitic.* **2000**, *51*, 233–248.
32. Mendes-Ferreira, A.; Mendes-Faia, A.; Leao, C. Survey of hydrogen sulphide production by wine yeasts. *J. Food Prot.* **2002**, *65*, 1033–1037. [CrossRef]
33. Swiegers, J.H.; Bartowsky, E.J.; Henschke, P.A.; Pretorius, I. Yeast and bacterial modulation of wine aroma and flavour. *Aust. J. Grape Wine Res.* **2005**, *11*, 139–173. [CrossRef]
34. Rauhut, D. Yeasts—production of sulfur compounds. In *Wine Microbiology and Biotechnology*; CRC Press: Chur, Switzerland, 1993; pp. 183–223. ISBN 978-0-415 27850-8.
35. Ugliano, M.; Henschke, P.A. Yeasts and wine flavour. In *Wine Chemistry and Biochemistry*; Moreno-Arribas, M.V., Polo, M.C., Eds.; Springer: New York, NY, USA, 2009; pp. 313–392. ISBN 978-0-387-74118-5.
36. Ugliano, M.; Fedrizzi, B.; Siebert, T.; Travis, B.; Magno, F.; Versini, G.; Henschke, P.A. Effect of nitrogen supplementation and saccharomyces species on hydrogen sulfide and other volatile sulfur compounds in shiraz fermentation and wine. *J. Agric. Food Chem.* **2009**, *57*, 4948–4955. [CrossRef] [PubMed]
37. Cordente, A.G.; Curtin, C.D.; Varela, C.; Pretorius, I.S. Flavour-active wine yeasts. *Appl. Microbiol. Biotechnol.* **2012**, *96*, 601–618. [CrossRef] [PubMed]
38. Guidi, I.D.; Farines, V.; Legras, J.-L.; Blondin, B. Development of a new assay for measuring H2S production during alcoholic fermentation: Application to the evaluation of the main factors impacting H2S production by three saccharomyces cerevisiae wine Strains. *Fermentation* **2021**, *7*, 213. [CrossRef]
39. Ferrer-Gallego, R.; Puxeu, M.; Martín, L.; Nart, E.; Hidalgo, C.; Andorrà, I. *Microbiological, physical, and chemical procedures to elaborate high-quality SO2-free wines. Grapes and Wines—Advances in Production, Processing, Analysis and Valorization*; InTech: Rijeka, Croatia, 2018; pp. 171–193.
40. Yamagata, S. Roles of O-Acetyl-l-Homoserine sulfhydrylases in microorganisms. *Biochimie* **1989**, *71*, 1125–1143. [CrossRef]
41. Henschke, P.A.; Jiranek, V. Yeast—Growth during Fermentation. In *Wine Microbiology and Biotechnology*; CRC Press: Boca Raton, FL, USA, 1993; pp. 27–54. ISBN 978-0-415-27850-8.
42. Park, S.K.; Boulton, R.B.; Noble, A.C. Formation of hydrogen sulfide and glutathione during fermentation of white grape musts. *Am. J. Enol. Vitic.* **2000**, *51*, 91–97.
43. Moreira, N.; Mendes, F.; Pereira, O.; Guedes de Pinho, P.; Hogg, T.; Vasconcelos, I. Volatile sulphur compounds in wines related to yeast metabolism and nitrogen composition of grape musts. *Anal. Chim. Acta* **2002**, *458*, 157–167. [CrossRef]

44. Vos, P.J.A.; Gray, R.S. The origin and control of hydrogen sulfide during fermentation of grape must. *Am. J. Enol. Vitic.* **1979**, *30*, 187–197.
45. Henschke, P.A.; Jiranek, V. Hydrogen sulfide formation during fermentation: Effect of nitrogen composition in model grape must. In Proceedings of the International Symposium on Nitrogen in Grapes and Wine, Seattle, WA, USA, 18–19 June 1991; American Society for Enology and Viticulture: Davis, CA, USA, 1991; pp. 172–184, ISBN 0-9630711-0-6.
46. Jiranek, V.; Langridge, P.; Henschke, P.A. Determination of Sulphite reductase activity and its response to assimilable nitrogen status in a commercial saccharomyces cerevisiae wine yeast. *J. Appl. Bacteriol.* **1996**, *81*, 329–336. [CrossRef]
47. Prokes, K.; Baron, M.; Mlcek, J.; Jurikova, T.; Adamkova, A.; Ercisli, S.; Sochor, J. The influence of traditional and immobilized yeast on the amino-acid content of sparkling wine. *Fermentation* **2022**, *8*, 36. [CrossRef]
48. Hallinan, C.P.; Saul, D.J.; Jiranek, V. Differential utilisation of sulfur compounds for H2S liberation by nitrogen-starved wine yeasts. *Aust. J. Grape Wine Res.* **1999**, *5*, 82–90. [CrossRef]
49. Jiranek, V.; Langridge, P.; Henschke, P. Regulation of hydrogen sulfide liberation in wine-producing saccharomyces cerevisiae strains by assimilable nitrogen. *Appl. Environ. Microbiol.* **1995**, *61*, 461–467. [CrossRef]
50. Kinzurik, M.I.; Herbst-Johnstone, M.; Gardner, R.C.; Fedrizzi, B. Hydrogen sulfide production during yeast fermentation causes the accumulation of ethanethiol, s-ethyl thioacetate and diethyl disulfide. *Food Chem.* **2016**, *209*, 341–347. [CrossRef]
51. Franco-Luesma, E.; Ferreira, V. Reductive off-odors in wines: Formation and release of H2S and methanethiol during the accelerated anoxic storage of wines. *Food Chem.* **2016**, *199*, 42–50. [CrossRef]
52. Perpète, P.; Duthoit, O.; De Maeyer, S.; Imray, L.; Lawton, A.I.; Stavropoulos, K.E.; Gitonga, V.W.; Hewlins, M.J.E.; Richard Dickinson, J. Methionine catabolism in saccharomyces cerevisiae. *FEMS Yeast Res.* **2006**, *6*, 48–56. [CrossRef]
53. Gruenwedel, D.W.; Patnaik, R.K. Release of hydrogen sulfide and methyl mercaptan from sulfur-containing amino acids. *J. Agric. Food Chem.* **1971**, *19*, 775–779. [CrossRef]
54. Ugliano, M.; Kwiatkowski, M.; Vidal, S.; Capone, D.; Siebert, T.; Dieval, J.-B.; Aagaard, O.; Waters, E.J. Evolution of 3-mercaptohexanol, hydrogen sulfide, and methyl mercaptan during bottle storage of sauvignon blanc wines. Effect of Glutathione, copper, oxygen exposure, and closure-derived oxygen. *J. Agric. Food Chem.* **2011**, *59*, 2564–2572. [CrossRef]
55. Bekker, M.Z.; Day, M.P.; Holt, H.; Wilkes, E.; Smith, P.A. Effect of oxygen exposure during fermentation on volatile sulfur compounds in S Hiraz wine and a comparison of strategies for remediation of reductive character. *Aust. J. Grape Wine Res.* **2016**, *22*, 24–35. [CrossRef]
56. Nguyen, D.-D.; Nicolau, L.; Dykes, S.I.; Kilmartin, P.A. Influence of microoxygenation on reductive sulfur off-odors and color development in a cabernet sauvignon wine. *Am. J. Enol. Vitic.* **2010**, *61*, 457–464. [CrossRef]
57. Guitart, A.; Orte, P.H.; Ferreira, V.; Peña, C.; Cacho, J. Some observations about the correlation between the amino acid content of musts and wines of the chardonnay variety and their fermentation aromas. *Am. J. Enol. Vitic.* **1999**, *50*, 253–258.
58. Kreitman, G.Y.; Danilewicz, J.C.; Jeffery, D.W.; Elias, R.J. Copper(II)-Mediated hydrogen sulfide and thiol oxidation to disulfides and organic polysulfanes and their reductive cleavage in wine: Mechanistic elucidation and potential applications. *J. Agric. Food Chem.* **2017**, *65*, 2564–2571. [CrossRef]
59. Kreitman, G.Y.; Elias, R.J.; Jeffery, D.W.; Sacks, G.L. Loss and formation of malodorous volatile sulfhydryl compounds during wine storage. *Crit. Rev. Food Sci. Nut.* **2019**, *59*, 1728–1752. [CrossRef]
60. Vela, E.; Hernández-Orte, P.; Franco-Luesma, E.; Ferreira, V. The effects of copper fining on the wine content in sulfur off-odors and on their evolution during accelerated anoxic storage. *Food Chem.* **2017**, *231*, 212–221. [CrossRef]
61. International Code of Oenological Practices. Available online: https://www.oiv.int/public/medias/3570/e-code-ii-358.pdf (accessed on 10 March 2022).
62. Franco-Luesma, E.; Ferreira, V. Formation and release of H2S, methanethiol, and dimethylsulfide during the anoxic storage of wines at room temperature. *J. Agric. Food Chem.* **2016**, *64*, 6317–6326. [CrossRef] [PubMed]
63. Huang, X.; Lu, Q.; Hao, H.; Wei, Q.; Shi, B.; Yu, J.; Wang, C.; Wang, Y. Evaluation of the treatability of various odor compounds by powdered activated carbon. *Water Res.* **2019**, *156*, 414–424. [CrossRef] [PubMed]
64. Ailer, Š. Vplyv Aplikácie Kontaktných Listových Hnojív Na Obsah hygienicky významných chemických prvkov v hroznovom mušte a víne. *Bull. Food Res.* **1999**, *38*, 183–193.
65. Dias, D.; Guarda, A.; Wiethan, B.A.; Claussen, L.E.; Bohrer, D.; De Carvalho, L.M.; Do Nascimento, P.C. Influence of ethanethiol in antioxidant activity and in total phenolics concentration of wines. Comparative study against control samples. *J. Food Qual.* **2013**, *36*, 432–440. [CrossRef]
66. Allison, R.B. Hydrogen Sulfide Development In Wine During Anoxic Storage. Ph.D. Thesis, Cornell University, Ithaca, NY, USA, 2021. [CrossRef]
67. Ribéreau-Gayon, P.; Dubourdieu, D.; Donèche, B.; Lonvaud, A. *Handbook of Enology, Volume 1: The Microbiology of Wine and Vinifications*; John Wiley & Sons: Hoboken, NJ, USA, 2006; p. 512. ISBN 978-0-470-01035-8.
68. Kraft, D.N. Impact of Lees Content, Nitrogen, and Elemental Sulfur on Volatile Sulfur Compound Formation in Vitis Vinifera L. cv. "Pinot noir" Wine. Master's Thesis, Oregon State University, Corvallis, OR, USA, 2015.
69. Goode, J. *Flawless: Understanding Faults in Wine*; University of California Press: Berkely, CA, USA, 2018; p. 237. ISBN 978-0-520-97131-8.
70. Viviers, M.Z.; Smith, M.E.; Wilkes, E.; Smith, P. Effects of five metals on the evolution of hydrogen sulfide, methanethiol, and dimethyl sulfide during anaerobic storage of chardonnay and shiraz wines. *J. Agric. Food Chem.* **2013**, *61*, 12385–12396. [CrossRef]

71. Clark, A.C.; Grant-Preece, P.; Cleghorn, N.; Scollary, G.R. Copper(II) addition to white wines containing hydrogen sulfide: Residual copper concentration and activity. *Aust. J. Grape Wine Res.* **2015**, *21*, 30–39. [CrossRef]
72. Bekker, M.Z.; Nandorfy, D.E.; Kulcsar, A.C.; Faucon, A.; Smith, P.A. Remediating 'Reductive'Characters in wine. In Proceedings of the 17th Australian Wine Industry Technical Conference, Adelaide, Australia, 21–24 July 2019; 2020; p. 97.
73. Pokrỳvková, J.; Ailer, Š.; Jedlička, J.; Chlebo, P.; Jurík, L. The use of a targeted must oxygenation method in the process of developing the archival potential of natural wine. *Appl. Sci.* **2020**, *10*, 4810. [CrossRef]
74. European Commission. Commission Delegated Regulation (EU) 2019/934—of 12 March 2019—Supplementing Regulation (EU) No 1308 /2013 of the European Parliament and of the Council as Regards Wine-Growing Areas Where the Alcoholic Strength May Be Increased, Authorised Oenological Practices and Restrictions Applicable to the Production and Conservation of Grapevine Products, the Minimum Percentage of Alcohol for by-Products and Their Disposal, and Publication of OIV Files. 52. *Off. J. Eur. Union L.* **2019**, *149*, 1–52.
75. Singleton, V.L.; Kramlinga, T.E. Browning of white wines and an accelerated test for browning capacity. *Am. J. Enol. Vitic.* **1976**, *27*, 157–160.
76. Salacha, M.-I.; Kallithraka, S.; Tzourou, I. Browning of white wines: Correlation with antioxidant characteristics, total polyphenolic composition and flavanol content. *IJFST* **2008**, *43*, 1073–1077. [CrossRef]
77. Karbowiak, T.; Mansfield, A.K.; Barrera-García, V.D.; Chassagne, D. Sorption and diffusion properties of volatile phenols into cork. *Food Chem.* **2010**, *122*, 1089–1094. [CrossRef]
78. Tarko, T.; Duda-Chodak, A.; Sroka, P.; Siuta, M. The impact of oxygen at various stages of vinification on the chemical composition and the antioxidant and sensory properties of white and red wines. *Int. J. Food Sci.* **2020**, *2020*, 7902974. [CrossRef]
79. Makhotkina, O.; Kilmartin, P.A. Uncovering the influence of antioxidants on polyphenol oxidation in wines using an electrochemical method: Cyclic voltammetry. *J. Electroanal. Chem.* **2009**, *633*, 165–174. [CrossRef]
80. Kilmartin, P.A. The oxidation of red and white wines and its impact on wine aroma. *Chem. New Zealand* **2009**, *73*, 79–83.
81. Michlovský, M. *Bobule*; Vinselekt Michlovský a.s.: Lučni, Rakvice, Czech, 2014.
82. Oliveira, C.M.; Ferreira, A.C.S.; De Freitas, V.; Silva, A.M. Oxidation mechanisms occurring in wines. *Food Res. Int.* **2011**, *44*, 1115–1126. [CrossRef]
83. Milat, A.M.; Boban, M.; Teissedre, P.-L.; Šešelja-Perišin, A.; Jurić, D.; Skroza, D.; Generalić-Mekinić, I.; Ljubenkov, I.; Volarević, J.; Rasines-Perea, Z. Effects of oxidation and browning of macerated white wine on its antioxidant and direct vasodilatory activity. *J. Funct. Foods* **2019**, *59*, 138–147. [CrossRef]
84. Ivanova, V.; Vojnoski, B.; Stefova, M. Effect of winemaking treatment and wine aging on phenolic content in vranec wines. *J. Food Sci. Technol.* **2012**, *49*, 161–172. [CrossRef] [PubMed]
85. Ivanova, V.; Dörnyei, Á.; Márk, L.; Vojnoski, B.; Stafilov, T.; Stefova, M.; Kilár, F. Polyphenolic content of vranec wines produced by different vinification conditions. *Food Chem.* **2011**, *124*, 316–325. [CrossRef]
86. Sam, F.E.; Ma, T.-Z.; Salifu, R.; Wang, J.; Jiang, Y.-M.; Zhang, B.; Han, S.-Y. Techniques for dealcoholization of wines: Their impact on wine phenolic composition, volatile composition and sensory characteristics. *Foods* **2021**, *10*, 2498. [CrossRef]
87. Gutiérrez-Escobar, R.; Aliaño-González, M.J.; Cantos-Villar, E. Wine polyphenol content and its influence on wine quality and properties: A review. *Molecules* **2021**, *26*, 718. [CrossRef]
88. Deshaies, S.; Cazals, G.; Enjalbal, C.; Constantin, T.; Garcia, F.; Mouls, L.; Saucier, C. Red wine oxidation: Accelerated ageing tests, possible reaction mechanisms and application to syrah red wines. *Antioxidants* **2020**, *9*, 663. [CrossRef]
89. Vlahou, E.; Christofi, S.; Roussis, I.G.; Kallithraka, S. Browning development and antioxidant compounds in white wines after selenium, iron, and peroxide addition. *Appl. Sci.* **2022**, *12*, 3834. [CrossRef]
90. Cacho, J.; Castells, J.E.; Esteban, A.; Laguna, B.; Sagristá, N. Iron, copper, and manganese influence on wine oxidation. *Am. J. Enol. Vitic.* **1995**, *46*, 380–384.
91. Moreno, J.; Peinado, R. *Enological Chemistry*; Academic Press: Cambridge, MA, USA, 2012; p. 443. ISBN 978-0-12-388439-8.
92. Li, H.; Guo, A.; Wang, H. Mechanisms of Oxidative Browning of Wine. *Food Chem.* **2008**, *108*, 1–13. [CrossRef]
93. Cheynier, V.F.; Van Hulst, M.W. Oxidation of trans-caftaric acid and 2-S-glutathionylcaftaric acid in model solutions. *J. Agric. Food Chem.* **1988**, *36*, 10–15. [CrossRef]
94. Ünal, M.Ü.; Şener, A. Correlation between browning degree and composition of importantturkish white wine grape varieties. *Turk. J. Agric. For.* **2016**, *40*, 62–67. [CrossRef]
95. Sartor, S.; Burin, V.M.; Ferreira-Lima, N.E.; Caliari, V.; Bordignon-Luiz, M.T. Polyphenolic profiling, browning, and glutathione content of sparkling wines produced with nontraditional grape varieties: Indicator of quality during the biological aging. *J. Food Sci.* **2019**, *84*, 3546–3554. [CrossRef]
96. Bradshaw, M.P.; Prenzler, P.D.; Scollary, G.R. Ascorbic acid-induced browning of (+)-catechin in a model wine system. *J. Agric. Food Chem.* **2001**, *49*, 934–939. [CrossRef]
97. Serra-Cayuela, A.; Jourdes, M.; Riu-Aumatell, M.; Buxaderas, S.; Teissedre, P.-L.; López-Tamames, E. Kinetics of browning, phenolics, and 5-hydroxymethylfurfural in commercial sparkling wines. *J. Agric. Food Chem.* **2014**, *62*, 1159–1166. [CrossRef]
98. Ailer, Š.; Serenčéš, R.; Kozelová, D.; Poláková, Z.; Jakabová, S. Possibilities for depleting the content of undesirable volatile phenolic compounds in white wine with the use of low-intervention and economically efficient grape processing technology. *Appl. Sci.* **2021**, *11*, 6735. [CrossRef]
99. Baroň, M. Možnosti snížení obsahu oxidu siřičitého ve víne. Část 1. *Vinařský Obzor.* **2013**, *7–8*, 406–409.

100. Baroň, M. Možnosti snížení obsahu oxidu siřičitého ve víne. Část 2. *Vinařský Obzor.* **2013**, *9*, 456–458.
101. Baroň, M. Možnosti snížení obsahu oxidu siřičitého ve víne. Část 3. *Vinařský Obzor.* **2013**, *10*, 508–511.
102. Day, M.P.; Schmidt, S.A.; Pearson, W.; Kolouchova, R.; Smith, P.A. Effect of passive oxygen exposure during pressing and handling on the chemical and sensory attributes of chardonnay wine. *Aust. J. Grape Wine Res.* **2019**, *25*, 185–200. [CrossRef]
103. Roure, F.; Anderson, G. Characteristics of oenotannins. In *The Australian & New Zealand Grapegrower and Winemaker*; Winetitles Media: Broadview, Australia, 2006; p. 48.
104. Nunes, P.; Muxagata, S.; Correia, A.C.; Nunes, F.M.; Cosme, F.; Jordão, A.M. Effect of oak wood barrel capacity and utilization time on phenolic and sensorial profile evolution of an encruzado white wine. *J. Sci. Food Agric.* **2017**, *97*, 4847–4856. [CrossRef]
105. Renaud, S.; Gueguen, R. The french paradox and wine drinking. *Novartis Found. Symp.* **1998**, *216*, 208–217.
106. Vidavalur, R.; Otani, H.; Singal, P.K.; Maulik, N. Significance of wine and resveratrol in cardiovascular disease: French Paradox revisited. *Exp. Clin. Cardiol.* **2006**, *11*, 217.
107. Razmkhab, S.; Lopez-Toledano, A.; Ortega, J.M.; Mayen, M.; Merida, J.; Medina, M. Adsorption of phenolic compounds and browning products in white wines by yeasts and their cell walls. *J. Agric. Food Chem.* **2002**, *50*, 7432–7437. [CrossRef]
108. Gil, M.; Avila-Salas, F.; Santos, L.S.; Iturmendi, N.; Moine, V.; Cheynier, V.; Saucier, C. Rosé wine fining using polyvinylpolypyrrolidone: Colorimetry, targeted polyphenomics, and molecular dynamics simulations. *J. Agric. Food Chem.* **2017**, *65*, 10591–10597. [CrossRef]
109. Gil, M.; Louazil, P.; Iturmendi, N.; Moine, V.; Cheynier, V.; Saucier, C. Effect of polyvinylpolypyrrolidone treatment on rosés wines during fermentation: Impact on color, polyphenols and thiol aromas. *Food Chem.* **2019**, *295*, 493–498. [CrossRef]
110. Bonilla, F.; Mayen, M.; Merida, J.; Medina, M. Yeasts used as fining treatment to correct browning in white wines. *J. Agric. Food Chem.* **2001**, *49*, 1928–1933. [CrossRef]
111. Alpeza, I.; Linke, I.; Maslov, L.; Jeromel, A. Atypical aging off-flavour and relation between sensory recognition and 2-aminoacetophenone in croatian wines. *J. Cent. Eur. Agric.* **2021**, *22*, 408–419. [CrossRef]
112. Christoph, N.; Bauer-Christoph, C.; Gessner, M.; Köhler, H.J. Die "Untypische Alterungsnote Im Wein", Teil I: Untersuchungen zum auftreten und zur sensorischen charakterisierung der "untypischen alterungsnote". *Rebe Wein* **1995**, *48*, 350–356.
113. Rapp, A.; Versini, G.; Ullemeyer, H. 2-Aminoacetophenone-causal component of untypical aging flavor (Naphthalene Note, Hybrid Note) of wine. *Vitis* **1993**, *32*, 61–62.
114. Schneider, V. UTA-Ein weltweites problem. *Der Winzer* **2013**, *7*, 24–28.
115. Linsenmeier, A.; Pfliehinger, M. Wirkungskette stress–Das Ungeklärte UTA-Rätsel. *Deut. Weinbau* **2009**, *24*, 18–21.
116. Linsenmeier, A.W.; Rauhut, D.; Sponholz, W.R. Aging and flavor deterioration in wine. In *Managing Wine Quality*; Elsevier: Amsterdam, The Netherlands, 2022; pp. 559–594.
117. Ailer, Š.; Valšíková, M.; Jedlička, J.; Mankovecký, J.; Baroň, M. Influence of sugar and ethanol content and color of wines on the sensory evaluation: From wine competition "Nemčiňany Wine Days" in Slovak Republic (2013–2016). *Erwerbs-Obstbau* **2020**, *62*, 9–16. [CrossRef]
118. Schneider, V. Atypical aging defect: Sensory discrimination, viticultural causes, and enological consequences. A review. *Am. J. Enol. Vitic.* **2014**, *65*, 277–284. [CrossRef]
119. Fischer, U.; Sponholz, R. Die sensorische beschreibung der untypischen alterungsnote. *Der Dtsch. Weinbau* **2000**, *3*, 16–21.
120. Henick-Kling, T.; Gerling, C.; Martinson, T.; Acree, T.; Lakso, A. Studies on the origin and sensory aspects of atypical aging in white wines. In Proceedings of the 15th International Enology Symposium; International Association of Enology, Management and Wine Marketing, Trier, Germany, 16 April 2008; p. 10.
121. Gessner, M.; Köhler, H.J.; Christoph, N. Die "Untypische Alterungsnote" in Wein, Teil VIII: Auswirkungen von Inhaltsstoffen Und Antioxidantien Auf Die Bildung von o-Aminoacetophenon. *Rebe Wein* **1999**, *51*, 264–267.
122. Christoph, N.; Bauer-Christoph, C.; Geßner, M.; Köhler, H.J.; Simat, T.; Hoenicke, K. Formation of 2-aminoacetophenone and formylaminoacetophenone in wine by reaction of sulfurous acid with indole-3-acetic acid. *Vitic. Enol. Sci.* **1998**, *53*, 79–86.
123. Linsenmeier, A.; Rauhut, D.; Kurbel, H.; Lohnertz, O.; Schubert, S. Untypical ageing off-flavour and masking effects due to long-term nitrogen fertilization. *Vitis-Geilweilerhof* **2007**, *46*, 33.
124. Maslov, L.; Jeromel, A.; Herjavec, S.; Korenika, A.-M.J.; Mihaljević, Ž.; Plavša, T. Indole-3-acetic acid and tryptophan in istrian malvasia grapes and wine. *J. Food Agric. Environ.* **2011**, *9*, 132–136.
125. Horlacher, N.; Link, K.; Schwack, W. Bacillus thuringiensis insecticides: Source of 2-aminoacetophenone formation in wine. *J. Exp. Food Chem.* **2016**, *2*, 113. [CrossRef]
126. Nardin, T.; Roman, T.; Dekker, S.; Nicolini, G.; Thei, F.; Masina, B.; Larcher, R. Evaluation of antioxidant supplementation in must on the development and potential reduction of different compounds involved in atypical ageing of wine using HPLC-HRMS. *LWT* **2022**, *154*, 112639. [CrossRef]
127. Facchini, P.J.; Huber-Allanach, K.L.; Tari, L.W. Plant Aromatic L-Amino acid decarboxylases: Evolution, biochemistry, regulation, and metabolic engineering applications. *Phytochemistry* **2000**, *54*, 121–138. [CrossRef]
128. Zhao, Y. Auxin biosynthesis: A simple two-step pathway converts tryptophan to indole-3-acetic acid in plants. *Mol. Plant* **2012**, *5*, 334–338. [CrossRef]
129. Hoenicke, K.; Simat, T.J.; Steinhart, H.; Christoph, N.; Geßner, M.; Köhler, H.-J. 'Untypical Aging off-Flavor' in Wine: Formation of 2-aminoacetophenone and evaluation of its influencing factors. *Anal. Chim. Acta* **2002**, *458*, 29–37. [CrossRef]

130. Linsenmeier, A.; Rauhut, D.; Kürbel, H.; Schubert, S.; Löhnertz, O. Ambivalence of the influence of nitrogen supply on O-aminoacetophenone in riesling wine. *Vitis* **2007**, *46*, 91–97.
131. Gessner, M.; Köhler, H.J.; Christoph, N.; Nagel-Derr, A.; Krell, U. Die "Untypische Alterungsnote" in Wein, Teil IX: Würzburger UTAFIX-Test: Ein einfaches diagnoseverfahren zur früherkennung von weinen Mit UTA-Neigung. *Rebe Wein* **1999**, *52*, 296–303.

Article

The Management of Dissolved Oxygen by a Polypropylene Hollow Fiber Membrane Contactor Affects Wine Aging

Francesco Errichiello, Luigi Picariello, Antonio Guerriero, Luigi Moio, Martino Forino and Angelita Gambuti *

Department of Agricultural Sciences, Section of Vine and Wine Sciences, University of Naples 'Federico II', Viale Italia, 83100 Avellino, Italy; francesco.errichiello@unina.it (F.E.); luigi.picariello@unina.it (L.P.); a.guerriero1385@gmail.com (A.G.); luigi.moio@unina.it (L.M.); forino@unina.it (M.F.)
* Correspondence: angelita.gambuti@unina.it; Tel.: +39-081-2532-605

Abstract: Background: Numerous oenological practices can cause an excess of dissolved oxygen in wine, thus determining sensory and chromatic defects in the short- to long-term. Hence, it is necessary to manage the excess of oxygen before bottling. Methods: In this study, the management of the dissolved oxygen content by a polypropylene hollow fiber membrane contactor apparatus was performed in two wines from different grape varieties (Aglianico and Falanghina). The wines were analyzed after an 11-month aging. Anthocyanins and acetaldehyde content were evaluated by HPLC. In addition, other phenolic compounds and chromatic characteristics were analyzed by spectrophotometric methods. NMR and HR ESIMS analyses were conducted to evaluate the amount of pyranoanthocyanins and polymeric pigments. Results: After 11 months of aging, in both wines a decrease of free and total SO_2 with respect to initial values was detected. In the wines with the highest dissolved oxygen levels, a more remarkable loss was observed. No significant differences in terms of color parameters were detected. In red wine with the highest oxygen content, a massive formation of polymeric pigments and BSA reactive tannins was observed, as opposed to wines with lower oxygen levels. Conclusion: The study demonstrated that the membrane contactor can prove a successful tool to manage dissolved oxygen in wines as to prevent their oxidative spoilage.

Keywords: oxidation; membrane contactor; wine; polymeric pigments

1. Introduction

Wine is a chemically dynamic system, and even after fermentation its composition continues to evolve during the storage. These post-fermentation changes are referred to as aging, but a distinction is to be made between the changes occurring during the maturation phase (e.g., bulk storage of wine in tank or barrel), when the winemaker's intervention is still allowed, and those taking place during the aging phase once wine has been sealed in bottles and the only possible intervention is limited to the selection of the most appropriate storage conditions.

Among wine compounds, phenolics are those mostly affected by aging. They originate from grapes (flavonoids and non-flavonoids) and constitute one of the most important wine quality parameters. During winemaking and aging, phenolics mainly undergo oxidation reactions, which not only affect the phenolic composition itself but also determine changes in terms of sensory characteristics, such as color and astringency.

Phenolic compounds are the primary reactants to be oxidized in presence of oxygen and metals (Fe^{3+}, Cu^{2+}), giving rise to a cascade of chemical transformations that may result in an excessive deterioration of wine [1]. Wine oxidation consists of a series of reactions: first, oxygen is reduced to hydrogen peroxide by interacting with transition metals, including iron and copper ions, in the presence of catechol subunits that are oxidized to quinones [2]. Quinones strongly react with nucleophilic compounds, such as antioxidants (sulfur dioxide, glutathione, ascorbic acid), desirable aroma volatile thiols (i.e., 3-sulfanylhexanol), undesirable aroma thiols (i.e., hydrogen sulfide), amino acids (i.e., phenylalanine, methionine) and numerous polyphenols (mainly flavanols). The products of these reactions may lead to the formation of condensed polymeric pigments—particularly important in red wines—or even to the loss of color and varietal characters [3]. In a subsequent step, ferrous or cuprous species react with hydrogen peroxide by the Fenton reaction to yield the hydroxyl radical, a strong oxidant, capable of reacting with all organic constituents in proportion to concentration [4]. The most abundant organic compound in wine is ethanol, which is converted into acetaldehyde once oxidized by the hydroxyl radical.

As a consequence of oxidation, in red wine native anthocyanin pigments are quickly transformed into more stable pigments via various types of reactions such as aldehyde-mediated condensation reactions with tannins and cyclo-addition reactions leading to the formation of pyranoanthocyanins [5]. Oxidation reactions also contribute to modifying the wine astringency by changing the tannin structure as a consequence of intra and inter molecular reactions mediated by oxygen [6]. These "stabilized products" anthocyanin or pigmented tannins persist much longer in wine than their initial forms [7]. Thus, low amounts of oxygen in red wine are important to stabilize either color or astringency. Pasteur himself, in his studies on wine, theorized that only when a wine is exposed to oxygen can it develop attributes that make it a finely aged high-quality product. During winemaking, oxygen plays a crucial role in the fermentation process. It promotes the yeast biomass synthesis and favors a sound fermentation. Several studies have shown that the risk of stuck and sluggish fermentations is reduced after oxygen additions of 10–20 mg/L [8].

In white wines oxidation is usually associated with important changes in color. A brown color is normally unwanted, because it is a sign of oxidation in table white wine. Brown coloration can be induced by enzymatic or chemical oxidation mediated by oxygen. The latter is slower than the enzymatic-induced oxidation. White wine is generally more sensible to O_2 than red wine. Even small additions of O_2 to white wine can lead to loss of aroma, especially fruitiness with the appearance of off-flavors described as caramel, rancid, farmed-feed, honey-like and cooked vegetables. The quinones generated from oxidation can react with thiols by the Michael addition reaction or generate H_2O_2, as reported above. The oxidative environment through all the phases of winemaking is positively correlated to the formation of these products. During the aging of red wine in oak barrel, the oxidative process also induces the formation of sotolone through the oxidation of threonine or by the reaction of acetaldehyde with α-ketobutyric acid. The oxidative degradation of phenylalanine and β-phenylethanol in a barrel also leads to higher concentrations of phenylacetaldehyde [9].

As described above, oxygen can have either beneficial or detrimental effects on the wine quality. The level of oxygen exposure of wine during winemaking or aging is crucial as it can affect the final product. Singleton [10] estimated the amount of oxygen that a white or a red wine could absorb before oxidative defects emerge. In white wine, the toleration is around 10 air saturations as opposed to red wine, which can tolerate more than 30 air saturations (180 mL O_2/L). He also recommended about 10 saturations to improve the quality of red wine.

Oxygen management during the phases of vinification and storage is therefore important and must be handled according to the knowledge acquired as to avoid the insurgence of oxidative characters. After the winemaking, wine usually undergoes a series of stabilization practices such as decanting, refrigeration and filtration that can determine an uncontrolled oxygen inlet. Additionally, new vinification methods that use stainless steel tanks and systems allowing a controlled oxygen micro-supply (micro-oxygenation) are now common. Even when specific efforts have been made to produce wines, which are as resistant as possible to further oxygen intakes, all the uncontrolled dissolved oxygen in wine can determine a further development of oxidize characters once bottled. In this context, the use of a membrane contactor to manage oxygen in wine before bottling might be a successful strategy in order to obtain the best possible wines.

Membrane contactors are among the most used industrial systems, and this technology has been proven useful in a range of liquid/liquid and gas/liquid applications in fermentation, pharmaceuticals, wastewater treatment, chiral separations, semiconductor manufacturing, carbonation of beverages, metal ion extraction, protein extraction, VOC removal from waste gas, osmotic distillation and wine dealcoholization [11,12]. It is a device that achieves gas/liquid or liquid/liquid mass transfer without dispersion of one phase within another. Although membrane contactor technology was introduced as a tool for gas management in wines [13,14], until now few studies have dealt with the effectiveness of its application before bottling to regulate the evolution of white and red wine during bottle aging.

In this study, a partial deoxygenation was performed on two monovarietal wines: Aglianico and Falanghina. The effect of oxygen removal on several wine parameters as free and bound SO_2 and acetaldehyde content was evaluated after 11 months of bottle aging. For red wine, the effect on chromatic characteristics and main phenolic compounds was also evaluated.

2. Results and Discussion

In this study two commercial wines (Aglianico (R) and Falanghina (W)) were submitted to a deoxygenation process by using membrane contactor technology to obtain three wines with decreasing levels of dissolved oxygen (high (H), medium (M) and low (L) for each wine) before the bottling phase. After 11 months of bottle aging, the effects on sulfur dioxide, acetaldehyde, chromatic characters, polymeric pigments, VRF, BSA-tannins and total phenolics were evaluated.

2.1. Sulfur Dioxide

The concentration of free and total SO_2 was monitored after 11 months of aging in the Aglianico wine (Table 1). For all samples, a loss of SO_2 with respect to bottling time (free SO_2 18 mg/L, tot SO_2 43 mg/L) was observed and the greatest decline in the values of total SO_2 was observed in the wines with higher oxygen contents at bottling (RHO_2 and RMO_2), as expected.

Table 1. Evolution of free sulfur dioxide and total sulfur dioxide after 11 months of aging of treated red Aglianico and white Falanghina wines.

	Aglianico		
	RHO_2	RMO_2	RLO_2
Free SO_2	1.28 ± 0.00 B	1.60 ± 0.45 B	3.84 ± 0.00 A
Total SO_2	33.92 ± 0.00 B	36.16 ± 0.45 AB	37.44 ± 1.36 A
	Falanghina		
	WHO_2	WMO_2	WLO_2
Free SO_2	4.80 ± 0.45 B	7.36 ± 0.45 A	8.64 ± 0.45 A
Total SO_2	67.2 ± 0.91 B	79.68 ± 0.45 A	78.08 ± 0.91 A

RHO_2 (red high oxygen), RMO_2 (red medium oxygen), RLO_2 (red low oxygen), WHO_2 (white high oxygen), WMO_2 (white medium oxygen), WLO_2 (white low oxygen). Different letters indicate a statistically significant difference among treated wines. All the data are expresses as means (mg/L) ± standard deviation, ($p < 0.05$).

During the aging process, the most abundant forms of free SO_2 at the pH of the wine, the bisulfite ion (HSO_3^- in equilibrium with molecular SO_2), is consumed by the reaction with hydrogen peroxide and several electrophilic wine components, such as those produced by the oxidation cascade, including quinone and acetaldehyde [15]. In winery settings, it is common practice measuring the so-called "free SO_2" that is the sum of molecular SO_2 and bisulfite ion. This latter compound can form covalent adducts with electrophiles, called SO_2 binders, that can be classified as weak or strong based on the dissociation constant of the sulfite adducts formed. These sulfite-adducts are called bound SO_2 and the sum of free and bound SO$_2$ gives the "total SO$_2$". The free and bound SO$_2$ are in equilibrium with each other in wine. Moreover, during oxidation, by the equilibrium between these two forms of SO$_2$ and by the consumption of free SO$_2$, the combined form is released to restore the equilibrium. In all treated red wines, values of free SO$_2$ below 3.84 mg/L were detected after 11 months of bottle aging. These values are well below the quantification limit of official methods of analysis of free SO$_2$ [16]; therefore, it is more correct to consider negligeable the value of free SO$_2$ while the values of total SO$_2$ were lower in samples bottled with higher content of dissolved oxygen (RHO$_2$). The shift between free and combined during wine aging might be the reason why decreasing levels of total SO$_2$ were detected in wines as dissolved oxygen at bottling increased. Apart from the consumption of SO$_2$ owing to oxidation reactions, part of it can be also lost during the aging due to the reactions of sulfur dioxide with flavanols. The mechanism of formation of 4ß-sulfonated products is still uncertain. It is hypothesized that monomeric 4ß-sulfonated derivatives are formed by the acid-catalyzed depolymerization of proanthocyanidins [17].

The concentrations of free and total SO$_2$ (Table 1) were monitored after 11 months of aging in white wine whose behavior turned out to be the same as that in red wine. Further, in this case a loss of free and total SO$_2$ with respect to the initial values was detected (free SO$_2$ 26 mg/L and tot. SO$_2$ 89 mg/L). In addition, wines with higher content of oxygen at bottling (WHO$_2$) showed a lower content of free and total SO$_2$.

Acetaldehyde, formed by the metal-catalyzed oxidation of ethanol during wine oxidation, was higher at 11 months of aging in sample WHO$_2$ with respect to sample WMO$_2$, and WLO$_2$ as expected given the lower content of free and total SO$_2$ in WHO$_2$ samples (Table 2). Because the weight ratio between acetaldehyde and sulfur dioxide is 1.4/1 (1.4 mg of SO$_2$ consumed per 1 mg of CH$_3$CHO), we could assess that the amount of acetaldehyde in sample WHO$_2$ is totally combined with SO$_2$ (50 mg of acetaldehyde combines with 70 mg of SO$_2$) and considering the negligible levels of free sulfur dioxide after 11 months of aging, it is expected that further exposures to oxygen may lead to the appearance of free acetaldehyde.

Table 2. Evolution of acetaldehyde after 11 months of aging of treated Falanghina white wines.

	Falanghina
	Acetaldehyde (mg/L)
WHO$_2$	45.85 ± 0.11 A
WMO$_2$	44.82 ± 0.43 B
WLO$_2$	44.60 ± 0.15 B

WHO$_2$ (white high oxygen), WMO$_2$ (white medium oxygen), WLO$_2$ (white low oxygen). All the data are expressed as means (mg/L) ± standard deviation. Different letters indicate a statistically significant difference among treated wines. All the data are expresses as means ± standard deviation, ($p < 0.05$).

In red wines, acetaldehyde concentration does not differ among the analyzed samples. This is probably due to the fact that, in presence of anthocyanins and higher concentration of flavanols, acetaldehyde is involved in a number of reactions with these phenolics during the aging. As discussed below, the most important reaction involving acetaldehyde, anthocyanins and flavanols is the formation of ethyl-bridged compounds [18,19] and ethyl-linked oligomers, which can further react with additional acetaldehyde, anthocyanins, and flavanols to generate a pyran ring, or other polymeric-type structures. Ultimately, these products can alter the wine sensory attributes [20] by affecting some key wine characteristics such as color, flavor and astringency.

2.2. Effect on Pigments and Chromatic Characters

Data on the content of monomeric anthocyanins in treated red wines after 11 months of bottle aging showed (Table 3 and Supplementary Materials Figure S1) a loss of Malvidin-3-glucoside in the RHO$_2$ sample with a higher concentration of oxygen than in RMO$_2$ and RLO$_2$. Consistently, differences in terms of total anthocyanins were detected among wines (Figure 1). Wine with a low concentration of oxygen at bottling showed a higher concentration of total native anthocyanins compared to the ones with higher concentrations of dissolved oxygen. The effect of the membrane contactor treatment on various pigment classes, determined by the Harbertson method, included an expected significant low concentration in SPP (short polymeric pigments) in samples RLO_2 compared to the samples RHO_2 and RMO_2 that showed the highest increase of these important stable compounds (Table 4). LPP (long polymeric pigments) were not significantly different in all samples. Polymeric pigments (SPP and LPP) are defined as pigments resistant to bisulfite bleaching. They are formed by the reaction between anthocyanins and tannins during the wine aging [21], leading to a stabilization of the color over time. The main difference between these two classes of pigments is that, unlike SPP, LPP tend to precipitate with protein [22]. As red wine ages, a greater formation of LPP compared to SPP is usually observed. Thus, the changes detected for SPP and not for LPP can reflect an early oxidative state of Aglianico wines after 11 months of aging. The involvement of native anthocyanins in reactions yielding new polymeric pigments is consistent with the decrease of total native anthocyanins shown in (Table 4), and with similar effects observed in red wines during micro-oxygenation [23].

Table 3. Native anthocyanins.

	Aglianico		
	RHO_2	RMO_2	RLO_2
Delf-3-gl	9.62 ± 0.02 A	9.98 ± 0.28 A	10.26 ± 0.40 A
Petu-3-gl	9.36 ± 4.47 A	13.40 ± 0.34 A	14.15 ± 0.47 A
Peon-3-gl	2.43 ± 0.26 A	2.66 ± 0.25 A	2.86 ± 0.22 A
Malv3-gl	56.94 ± 0.18 B	59.08 ± 0.46 AB	60.64 ± 1.54 A
Mal-3-Acgl	8.74 ± 0.05 A	9.29 ± 0.07 A	9.47 ± 0.38 A
Mal-3-Cumgl	2.44 ± 0.11 A	2.58 ± 0.20 A	3.08 ± 0.48 A

RHO_2 (red high oxygen), RMO_2 (red medium oxygen), RLO_2 (red low oxygen). Delf-3-gl (Delphinidin-3-glucoside), Cyan-3-gl (Cyanidin-3-glucoside), Petu-3-gl (Petunidin-3-glucoside), Peon-3-gl (Peonidin-3-glucoside), Malv-3-gl (Malvidin-3-glucoside), Mal-3-Acgl (Malvidin-3-acetyglucoside), Mal-3-Cumgl (Malvidin-3-coumarylglucoside). All the data are expressed as means (mg/L) ± standard deviation. Different letters indicate a statistically significant difference among treated wines. All the data are expresses as means ± standard deviation, ($p < 0.05$).

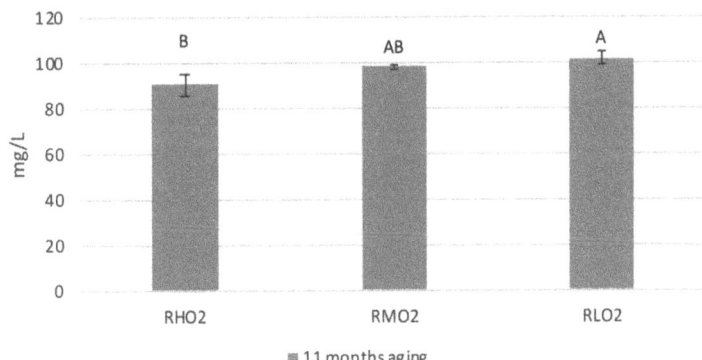

Figure 1. Total native anthocyanin. RHO_2 (red high oxygen), RMO_2 (red medium oxygen), RLO_2 (red low oxygen). All the data are expressed as means ± standard deviation. Different letters indicate a statistically significant difference among treated wines. All the data are expressed as means ± standard deviation, ($p < 0.05$).

Table 4. Polymeric pigments (SPP) Short polymeric pigments and (LPP) Long polymeric pigments.

	Aglianico		
	RHO$_2$	RMO$_2$	RLO$_2$
SPP	0.68 ± 0.02 A	0.65 ± 0.01 AB	0.64 ± 0.00 B
LPP	0.52 ± 0.03 A	0.52 ± 0.01 A	0.50 ± 0.02 A

RHO$_2$ (red high oxygen), RMO$_2$ (red medium oxygen), RLO$_2$ (red low oxygen). All the data are expressed as means (Abs Units) ± standard deviation. Different letters indicate a statistically significant difference among treated wines. All the data are expresses as means ± standard deviation, ($p < 0.05$).

As wine ages and through different oxygen exposures, these polymeric pigments become of crucial importance for the wine color and small amounts of acetaldehyde can react with anthocyanins to produce new stable red pigments [19,24]. The fact that significant differences in terms of color intensity and hue were not detected (Table 5) could be probably due, as already observed for LPP, to the relatively short time of aging.

Table 5. Chromatic Characteristics.

	Aglianico	
	Color Intensity	Hue
RHO$_2$	8.09 ± 0.14 A	0.73 ± 0.00 A
RMO$_2$	7.83 ± 0.16 A	0.73 ± 0.01 A
RLO$_2$	7.84 ± 0.27 A	0.74 ± 0.01 A

RHO$_2$ (red high oxygen), RMO$_2$ (red medium oxygen), RLO$_2$ (red low oxygen). All the data are expressed as means (Abs Units) ± standard deviation. Different letters indicate a statistically significant difference among treated wines. All the data are expresses as means ± standard deviation, ($p < 0.05$).

2.3. Effect on Red Wine Pigments: NMR and HR ESIMS Analyses

To the aim of understanding the molecular basis of the observed changes in the wines treated with different levels of oxygen, samples of RHO$_2$, RMO$_2$, and RLO$_2$ were subjected to NMR-based analysis, as described in the Section 3. A careful inspection of the obtained ^{1}H-NMR spectra of the three samples did not bring to light any discernible difference among the three compared wines (Supplementary Materials Figure S2). Hence, we decided to investigate the same wines by means of HR ESIMS given the intrinsic higher sensitivity of the technique when compared to NMR spectroscopy. The three wines were fractionated by HPLC/Vis by using a C-18 column. Three fractions were obtained for each wine: the first fraction was collected from 15 to 25 min (fraction 1), the second one from 25 to 30 min (fraction 2) and, finally, a third fraction (fraction 3) from 30 min through the end of the chromatographic run. Consistently with data reported in literature [25], in the first fraction non-acetylated anthocyanins were expected to be occurring, while potential pyranoanthocyanins were to be collected in the second fraction. The three obtained fractions (1–3) for each of the three analyzed wines (RHO$_2$, RMO$_2$ and RLO$_2$) were all subjected to full scan HR ESIMS analysis in the positive ion mode. In fraction 1 of all wines, we detected an ion peak that was assigned to malvidin-3-O-glucoside (493.1332; $\Delta = -1.648$; corresponding to C$_{23}$H$_{25}$O$_{12}^+$), whilst ion peaks related to the other common wine anthocyanins, when detected, presented errors above 10 ppm and were not considered reliable (Supplementary Materials Figure S3). In regard to fractions 2 and 3 of RMO$_2$ and RLO$_2$, they turned out to be basically superimposable with each other, while the mass spectra of fractions 2 and 3 of the RHO$_2$ wine showed some interesting peculiarities. More specifically, in RHO$_2$ fraction 2 an ion peak centered at m/z 517.1317 ($\Delta = -4.569$; corresponding to C$_{25}$H$_{25}$O$_{12}^+$) was observed. This peak was attributed to Vitisin B (Supplementary Materials Figure S4) [26]. In the mass spectrum of fraction 3 of RHO$_2$ two ion peaks at m/z 809.2294 ($\Delta = 0.827$; corresponding to C$_{40}$H$_{41}$O$_{18}^+$) (Supplementary Materials Figure S5) and 1029.2871 ($\Delta = -0.065$; corresponding to C$_{48}$H$_{53}$O$_{25}^+$) (Supplementary Materials Figure S6), respectively, were contained. These ion peaks were indicative of the occurrence of polymeric pigments. The peak at m/z 809 was attributed to ethylidene-bridged dimers constituted by one malvidin-3-O-glucoside unit and one (epi)catechin moiety [27], and the peak at m/z 1029 was assigned to the ethylidene-bridged dimer constituted by two malvidin-3-O-glucoside unit, of which one occurred in its flavylium form and the other one in its pseudobase form [28,29].

Vitisin B and ethylidene-bridged pigments (m/z 809 and 1029) are the result of the chemical reaction between acetaldehyde and anthocyanins or flavan-3-ols. Acetaldehyde can act either as a nucleophile at its alpha position or as an electrophile at the carbonyl functionality. The reaction between the nucleophile acetaldehyde and the electrophile C4 position of anthocyanins leads to the formation of Vitisin B, a quite stable compound classified as a pyranoanthocyanin. Conversely, when acetaldehyde acts as an electrophile by undergoing a nucleophilic attack by the C8, and to a lesser extent even by the C6 positions of either flavan-3-ols or anthocyanins, ethylidene-bridged dimers are formed. It is not surprising that we observed Vitisin B and red pigments only in the RHO_2 wine, since such products, as discussed above, are formed by the reaction of acetaldehyde with anthocyanins and flavan-3-ols, and acetaldehyde is a molecule mainly resulting from the oxidative process undergone by wines over time by means of exposure to atmospheric oxygen. Hence, higher quantities of acetaldehyde are certainly present in RHO_2 than in RMO_2 and RLO_2 wines that appear to have been protected from oxygen to a greater extent than RHO_2.

2.4. Effect on VRF, BSA-Tannins and Total Phenols

Although soon after the membrane contactor treatment, the levels of total phenolics were similar among the treated wines, after 11 months of bottle aging, the amount of total phenols was higher in the sample with higher concentrations of oxygen at bottling, as shown in Table 6. This could be probably due to the role of oxygen played in the formation of phenolic compounds more reactive to iron and in variation of the molecular structure of monomeric and polymeric phenolic structures as already shown in wines undergoing different oxygen uptakes during the aging [6]. This is confirmed by the trend observed in Table 6 for BSA-reactive to tannins and flavans reactive to vanillin and in Table 4 for SPP, which showed a statistical difference after the 11 months of aging.

Table 6. Total phenols, flavans reactive to vanillin and tannins reactive to BSA.

	Aglianico		
	RHO_2	RMO_2	RLO_2
BSA Reactive Tannins	84.43 ± 19.14 A	41.51 ± 11.98 B	37.74 ± 26.39 B
FRV (mg/L)	1064.84 ± 66.66 A	931.69 ± 52.26 B	1001.21 ± 3.33 A
Total Phenols (mg/L)	534.85 ± 27.81 A	494.28 ± 6.75 B	474.79 ± 4.75 C

RHO_2 (red high oxygen), RMO_2 (red medium oxygen), RLO_2 (red low oxygen). All the data are expressed as means ± standard deviation. Different letters indicate a statistically significant difference among treated wines. All the data are expresses as means (mg/L) ± standard deviation, ($p < 0.05$).

Tannin concentrations reactive to BSA-proteins were determined by the Habertson method in RHO_2, RMO_2 and RLO_2. During the aging, an increase of the level of tannins reactive to BSA in RHO_2 was observed, consistently with a possible polymerization of tannin structures [27]. Indeed, Harbertson [30] showed that the BSA precipitation increased as a function of the increasing degree of polymerization (or size) from trimers to octamers. As a consequence, every change in tannin composition and size can affect their capability of reacting with BSA. The oxidation of tannins causes the formation of intramolecular as well as intermolecular bonds between flavonoids. The latter cause the polymers to elongate and to become more reactive to salivary proteins [31]. These types of reactions can in fact modify the tannin structure and, thus, the hydrogen bonds and hydrophobic interactions with proteins [32].

As astringency is caused by the tannin-induced aggregation and the precipitation of salivary proteins [33], the increase of BSA-tannins in RHO_2 suggests that the changes undergone by tannins by means of these reactions during the aging could contribute to modify the astringency perception.

Vanillin-reactive flavans (VRF) can instead provide further information related to the size of condensed tannins. In fact, vanillin reacts with the A ring of flavanols at either position 6 or 8 but also acetaldehyde reacts with the same positions of the A ring of flavanols. Therefore, a decrease of VRF may be regarded as an indirect measure of the oxidative polymerization of flavanols linked to reactions triggered by acetaldehyde and involving tannins and anthocyanins [34].

In our study, a slightly lower concentration of VRF was observed in the RMO_2 sample. The lack of a clear trend as a function of the oxygen amount could be due to the fact that these molecules can also undergo hydrolytic cleavage in presence of oxygen by new molecular rearrangements with the formation of new intra and intermolecular bonds [6].

3. Materials and Methods

3.1. Wines

Two commercial red Aglianico and white Falanghina wines produced in Southern Italy by Cantina del Taburno winery were used. Details of the samples together with some base parameters are shown in Table 7. Base parameters were determined according to the OIV compendium of international methods of wine and must analysis (2007).

Table 7. Wine chemical parameters.

	Falanghina	Aglianico
Alcohol ($v/v\%$)	13.30	13.85
Sugars (g/L)	3.03	1.14
pH	3.21	3.45
Total acidity (g/L)	5.68	5.44
Volative acidity (g/L)	0.21	0.32
Malic acid (g/L)	1.78	0.75
Dry extract (g/L)	21.57	27.82

3.2. Wine Oxygen Management

An industrial system ISIOX (Tebaldi s.r.l) equipped with a gas-liquid membrane contactor (Liqui-Cel®, Transverse-flow, South Lakes Dr. Charlotte, NC, USA, cut off 50 g/mol) and a centrifugal pump in stainless steel was used. The membrane provides a fixed and well-determined interface for gas/liquid mass transfer without dispending one phase into another. The structure of the membrane contactor (hydrophobic hollow fiber) is made of polypropylene (PP) and offers a very large contact area (gas/liquid) of 20 m².

The deoxygenation process consists of continuous cycles in which N_2, vacuum or a combination of the first two processes (mix) circulating on one surface of polypropylene membrane is gradually enriched by oxygen deriving from wine circulating on the other side of the membrane. The driving force for the process is the difference in partial pressure of oxygen across the membrane. During the process, wine continuously circulates from a closed tank to the deoxygenation apparatus.

In order to achieve the required level of oxygen, the oxygen content in the wine was monitored through all the processes until the target level was reached.

The process control was carried out by monitoring O_2 level in wines by using a PC incorporated with a very simple programming logic and specific sensors, which monitor the temperature and the oxygen content (luminescence system, Hach Lange, measurement range 0 to 20 mg/L O_2, resolution: 0.01 mg/L O_2, accuracy: 0–5 mg/L $O_2 \pm 0.1$). To control the inlet and outlet pressure, electronic pressure switches were used, as well as the pressure of the gas in the process and the level of any vacuum.

Different deoxygenation treatments were applied to wine in order to obtain different samples at different oxygen content. The amount of oxygen in white wine was respectively white High O_2, WHO_2 = 2.7 mg/L, white Medium O_2, WMO_2 = 1 mg/L, white Low O_2, WLO_2 = 0.25 mg/L.

The amount of oxygen in red wine was respectively red high O_2, RHO_2 = 1 mg/L, red Medium O_2, $RMO2$ = 0.5 mg/L, red Low O_2, $RLO2$ = 0.2 mg/L. Both sets of wines were filtered at 1 µm before being injected into the machine, with the purpose of avoiding fouling and wetting membrane phenomenon [35].

3.3. Samples Bottling and Aging

After the treatments, wines were bottled and all bottles were sealed using Nomacorc coextruded synthetic closures (Nomacorc SA, Thimister Clermont, Belgium) select green 100, which allow oxygen to pass through the cork in a controlled manner (0.4 mg O_2 after 3 months, 0.7 mg O_2 after 6 months, 1.2 mg O_2 after 12 months and 1.1 mg O_2 year after the first year of aging). Bottles were aged for 11 months at 15 °C.

3.4. Methods of Analysis

3.4.1. High-Performance Liquid Chromatography Determination of Acetaldehyde

Analyses of acetaldehyde were performed by derivatizing experimental sample wine and HPLC.

The derivatization analysis was as follows: 100 µL of wine sample was dispensed into a vial, followed by the addition of 20 µL of freshly prepared 1.120 mg/L SO_2 solution, 20 µL of 25% sulfuric acid (Carlo Erba reagent 96%), and 140 µL of 2 g/L 2,4-dinitrophenylhydrazine reagent. After mixing, the solution was allowed to react for 15 min at 65 °C and then promptly cooled to room temperature [36].

The HPLC used was a Shimadzu LC10 ADVP apparatus (Shimadzu Italy, Milan, Italy) equipped with a SCL-10AVP system controller, two LC-10ADVP pumps to create the needed solvent gradient, a SPD-M 10 AVP detector and an injection system full Rheodyne model 7725 (Rheodyne, Cotati, CA, USA). The separation was carried out on a Waters Spherisorb column (C 18, Silica particle substrate, ODS2 250 × 4.6 mm, 5 µm particles diameter, 80 Å pore size) equipped with a guard column. Optimum efficiency of separation was obtained using a flow rate of 0.75 mL/min, column temperature of 35 °C; mobile phase solvents were: (A) 0.5% formic acid (Sigma Aldrich ≥ 95%) in water milli-Q (Sigma Aldrich) and (B) acetonitrile (Sigma Aldrich ≥ 99,9%); gradient elution protocol was: 35% B to 60% B (t = 8 min), 60% B to 90% B (t = 13 min), 90% B to 95% B (t = 15 min, 2-min hold), 95% B to 35% B (t = 17 min, 4-min hold), total run time, 21 min, samples injections 50 µL and the detection was performed by monitoring the absorbance signals at 365 nm.

The calibration curves were made up by injecting 5 solutions (in triplicate) containing their respective standards covering the range of linearity 10–120mg/L and were characterized by a correlation coefficient (R^2) > 0.976. Three analytical replicates were carried out for each experimental replicate.

3.4.2. High-Performance Liquid Chromatography Analyses of Anthocyanins

Analyses of native anthocyanins were performed by a HPLC Shimadzu LC10 ADVP apparatus (Shimadzu Italy, Milan, Italy) equipped with a SCL-10AVP system controller, two LC-10ADVP pumps to create the needed solvent gradient, an SPD-M 10 AVP detector and an injection system full Rheodyne model 7725 (Rheodyne, Cotati, CA, USA). According to the method described in the OIV Compendium of International Methods of Analysis of Wine and Musts [37].

The HPLC solvents were the following: solvent A: water milli-Q (Sigma-Aldrich, Milan, Italy)/formic acid (Sigma-Aldrich ≥ 95%)/acetonitrile (SigmaAldrich ≥ 99.9%) (87:10:3) v/v; solvent B: water/formic acid/acetonitrile (40:10:50) v/v. The gradient was: zero-time conditions 94% A and 6% B; after 15 min, the pumps were adjusted to 70% A and 30% B, at 30 min to 50% A and 50%B, at 35 min to 40% A and 60% B, at 41 min, end of analysis, to 94% A and 6% B. A 5 min re-equilibration time was applied before the successive analysis as reported by [38]. The column used for the analyses was a waters spherisorb column (C 18, Silica particle substrate, ODS2 250 × 4.6 mm, 5 µm particles diameter, 80 Å pore size) with a precolumn was used. An amount of 50 µL of calibration standards or wine was injected onto the column. The absorbance signals at 520 nm were detected. Detector sensitivity was 0.01 Absorbance units full scale (AUFS). All the samples were filtered through 0.45 µm Durapore membrane filters (Millipore-Ireland) into glass vials and immediately injected into the HPLC system.

The calibration curve was obtained by injecting 5 solutions (in triplicate) containing increasing concentrations of malvidin-3-monoglucoside (Extrasynthese, Lyon, France). The calibration was characterized by a correlation coefficient (R^2) = 0.996. The linearity range of the calibration curve was 2–200 mg/L. The precision of the method used was tested by six replicate analyses of a red wine sample containing 118.4 mg/L of total monomeric anthocyanins. The coefficient of variation was included between 1.1% (for malvidin 3-monoglucoside) and 9.1% (for malvidin 3-(6II-coumaroyl)-glucoside) and demonstrated the good reproducibility of the HPLC analysis. The monomeric anthocyanins concentrations were expressed as mg/L of malvidin-3-monoglucoside.

Fractionation of RLO_2, RMO_2 and RHO_2 wines was performed according to the OIV method of analysis by using the same HPLC Shimadzu LC10 ADVP as reported above. Three analytical replicates were carried out for each experimental replicate.

3.4.3. High Resolution Electrospray Ionization Mass Spectrometry (HR ESIMS) Analyses of Red Wines

The HPLC-separation of red wines conducted as described above afforded three fractions for each wine. The first fraction was collected from 15 to 25 min (fraction 1), the second one from 25 to 30 min (fraction 2) and the third fraction (fraction 3) from 30 min through the end of the chromatographic run (45 min). Each collected fraction was dried, solubilized in methanol and analyzed by HR-ESIMS in continuous flow injection in the positive ion mode. HR ESIMS experiments were performed on an Agilent 1260 Infinity II HPLC quaternary system coupled to a linear ion trap LTQ Orbitrap XL hybrid Fourier transform MS (FTMS) instrument equipped with an ESI ION MAX source (Thermo-Fisher). The following source settings were used (mass range m/z 100–2000): spray voltage 4.5 kV, capillary temperature 300 °C, capillary voltage 15 V, sheath gas 20 and auxiliary gas 21 (arbitrary units), tube lens voltage 140 V, and 25% collision energy. Calculation of elemental formulae was conducted by using Xcalibur software v 2.0.7 with a mass tolerance constrain of 5 ppm.

3.4.4. NMR Experiments

An amount of 2 mL of each wine sample (RHO_2, RMO_2 and RLO_2) was lyophilized and solubilized in 700 µL of CD_3OD (δ_H 3.31; δ_C 49.0 ppm). NMR experiments were run on a Varian Unity Inova 700 spectrometer equipped with a 13C Enhanced HCN Cold Probe and by using a Shigemi 5 mm NMR tubes. The ^{1}H-NMR standard Varian pulse sequence was employed.

3.4.5. Standard Chemical Analyses and Spectrophotometric Measurement

According to "OIV Compendium of International Wine and Must Analysis of Wine and Musts Analysis 2007" [37], standard chemical analysis (alcohol strength calculated by volume, reducing sugar, total acidity, pH, volatile acidity, malic acidity and total dry matter) was measured.

Spectrophotometric analyses were performed by a Jenway 7305 spectrophotometer. Chromatic characteristics, color intensity and hue was determined according to OIV methods [37]. The color intensity was determined as the sum of abs 420 nm, abs 520 nm and abs 620 nm and hue as abs 420 nm/abs 520 nm ratio.

BSA-reactive tannins, short polymeric pigments (SPP), large polymeric pigments (LPP) and total phenols were determined by the Harbertson et al. assay [22]. Short polymeric pigments (SPP) and large polymeric pigments (LPP) were obtained by combining analysis of supernatant obtained after protein precipitation using bovine serum albumin BSA

(Sigma-Aldrich) with the bisulphite bleaching of pigments in wine. BSA-tannin complex in the pellet was redissolved, added with ferric chloride and read at 510 nm. Total phenols were quantified by reading at 510 nm the sample as follows: 50 µL of wine was added to 825 µL of buffer C and read at the spectrophotometer as a blank solution. After the addition of 125 µL of ferric chloride, the sample was read again to quantify the amount of iron-reactive phenols.

Vanillin-reactive flavans (VRF) were determined according to Gambuti et al. [39]. Briefly, 750 µL of a solution of vanillin (4% in methanol) was added to 125 µL of diluted wine and, after 5 min, 375 µL of concentrated hydrochloric acid was added. After a 15-min incubation of the mixture at 20 °C, the absorbance was determined at 500 nm and read against a blank solution in which pure methanol was used instead of the solution of the solution of vanillin. Concentrations were calculated as (+)-catechin (mg/L).

3.4.6. Statistical Analysis

Quantitative data were compared using Tukey's least significant differences procedure, all the variance resulted homogeneous. When the variances were not homogeneous, data were analyzed using Kruskal–Wallis test. When results of the Kruskal–Wallis test were significant ($p < 0.05$), the significance of between-group differences was determined by the Bonferroni–Dunn test (5% significance level). These analyses were performed using XLSTAT (version 2013.6.04; Addinsoft, Paris, France). All data are means of three values.

4. Conclusions

Results obtained in this study confirmed that different contents of oxygen in bottled wine have an impact on wine aging and oxidation.

The oxygen management in the Aglianico wine by means of a polypropylene hollow fiber membrane contractor determined, after 11 months of aging, a lower content of free and total SO_2 in sample with higher level of dissolved oxygen. The same behavior was observed in the Falanghina wine with an increase in terms of acetaldehyde in the sample with higher levels of oxygen.

In regard to phenolic compounds, a greater loss of total native anthocyanins was observed in red wines. Their content was higher in the wine with lower concentration of oxygen. However, the loss of total native anthocyanins did not affect the color parameters of wines, such as color intensity and tonality.

BSA-reactive tannins and vanillin-reactive flavans were lower in samples containing medium and low level of oxygen with respect to the samples with higher contents of oxygen. This is due to the fact that oxygen, by participating to oxidation reactions, concurs to the formation of polymeric pigments. As expected, in red wines with the highest content of dissolved oxygen after the membrane contactor treatment, Vitisin B, ethylidene-bridged dimers and acetaldehyde were more abundant in comparison to wines treated as to have a lower content of oxygen in pre-bottling phase.

In conclusion, the deoxygenation of wine by membrane contractor could be a suitable technique for wine industry to prevent all those oxidative phenomena that could change the final quality of red and white wines affecting the content of sulfur dioxide and acetaldehyde (especially in white wines). Nevertheless, oenologists have to consider that the decrease of oxygen content may affect the color stability in red wines.

Supplementary Materials: The following are available online. Figure S1: Enlargements of ^{1}H-NMR spectra of wines RHO_2, RMO_2 and RLO_2 registered in CD_3OD. Figure S2: Structure of malvidin 3-O-glucoside and relative HR ESIMS in the positive ion mode. Figure S3: Structure of Vitisin B and relative HR ESIMS in the positive ion mode. Figure S4: Structure of ethylidene-bridged dimer constituted by one unit of malvidin 3-O-glucoside (bottom) and one unit of (epi)catechin (top) with the relative HR ESIMS in the positive ion mode. Figure S5: Structure of ethylidene-bridged dimer constituted by two units of malvidin 3-O-glucoside, of which the one at the bottom is in its flavylium form and the one on the top in its pseudobase form, along with the relative HR ESIMS in positive ion mode.

Author Contributions: Conceptualization, A.G. (Angelita Gambuti) and F.E.; methodology, L.P. and M.F.; software, L.P.; validation, A.G. (Angelita Gambuti) and M.F.; formal analysis, A.G. (Antonio Guerriero) and F.E.; investigation, F.E. and A.G. (Antonio Guerriero); resources, L.M.; data curation, L.P.; writing—original draft preparation, F.E.; writing—review and editing, A.G. (Angelita Gambuti); visualization, M.F.; supervision, A.G. (Angelita Gambuti) and L.M. All authors have read and agreed to the published version of the manuscript.

Funding: This research received no external funding.

Institutional Review Board Statement: Not applicable.

Informed Consent Statement: Not applicable.

Data Availability Statement: Not applicable.

Acknowledgments: The authors would like to thank Tebaldi srl for providing the membrane contactor machine and instrumental support (HPLC-DAD autosampler).

Conflicts of Interest: The authors declare no conflict of interest.

References

1. Waterhouse, A.L.; Laurie, V.F. Oxidation of wine phenolics: A critical evaluation and hypotheses. *Am. J. Enol. Vitic.* **2006**, *57*, 306–313.
2. Danilewicz, J.C. Mechanism of autoxidation of polyphenols and participation of sulfite in wine: Key role of iron. *Am. J. Enol. Vitic.* **2011**, *62*, 319–328. [CrossRef]
3. Nikolantonaki, M.; Waterhouse, A.L. A method to quantify quinone reaction rates with wine relevant nucleophiles: A key to the understanding of oxidative loss of varietal thiols. *J. Agric. Food Chem.* **2012**, *60*, 8484–8491. [CrossRef] [PubMed]
4. Elias, R.J.; Andersen, M.L.; Skibsted, L.H.; Waterhouse, A.L. Identification of free radical intermediates in oxidized wine using electron paramagnetic resonance spin trapping. *J. Agric. Food Chem.* **2009**, *57*, 4359–4365. [CrossRef] [PubMed]
5. Oliveira, C.M.; Ferreira, A.C.S.; De Freitas, V.; Silva, A.M. Oxidation mechanisms occurring in wines. *Food Res. Int.* **2011**, *44*, 1115–1126. [CrossRef]
6. Mouls, L.; Fulcrand, H. UPLC-ESI-MS study of the oxidation markers released from tannin depolymerization: Toward a better characterization of the tannin evolution over food and beverage processing. *J. Mass Spectrom.* **2012**, *47*, 1450–1457. [CrossRef] [PubMed]
7. Waterhouse, A.L. Wine phenolics. *Ann. N. Y. Acad. Sci.* **2002**, *957*, 21–36. [CrossRef]
8. Rosenfeld, E.; Schaeffer, J.; Beauvoit, B.; Salmon, J.M. Isolation and properties of promitochondria from anaerobic stationary-phase yeast cells. *Antonie Van Leeuwenhoek* **2004**, *85*, 9–21. [CrossRef]
9. Jarauta, I.; Cacho, J.; Ferreira, V. Concurrent phenomena contributing to the formation of the aroma of wine during aging in oak wood: An analytical study. *J. Agric. Food Chem.* **2005**, *53*, 4166–4177. [CrossRef]
10. Singleton, V.L.; Trousdale, E.; Zaya, J. Oxidation of wines. 1. Young white wines periodically exposed to air. *Am. J. Enol. Vitic.* **1979**, *30*, 49–54.
11. Gabelman, A.; Hwang, S.T. Hollow fiber membrane contactors. *J. Membr. Sci.* **1999**, *159*, 61–106. [CrossRef]
12. Gambuti, A.; Rinaldi, A.; Lisanti, M.T.; Pessina, R.; Moio, L. Partial dealcoholisation of red wines by membrane contactor technique: Influence on colour, phenolic compounds and saliva precipitation index. *Eur. Food Res. Technol.* **2011**, *233*, 647–655. [CrossRef]
13. Blank, A.; Vidal, J.C. Gas management by membrane contactor: Ester and higher alcohol losses, and comparison with porous injector. *Bull. l'OIV* **2012**, *85*, 5–14.
14. Schonenberger, P.; Baumann, I.; Jaquerod, A.; Ducruet, J. Membrane Contactor: A Nondispersive and Precise Method to Control CO2 and O2 Concentrations in Wine. *Am. J. Enol. Vitic.* **2014**, *65*, 510–513. [CrossRef]
15. Waterhouse, A.L.; Sacks, G.L.; Jeffery, D.W. *Understanding Wine Chemistry*; John Wiley & Sons: Hoboken, NJ, USA, 2016.
16. Jenkins, T.W.; Howe, P.A.; Sacks, G.L.; Waterhouse, A.L. Determination of Molecular and "Truly" Free Sulfur Dioxide in Wine: A Comparison of Headspace and Conventional Methods. *Am. J. Enol. Vitic.* **2020**, *71*, 222–230. [CrossRef]
17. Bonaldo, F.; Guella, G.; Mattivi, F.; Catorci, D.; Arapitsas, P. Kinetic investigations of sulfite addition to flavanols. *Sci. Rep.* **2020**, *10*, 12792. [CrossRef]
18. Schmidtke, L.M.; Clark, A.C.; Scollary, G.R. Micro-oxygenation of red wine: Techniques, applications, and outcomes. *Crit. Rev. Food Sci. Nutr.* **2011**, *51*, 115–131. [CrossRef]
19. Timberlake, C.F.; Bridle, P. Interactions between anthocyanins, phenolic compounds, and acetaldehyde and their significance in red wines. *Am. J. Enol. Vitic.* **1976**, *27*, 97–105.
20. He, F.; Liang, N.N.; Mu, L.; Pan, Q.H.; Wang, J.; Reeves, M.J.; Duan, C.Q. Anthocyanins and their variation in red wines I. Monomeric anthocyanins and their color expression. *Molecules* **2012**, *17*, 1571–1601. [CrossRef]
21. Fulcrand, H.; Dueñas, M.; Salas, E.; Cheynier, V. Phenolic reactions during winemaking and aging. *Am. J. Enol. Vitic.* **2006**, *57*, 289–297.

22. Harbertson, J.F.; Picciotto, E.A.; Adams, D.O. Measurement of polymeric pigments in grape berry extract sand wines using a protein precipitation assay combined with bisulfite bleaching. *Am. J. Enol. Vitic.* **2003**, *54*, 301–306.
23. Picariello, L.; Rinaldi, A.; Forino, M.; Errichiello, F.; Moio, L.; Gambuti, A. Effect of Different Enological Tannins on Oxygen Consumption, Phenolic Compounds, Color and Astringency Evolution of Aglianico Wine. *Molecules* **2020**, *25*, 4607. [CrossRef]
24. Fulcrand, H.; dos Santos, P.J.C.; Sarni-Manchado, P.; Cheynier, V.; Favre-Bonvin, J. Structure of new anthocyanin-derived wine pigments. *J. Chem. Soc. Perkin Trans.* **1996**, *1*, 735–739. [CrossRef]
25. Rentzsch, M.; Schwarz, M.; Winterhalter, P.; Blanco-Vega, D.; Hermosín-Gutiérrez, I. Survey on the content of vitisin A and hydroxyphenyl-pyranoanthocyanins in Tempranillo wines. *Food Chem.* **2010**, *119*, 1426–1434. [CrossRef]
26. Bakker, J.; Timberlake, C.F. Isolation, identification, and characterization of new color-stable anthocyanins occurring in some red wines. *J. Agric. Food Chem.* **1997**, *45*, 35–43. [CrossRef]
27. Es-Safi, N.E.; Fulcrand, H.; Cheynier, V.; Moutounet, M. Studies on the acetaldehyde-induced condensation of (−)-epicatechin and malvidin 3-*O*-glucoside in a model solution system. *J. Agric. Food Chem.* **1999**, *47*, 2096–2102. [CrossRef]
28. Atanasova, V.; Fulcrand, H.; Le Guernevé, C.; Cheynier, V.; Moutounet, M. Structure of a new dimeric acetaldehyde malvidin 3-glucoside condensation product. *Tetrahedron Lett.* **2002**, *43*, 6151–6153. [CrossRef]
29. Forino, M.; Picariello, L.; Lopatriello, A.; Moio, L.; Gambuti, A. New insights into the chemical bases of wine color evolution and stability: The key role of acetaldehyde. *Eur. Food Res. Tech.* **2020**, *246*, 733–743. [CrossRef]
30. Harbertson, J.F.; Mireles, M.; Yu, Y. Improvement of BSA tannin precipitation assay by reformulation of resuspension buffer. *Am. J. Enol. Vitic.* **2014**, *10*, 16–21. [CrossRef]
31. Poncet-Legrand, C.; Cabane, B.; Bautista-Ortín, A.B.; Carrillo, S.; Fulcrand, H.; Perez, J.; Vernhet, A. Tannin oxidation: Intra-versus intermolecular reactions. *Biomacromolecules* **2010**, *11*, 2376–2386. [CrossRef]
32. Watrelot, A.A.; Day, M.P.; Schulkin, A.; Falconer, R.J.; Smith, P.; Waterhouse, A.L.; Bindon, K.A. Oxygen exposure during red wine fermentation modifies tannin reactivity with poly-L-proline. *Food Chem.* **2019**, *297*, 124923. [CrossRef] [PubMed]
33. Soares, S.; Sousa, A.; Mateus, N.; de Freitas, V. Effect of condensed tannins addition on the astringency of red wines. *Chem. Sens.* **2012**, *37*, 191–198. [CrossRef] [PubMed]
34. Gambuti, A.; Rinaldi, A.; Ugliano, M.; Moio, L. Evolution of phenolic compounds and astringency during aging of red wine: Effect of oxygen exposure before and after bottling. *J. Agric. Food Chem.* **2012**, *61*, 1618–1627. [CrossRef]
35. Breniaux, M.; Zeng, L.; Bayrounat, F.; Ghidossi, R. Gas transfer management by membrane contactors in an oenological context: Influence of operating parameters and membrane materials. *Sep. Purif. Technol.* **2019**, *227*, 115733. [CrossRef]
36. Han, G.; Wang, H.; Webb, M.R.; Waterhouse, A.L. A rapid, one step preparation for measuring selected free plus SO2- bound wine carbonyls by HPLC-DAD/MS. *Talanta* **2015**, *134*, 596–602. [CrossRef] [PubMed]
37. OIV. *Compendium of International Methods of Wine and must Analysis*; Office International de la Vigne et du Vin: Paris, France, 2017.
38. Picariello, L.; Gambuti, A.; Picariello, B.; Moio, L. Evolution of pigments, tannins and acetaldehyde during forced oxidation of red wine: Effect of tannins addition. *LWT Food Sci. Technol.* **2017**, *77*, 370–375. [CrossRef]
39. Gambuti, A.; Han, G.; Peterson, A.L.; Waterhouse, A.L. Sulfur dioxide and glutathione alter the outcome of microoxygenation. *Am. J. Enol. Vitic.* **2015**, *66*, 411–423. [CrossRef]

Article

Comparison of a Rapid Light-Induced and Forced Test to Study the Oxidative Stability of White Wines

Emilio Celotti [1], Georgios Lazaridis [1], Jakob Figelj [1], Yuri Scutaru [2] and Andrea Natolino [1,*]

[1] Department of Agricultural, Food, Environmental and Animal Sciences, University of Udine, 33100 Udine, Italy; emilio.celotti@uniud.it (E.C.); lazaridis.georgios@spes.uniud.it (G.L.); figelj.jakob@spes.uniud.it (J.F.)
[2] Department of Oenology and Chemistry, Technical University of Moldova, MD-2004 Chisinau, Moldova; iurie.scutaru@enl.utm.md
* Correspondence: andrea.natolino@uniud.it; Tel.: +39-0432-558376

Abstract: The oxidation processes of white wines can occur during storage and commercialization due to several factors, and these can negatively affect the color, aroma, and quality of the wine. Wineries should have faster and simpler methods that provide valuable information on oxidation stability of wines and allow fast decision-making procedures, able to trigger suitable technological interventions. Using a portable prototype instrument for light irradiations at different wavelengths and times was considered and evaluated on sensorial, spectrophotometric, and colorimetric parameters of white wines. The sensorial analysis revealed that white and light blue were the most significant, after only 1 h of irradiation. The experimental results showed that hydrogen peroxide could enhance the effect of light treatment, allowing a contemporary evaluation of the oxidation stability of wine against light and chemical stresses. As expected, a good correlation ($R^2 > 0.89$) between optical density at 420 nm and b* parameter was highlighted. The synergic effect of light and H_2O_2 was also studied on the hydrolyzable and condensed tannins' additions to white wine. The proposed methodology could be used to evaluate the oxidative stability of white wines, but also to evaluate the effect of some oenological adjuvants on wine stability.

Keywords: white wine; wine oxidation; browning; light exposure; tannins

1. Introduction

Storage and commercialization are crucial steps for both wine producers and consumers, as the wine quality needs to be ensured and maintained. White wines are usually consumed young within a year of production, to maintain their color, and fresh, fruity, and floral aroma. During storage, uncontrolled oxidation reactions can occur and cause several faults: volatile acidity increases [1], color changes from green and light yellow to brown and dark hues [2], flavor decay with many olfactory notes' losses, and off-flavors' formation [3,4].

Wine oxidation is a complex phenomenon since wine contains several organic and inorganic compounds, and it can be divided into enzymatic and non-enzymatic oxidation. The non-enzymatic process, also called chemical oxidation, has been studied in the last decades and begins with the oxidation of polyphenols containing a catechol or a galloyl group, such as catechin, epicatechin, gallocatechin, gallic acid and its esters, and caffeic acid [5]. White wines contain lower polyphenol concentrations (0.2 to 0.5 g/L), mainly hydroxycinnamic acids, but they remain crucial for oxidation-related issues in wine browning and aroma changes. The oxidation of polyphenols can lead to the formation of o-quinones with different degrees of polymerization and, due to their instability, further reactions can happen, and brown pigments can be formed [6]. The phenol oxidation induces, contemporary, the oxygen reduction to hydrogen peroxide, which is a potent oxidative compound that can also oxidize ethanol to acetaldehyde in the presence of transition metals [7]. Besides transition metals (i.e., Fe, Cu), other factors could affect the non-enzymatic oxidation: temperature, pH, and light exposure [8].

Light exposure can cause sensorial changes in wine, with the formation of volatile sulfur compounds and the oxidative browning spoilage. The light-induced off-flavors are mainly due to the riboflavin and are associated with the so-called light-struck taste (LST), characterized by unpleasant cabbage- and onion-like odors. The riboflavin is a highly photosensitive vitamin that induces the photooxidation of methionine generating methanethiol and dimethyl disulfide [9].

Light can induce the photodegradation of tartaric acid and the formation of glyoxal and glyoxylic acid. These two compounds can bin two flavanol units, forming a dimer that undergoes a dehydration and oxidation leading to a formation of yellow pigments and contributing to the oxidative browning of white wines [2,6,10]. These mechanisms of photochemical oxidation are favored by the dissolved oxygen and the transition metal ions, such as iron [9].

Several analytical methods have been proposed to evaluate the oxidative susceptibility or oxidation status of white wines. The cyclic voltammetry is a useful fingerprint technology for quantifying dynamic changes in wines' composition related to their redox state [11], or investigating redox potentials of various wine compounds measuring the anodic peak intensity [12–14]. The oxidative mechanism can be also investigated through the detection of radical species with the so-called Electron Paramagnetic Resonance spectroscopy (EPR) [15,16]. Another approach comprises the metabolomics analysis by UHPLC/QTOF-MS systems that revealed some specific compounds able to discriminate the different oxidative statuses of white wines [17].

All these aforementioned analytical techniques are complex, expensive, and they require high knowledge and highly-skilled technicians. Wine quality control requires the availability of simple and rapid analytical methods, allowing regular and punctual monitoring of the different production steps and a fast decision-making procedure, able to trigger suitable technological interventions, in case of deviations from the normal winemaking conditions.

Due to the complexity of wine oxidation phenomena and the current analytical methods to evaluate the stability and shelf-life of white wines, it is necessary to have new approaches that allow more rapid, complete, and reliable evaluations. The initial aim of the present work is to evaluate, by sensorial analysis, the effect of light exposure at different wavelengths and times, using a portable prototype instrument. Moreover, the evaluation of light, chemical stresses, and their combination was carried out through spectrophotometric and colorimetric indices: optical density at 420 nm, catechins content, L*, a*, and b* parameters. The addition of hydrolyzable and grape tannins at different concentrations (50, 200, and 500 mg/L) were also considered and studied on the oxidation stability of Pinot Gris wine.

2. Results and Discussion

The chemical properties of Pinot Gris and Chardonnay wines are shown in Table 1. All the experimental values are in the common ranges related to the cultivation area and grape variety [18]. The oxidation stability was evaluated by the Polyphenols Oxidative Medium (POM) test, and the results revealed initial technical stability, which allow a better evaluation of chemical and light stresses affecting wine quality during storage and commercialization stages.

Table 1. Chemical properties of Pinot Gris and Chardonnay.

Chemical Parameter	Pinot Gris	Chardonnay
Alcohol (% v/v)	12.95	12.48
pH	3.35	3.29
Total acidity (g/L)	5.88	6.42
Free SO_2 (mg/L)	20	27
Total SO_2 (mg/L)	90	81
Reducing sugars (g/L)	2.16	5.74
POM test (%)	15	37

2.1. Light Exposures and Sensory Analysis

Figure 1 shows the results of sensorial analysis carried out on a Pinot Gris wine before (Control) and after 7 h of irradiation with different light colors: violet, blue, light blue, green, yellow, red, and white.

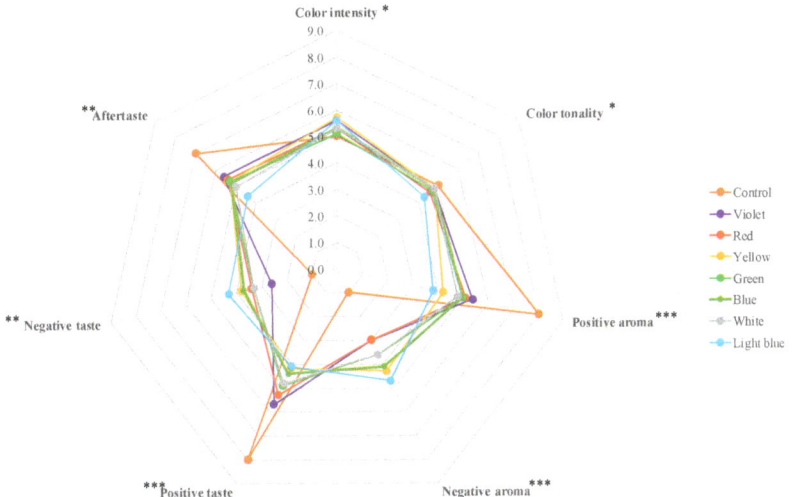

Figure 1. Sensorial analysis of Pinot Gris after 7 h and different light color exposures. * $p < 0.05$; ** $p < 0.01$; *** $p < 0.001$.

The radar chart reported in Figure 1 shows that the light irradiations at different wavelengths slightly affect the visual characteristics of the Pinot Gris wine. The irradiations induced a general increase of color intensity, due to an increase of amber yellow/brown color shades, and a contemporary decrease of yellow straw color tonality. The higher variation was detected after the light blue treatment with a wavelength range between 476 and 495 nm.

The panel group of trained judges highlighted the remarkable effects of light exposures over aroma and taste perceptions. The light irradiations induced a significant decrease of positive aroma and taste perceptions, and a contemporary increase of negative sensory notes. The wine tasting after light blue treatment pointed out the highest variation on aroma, taste, and aftertaste perceptions.

Experimental trials on Chardonnay wine pointed out that greater significant modifications on chemical composition occurred after light exposure at low wavelengths in the visible spectrum range or in the ultraviolet spectrum. Specifically, blue and violet light tonality allowed more significant changes on analytical indices, compared to red, orange and yellow light irradiation [19].

Considering the sensorial evaluation depicted in Figure 1, subsequently, a shorter exposure (1 h) with light blue irradiation was carried out to evaluate if shorter times are enough to induce significant effects on sensory indices.

A Chardonnay wine was also considered over Pinot Gris variety. After 1 h of light blue exposure, a sensorial analysis was carried out by the same trained panel group, evaluating more specific sensory descriptors on wine color, aroma, and taste. The results are reported in Figure 2.

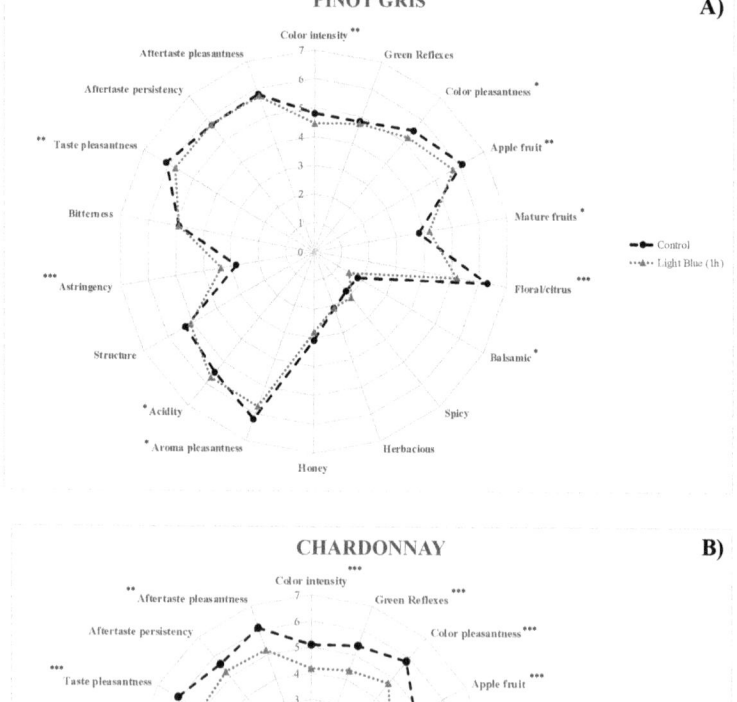

* $p<0.05$; ** $p<0.01$; *** $p<0.001$

Figure 2. Sensorial analysis of Pinot Gris (**A**) and Chardonnay (**B**) wine before (Control) and after 1 h exposure to blue light.

The sensory analysis of Pinot Gris highlighted that the light blue treatment induced a common decrease of chromatic, aroma, and taste pleasantness, compared to the untreated sample. The judges evaluated a decrease of apple, citrus, and honey notes, but an increase of mature fruits and spicy perceptions. Taste evaluation pointed out a potential increase of acidity and astringency descriptors. Instead, no relevant effects were revealed about the aftertaste. Moreover, the comparison of untreated and irradiated Chardonnay wine revealed a common decrease of the considered descriptors, with a decrease of sensory quality after light blue exposure. It is remarkable, again, the significant increase of the astringency perceptions after light treatment. Polyphenols are sensitive to various physical and chemical agents, such as temperature, light, oxygen, and others [20], which can significantly affect some of their chemical and sensory properties [21].

The aforementioned effect is also reported for other food matrices, such as milk [22], highlighting a potential effect of light exposures on astringency perception.

Figure 3 shows the POM test results of Pinot Gris and Chardonnay wine before and after exposure to different color lights (light blue and white) at different times (10 and 20 min).

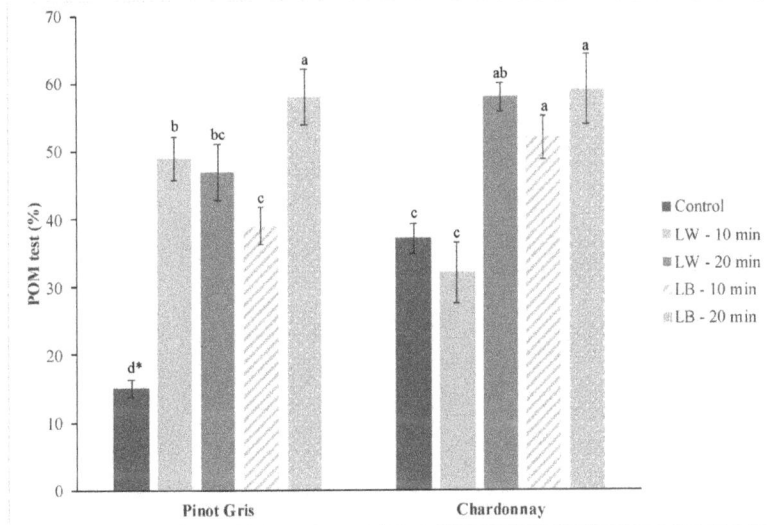

*Different letters indicate significant differences ($p < 0.05$).

Figure 3. POM test value of Pinot Gris and Chardonnay wine before (Control) and after different light color (white and light blue) exposures at 10 and 20 min.

Shorter exposure times below 1 h were adopted to define a potential analysis protocol suitable to be directly applied in wineries, considering their conventional production times. Accelerating decision-making procedures become fundamental for wineries to evaluate the oxidation stability and the shelf-life of the wines, allowing reduced sampling time, avoiding sample storage and transport, and reducing environmental risks [23].

The light exposure significantly increased the oxidability of the wines, independently of wavelength and exposure time, as pointed out by experimental results (Figure 3). The light blue exposure for 20 min allowed the highest variations of POM test results: from 15% to 58% for the Pinot Gris, and from 37% to 59% for the Chardonnay wine. The increase of oxidability could be due to the photosensitivity induced by light exposures on riboflavin in the wine, a photosensitizing agent that promotes the oxidation phenomena. When riboflavin is exposed to light, it reaches the singlet state that is converted to the triplet state with an intercrossing system. Riboflavin is reduced by the acquisition of two electrons from a donor compound, that consequently undergoes oxidation [8].

2.2. Individual and Combined Stress Trials

Figure 4 shows the experimental results of optical densities at 420 nm (O.D. 420), catechins contents (C.C.), a*, and b* parameters of the Pinot Gris samples before (Control) and after different treatments: white (LW) and light blue (LB) exposures, hydrogen peroxide (H_2O_2), and ascorbic acid (Asc.Ac.). Moreover, also, some treatment combinations were considered to evaluate possible synergic effects: LW + H_2O_2, LW + H_2O_2, LB + H_2O_2, LW + Asc.Ac., LB + Asc.Ac., LW + H_2O_2 + Asc.Ac., and LB + H_2O_2 + Asc.Ac. The O.D. 420, catechins content, L*, a*, and b* parameters were adopted as response variables and used to determine the coefficient of the third-order polynomial model used for ANOVA. The estimated coefficients are given in Table 2.

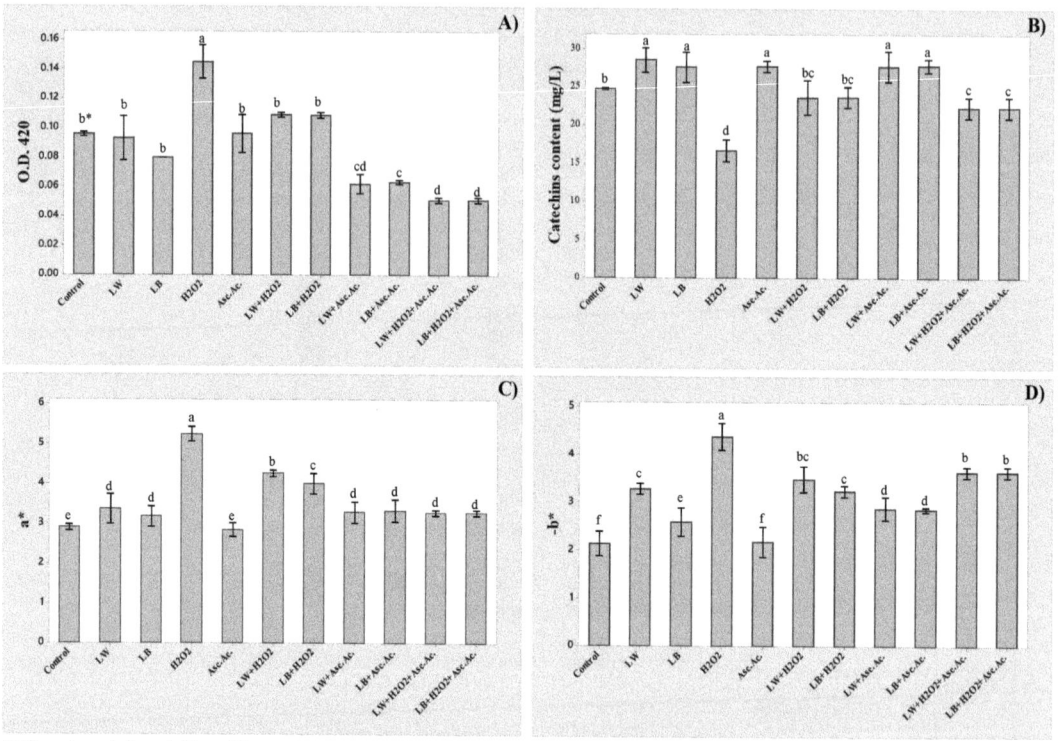

*Different letters indicate significant differences ($p < 0.05$)

Figure 4. Results of OD420 (**A**), catechins (**B**), a* (**C**), and b* (**D**) before and after several chemical and light stresses.

Third-order polynomial equations were found well to represent the experimental data, as indicated by the estimated coefficients of the determination R^2 and R^2-adj. The L* parameter was considered not suitable as response variables, considering the lowest values of R^2 (<86%) and R^2-adj (<80%).

The light exposures (LW and LB) affected significantly the catechins content, a*, and b* parameter. The magnitude of regression coefficients revealed that the white light (LW) induced higher changes on analytical parameters, compared to light blue treatment (LB). Instead, the light exposures did not affect the optical densities at 420 nm, which remain the same as untreated wine (Control).

The oxidation level of white wine in the bottle is commonly estimated by color with the extent of browning at 420 nm [8]. As reported in Figure 4, the OD at 420 nm increased significantly only after the addition of hydrogen peroxide, which is the most significant factor for all the analytical parameters ($p < 0.001$).

Table 2. Regression coefficients of the ANOVA of third-order polynomial model for OD 420 nm, catechins, L*, a* and b* parameters after light exposure (LB and LW), hydrogen peroxide, and ascorbic acid addition.

Terms	Coefficients				
	O.D. 420 nm	Catechins	L*	a*	b*
Constant	0.09657	24.700	18.680	2.9033	−2.1400
Light					
LB	−0.01567	2.890 ***	0.703 *	0.2700 **	−0.4533 ***
LW	−0.00600	3.800 ***	−0.540 *	0.4600 ***	−1.1400 ***
H_2O_2	0.04933 ***	−8.037 ***	−1.193***	2.3400 ***	−2.2400 ***
Asc.Ac.	0.00033	2.99	0.170	−0.0633	−0.0400
Light·H_2O_2					
LB·H_2O_2	−0.0433 ***	4.124 ***	−0.480	−1.5167 ***	1.597 ***
LW·H_2O_2	−0.0497 ***	3.160 ***	0.733 *	−1.4433 ***	2.033 ***
Light·Asc.Ac.					
LB·Asc.Ac.	−0.017	−2.805 ***	−1.257 ***	0.2133 *	−0.227
LW·Asc.Ac.	−0.028	−3.802 ***	0.267	−0.0200	0.440
Light H_2O_2·Asc.Ac.					
LB H_2O_2·Asc.Ac.	−0.0182	−1.554	1.723	−0.870 ***	−0.130
LW H_2O_2·Asc.Ac.	−0.0105	−0.503	0.230	−0.900 ***	−0.550 ***
R^2	90.10%	97.98%	85.97%	98.93%	98.79%
R^2-adj.	85.60%	97.06%	79.59%	98.44%	98.24%

* $p < 0.05$; ** $p < 0.01$; *** $p < 0.001$.

The ascorbic acid is commonly employed during winemaking processes due to its antioxidant properties in a complementary role with sulfur dioxide [24]. Despite the antioxidant role, the ascorbic acid showed some pro-oxidant effects, such as its oxidative degradation to dehydroascorbic acid and hydrogen peroxide [25]. The only addition of ascorbic acid resulted statistically significant only for catechins content; no significant changes were highlighted for OD 420 nm, a*, and b* parameter.

It is notable that the interaction between light and hydrogen peroxide addition (LW*H_2O_2, LB*H_2O_2), resulted highly significant ($p < 0.001$) for all the analytical indexes. The addition of H_2O_2 allowed the increase of light exposures effect, particularly when a* and b* parameters were considered.

The hydrogen peroxide addition, also implemented in the conventional POM test, enhanced the light-induced effect on spectrophotometric and colorimetric measurements. The combination of light and H_2O_2 addition allowed better detection of the oxidative status of white wines and it could represent the simplest, fastest, and more complete approach than what is currently adopted by wineries.

The combination of light and acid ascorbic addition did not highlight significant changes, particularly on OD 420 nm, catechins, and a* parameter. The ascorbic acid is not significant, as reported by the significance of regression coefficients in Table 2.

As depicted in Figure 5, it is remarkable that b* parameter was the most sensitive analytical index, and it allowed the detection of single and combined stress treatments. The light exposure, specifically with white light for 20 min, increased significantly the b* parameter from 2.14 ± 0.10 (Control) to 3.28 ± 0.05 (LW). Moreover, the combination with H_2O_2 addition amplified the effect of white light exposure from 3.28 ± 0.05 (LW) to 3.69 ± 0.11 (LW + H_2O_2), and light blue exposure from 2.59 ± 0.12 (LB) to 3.24 ± 0.05 (LB + H_2O_2).

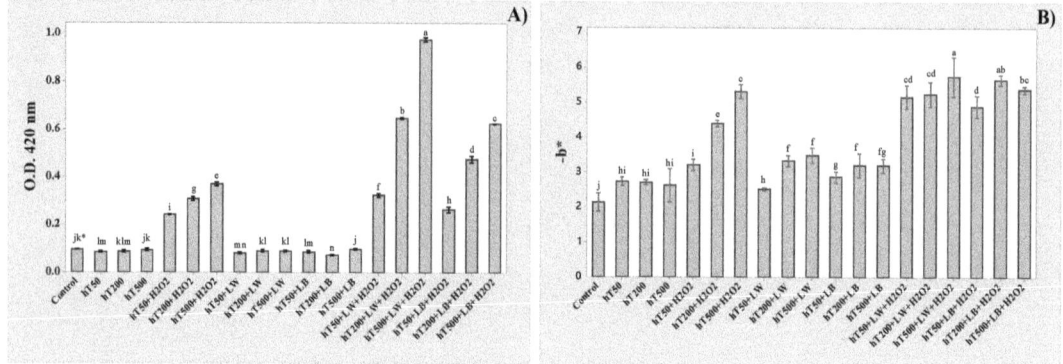

*Different letters indicate significant differences ($p < 0.05$)

Figure 5. Effect of hydrolyzable tannins' (hT) addition at different concentration (50, 200, and 500 mg/L), light exposure (LW, LB), and hydrogen peroxide addition (H_2O_2) on optical densities at 420 nm (**A**) and b* parameter (**B**).

Additionally, a good correlation between spectrophotometric and colorimetric measurements was found, specifically an inverse correlation between optical density at 420 nm and b* parameter ($R^2 > 0.89$).

2.3. Effect of Tannin Adjuvants

The use of phenols, both condensed and hydrolyzable tannins, is an acknowledged approach to limit some wine faults, such as the appearance of LST [9], thanks to their antioxidant properties as well as their ability in quenching the singlet oxygen [26,27]. During winemaking, tannins can be added on grape musts during pre-fermentative step or on finished wines as clarifying or stabilization agents.

Different concentrations of hydrolyzable (hT) and grape tannins (gT) were added to Pinot Gris wine, and combined with light exposures (LW, LB) and hydrogen peroxide addition (H_2O_2). The effect of tannins' addition and their combination with light and chemical stresses were studied on OD 420 nm (Figures 5A and 6A) and b* parameters (Figures 5B and 6B). As depicted in Figure 5A, the browning at 420 nm was not affected by the hT addition or by the concentration increase from 50 to 500 mg/L. An increase of b* parameter was pointed out from -2.14 ± 0.10 (Control) to 2.69 ± 0.09 (as mean value of hT50, hT200, and hT500), and no significant differences are shown due to the increase of hT concentration (Figure 5B).

Instead, the addition of grape tannins (gT) above 200 mg/L induced an increase of optical density at 420 nm from 0.096 ± 0.001 (Control) to 0.208 ± 0.001 (gT500) (Figure 6A).

As expected, the hydrogen peroxide induced a common increase of browning sample, and its effect was enhanced by the increase of tannin concentration.

It is remarkable to note the effect of light exposures (LW and LB), specifically comparing hydrolyzable (hT) and grape tannins (gT). The light exposure after hT addition did not affect the OD 420 nm, and b* parameter significantly enhanced at concentration above 200 mg/L. Instead, the light exposure showed a significant effect on both analytical parameters when grape tannins (gT) were added to Pinot Gris wine (Figure 6). As already reported in the literature [9,28], the tannins should be adequately chosen and their addition should be thoroughly evaluated in order not to alter the sensory properties of wine. Furthermore, as suggested by the experimental results, their choosing and addition should preserve or better enhance the oxidative stability of wines.

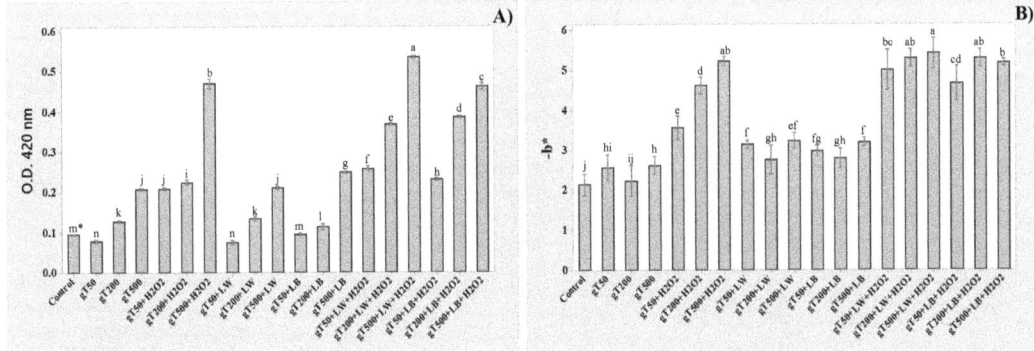

Figure 6. Effect of grape tannins' (gT) addition at different concentration (50, 200, and 500 mg/L), light exposure (LW, LB), and hydrogen peroxide addition (H_2O_2) on optical densities at 420 nm (**A**) and b* parameter (**B**).

A potential synergy behavior between light exposure and hydrogen peroxide can be highlighted for both tannins' categories, but a higher increase of analytical parameters was detected with hydrolyzable tannins (hT). Therefore, hydrolyzable tannins should be added carefully before bottling to prevent oxidation during the storage and commercialization stages.

3. Materials and Methods

3.1. Reagents and Solvents

All the solvents were of analytical grade (purity > 99%) and purchased from Sigma-Aldrich Co. (Milan, Italy). The chemical used, which include 4-(dimethylamino)-cinnamaldehyde (DAC) and (+)-catechin, were of analytical grade and purchased from Sigma-Aldrich Co. (Milan, Italy).

3.2. Wine Sample

A Pinot Gris wine from Friuli Venezia Giulia region (Italy) and 2018 vintage, was used for all the experimental trials. A Chardonnay wine from the same region and vintage was also considered. Grapes of Vitis vinifera var. Pinot Gris and Chardonnay were harvested by hand at technical maturity, and were transported immediately to the winery where they were destemmed and crushed. Grape marc was immediately separated and no maceration period was carried out. An SO_2 addition at 20 mg/L was made, and fermentation was carried out at 18 °C for 8 days and 6 days for Pinot Gris and Chardonnay, respectively. At the end of alcoholic fermentation, the wine was stored in a stainless-steel tank at 12 °C. The wine was collected in 0.75 L flint bottles, and stored in dark and fresh conditions until use. The chemical properties of Pinot Gris and Chardonnay wine are reported in Table 1.

3.3. Preliminary Stress Trials

The Pinot Gris wine was transferred in transparent glass flasks and irradiated for 4 h with 7 different color lights to evaluate the complete visible spectrum and its single ranges: violet (380–450 nm), blue (450–475 nm), light blue (476–495 nm), green (495–570 nm), yellow (570–600 nm), red (600–780 nm) and white (380–780 nm). Subsequently, the Pinot gris wine was exposed only to light blue (476–495 nm) for 1 h. A Chardonnay wine was also considered to evaluate the effect of light blue on another wine variety. All the light exposures were carried out in a portable instrument: a portable and closed case (Figure 7) constituted by multiple light-emitting diodes controlled by three switches, which allow the use of 7 different colors. The instrument was built to treat 0.75 L bottles. The case has also an air system to prevent an excessive increase of samples' temperature due to the light exposure, and a temperature probe [29].

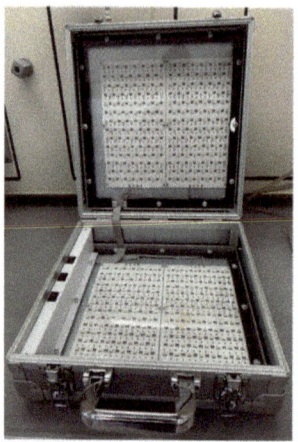

Figure 7. Protype instrument for light exposures.

3.4. Individual and Combined Stress Trials

The Pinot Gris wine was stressed by different light and chemical treatments:
- Light Exposure: light blue for 20 min; (BL)
- Light Exposure: white for 20 min; (WL)
- Hydrogen peroxide: 10 mL/L of H_2O_2 solution (3% v/v); (H_2O_2)
- Ascorbic acid: 100 mg/L; (Asc.Ac.)
- Hydrolyzable tannins: 50, 200, and 500 mg/L; (hT)
- Condensed tannins: 50, 200, and 500 mg/L; (gT)

The tannins' additions were carried out using commercial products: a mixture of ellagic tannins from oak as hydrolyzable tannins, and a mixture of grape skins' and seeds' tannins as condensed ones. Besides the single light or chemical treatment, it was some possible combinations between them were also considered.

Aliquots of 50 mL of wine samples were centrifugated at 3000 rpm for 5 min. The supernatant was transferred in 40 mL glass tubes and the chemical reagents were properly added. The light exposures were carried out in the same portable instrument previously described.

All the treatments were carried out in triplicate.

3.5. Sensorial Analysis

Pinot Gris and Chardonnay wines have been treated with different color lights, and evaluated by a sensorial analysis carried out by a panel group of 20 trained judges. The judges are researchers or oenologists experienced in wine tasting. The training of the panel group was carried out tasting wines of the same variety and category. Analysis focused on 7 general descriptors about color, aroma, and taste. Considering the results of the first evaluation, a second more detailed sensorial analysis was done considering more specific descriptors on wine color, aroma, and taste (Table 3). The judges scored the magnitude of each attribute from 1 to 9 where 1 was "low" and 9 was "high".

Table 3. Descriptors of sensorial analysis.

Descriptor Categories	Specific Descriptors	
	1st Sensorial Evaluation	2nd Sensorial Evaluation
COLOR	Intensity Tonality	Color intensity Green reflex Color preference
AROMA	Positive aroma Negative aroma	Apple Ripened fruit Citrus fruits Balsamic Spicy Vegetal Honey Aroma preference
TASTE	Positive taste Negative taste	Acidity Structure Astringency Bitterness Taste preference
AFTERTASTE	Aftertaste (general)	Persistency Pleasantness

3.6. Analytical Determination

3.6.1. Spectrophotometer Measurements

The Total Phenolic Indices (TPI) were calculated by measuring wine absorbance at 280 nm, according to Ribéreau-Gayon et al. [19]. The optical densities at 320 nm, related to hydroxycinnamic acid–tartaric acid esters (HCAs), and 420 nm, related to wine yellow color, were measured. All the determinations were carried out in a UV-Vis spectrophotometer (Shimadzu UV 1650, Milano, Italy), using distilled water as a control. All the measurements were performed in triplicate.

3.6.2. Flavan-3-ols' Content

Flavan-3-ols' content was determined according to the method proposed by Zironi et al. [30]. The chromogen reagent was prepared with 1 g of 4-(dimethylamino)-cinnamaldehyde (DAC) dissolved into 250 mL of 37% HCl and 750 mL of methanol. Next, 1 mL of diluted sample (1:25 v/v) was added to 5 mL of DAC solution. Then, absorbance was read at 640 nm in a UV-Vis spectrophotometer (Shimadzu UV 1650, Tokyo, Japan) against a blank prepared by substituting the sample with 1 mL of 10% ethanol solution. A calibration curve was made with several standard solutions of (+)-catechin, and measurements were carried out at 640 nm. All analyses were performed in triplicate. Results were expressed as milligrams of (+)-catechin equivalents per liter (mg/L).

3.6.3. Colorimetric Measurements

The chromatic characteristics and CIELAB parameters (L, a, b) were determined by measuring the transmittance of the wine every 10 nm over the visible spectrum (from 380 to 780 nm) using the illuminant D65 and 10° standard observer, following the official method published by International Organization of Vine and Wine [31]. All the measurements were carried out in a colorimeter Konica Minolta CR 300 (Tokyo, Japan).

3.6.4. Polyphenol Oxidative Medium (POM) Test

The predisposition of wines towards browning was determined by the so-called POM-test proposed by Müller-Späth (1992), with slight modification. Briefly, 5 mL of wine was heated at 60 °C for one hour, after, 25 µL of a 3% (v/v) hydrogen peroxide solution was added. The browning was estimated based on the percent increase of the absorbance at 420 nm, using the following Equation (1):

$$\text{POM (\%)} = \frac{\text{O.D.420}_{H_2O_2} - \text{O.D.420}_{H_2O}}{\text{O.D.420}_{H_2O}} \times 100 \qquad (1)$$

where O.D.420$_{H_2O_2}$ is the absorbance with H_2O_2 addition, and O.D.420$_{H_2O}$ with H_2O addition.

All the analyses were carried out with a UV-Vis spectrophotometer (Shimadzu UV 1650, Tokyo, Japan).

3.7. Statistical Analysis

All experiments and analysis were performed in triplicate and results are expressed as mean ± standard deviation. Minitab (version 17) was used for statistical analysis by one-way analysis of variance (ANOVA, with Tukey's honest significant difference (HSD) multiple comparison test) with the level of significance set at $p < 0.05$.

A third-order polynomial equation was used to express the response variable as a function of independent variable. The coefficients of the equation were determined by using Minitab 17 software (Minitab Inc., State College, PA, USA). The goodness of fit of the model was evaluated by the coefficients of determination R2 and R2-adj, and the analysis of variance (ANOVA, with Tukey's HSD multiple comparison).

4. Conclusions

White wine can be affected by light-dependent spoilages due to several factors, such as chemical composition, time and duration of light exposures, bottle shape, and color. Wineries can adopt several microbiological and technological approaches to prevent photochemical oxidation processes. However, the application of any oenological strategy needs analytical methods that provide valuable information on the oxidative stability of the wine.

A smart and portable prototype analytical instrument was adopted to expose a white wine at different light wavelengths. The sensorial analysis of untreated and treated samples showed significant changes after light treatments and some sensorial changes were detected, including an astringency increase. Time exposure of 20 min was considered enough to induce significant sensorial alterations. Treatments with white and light blue lights showed the highest significant changes on spectrophotometric and colorimetric determinations, specifically on optical densities at 420 nm and b* parameter. A good correlation was highlighted between OD420 nm and b* parameter ($R^2 > 0.89$).

The hydrogen peroxide addition allowed an increase of light-related effects, and a synergic effect could be pointed out between these two oxidative stresses.

The oenological strategy of tannins' addition was also considered and evaluated on the oxidation stability of white wines. The hydrolyzable tannins are more sensitive to the combination of light and hydrogen peroxide, compared to condensed ones. The use of tannins as clarifying and stabilizing agents should be done carefully so as not to alter the sensory quality and to promote oxidation processes.

The combination of multiple stresses using a portable instrument, associated with colorimetric measurements, could represent a valuable and fast approach that can be adopted by wineries to obtain useful information on the oxidative stability of white wines. Simple and fast methodologies are needed to accelerate decision-making procedures in order to maintain and ensure the quality of the wine until the bottle opening.

Author Contributions: Conceptualization, E.C.; methodology, E.C.; validation, E.C. and Y.S.; formal analysis, G.L. and J.F.; investigation, A.N.; resources, E.C.; data curation, A.N.; writing—original draft preparation, A.N.; writing—review and editing, A.N. and E.C..; supervision, E.C. and Y.S.; funding acquisition, E.C. All authors have read and agreed to the published version of the manuscript.

Funding: This research received no external funding.

Institutional Review Board Statement: Not applicable.

Informed Consent Statement: Not applicable.

Data Availability Statement: All data has been made available through the manuscript itself.

Acknowledgments: The authors are grateful to Ever s.r.l., especially to Giovanni Enrico Branca and Elisabetta Bellantuono, for economic and technical support. The authors are grateful to Marco D'Andrea and Lara Malgarin for their support on analytical determinations.

Conflicts of Interest: The authors declare no conflict of interest.

Sample Availability: Samples of the compounds are not available from the authors.

References

1. Waterhouse, A.L.; Sacks, G.L.; Jeffery, D.W. *Understanding Wine Chemistry*; John Wiley & Sons, Ltd.: Chichester, UK, 2016; ISBN 978-1-118-73072-0.
2. Grant-Preece, P.; Barril, C.; Schmidtke, L.M.; Scollary, G.R.; Clark, A.C. Light-Induced Changes in Bottled White Wine and Underlying Photochemical Mechanisms. *Crit. Rev. Food Sci. Nutr.* **2017**, *57*, 743–754. [CrossRef]
3. Pons, A.; Nikolantonaki, M.; Lavigne, V.; Shinoda, K.; Dubourdieu, D.; Darriet, P. New Insights into Intrinsic and Extrinsic Factors Triggering Premature Aging in White Wines. In *ACS Symposium Series*; Ebeler, S.B., Sacks, G., Vidal, S., Winterhalter, P., Eds.; American Chemical Society: Washington, DC, USA, 2015; Volume 1203, pp. 229–251. ISBN 978-0-8412-3010-1.
4. Freitas, A.I.; Pereira, V.; Leça, J.M.; Pereira, A.C.; Albuquerque, F.; Marques, J.C. A Simple Emulsification-Assisted Extraction Method for the GC–MS/SIM Analysis of Wine Markers of Aging and Oxidation: Application for Studying Micro-Oxygenation in Madeira Wine. *Food Anal. Methods* **2018**, *11*, 2056–2065. [CrossRef]
5. Oliveira, C.M.; Ferreira, A.C.S.; De Freitas, V.; Silva, A.M.S. Oxidation Mechanisms Occurring in Wines. *Food Res. Int.* **2011**, *44*, 1115–1126. [CrossRef]
6. Li, H.; Guo, A.; Wang, H. Mechanisms of Oxidative Browning of Wine. *Food Chem.* **2008**, *108*, 1–13. [CrossRef]
7. Guo, A.; Kontoudakis, N.; Scollary, G.R.; Clark, A.C. Production and Isomeric Distribution of Xanthylium Cation Pigments and Their Precursors in Wine-like Conditions: Impact of Cu(II), Fe(II), Fe(III), Mn(II), Zn(II), and Al(III). *J. Agric. Food Chem.* **2017**, *65*, 2414–2425. [CrossRef] [PubMed]
8. Gabrielli, M.; Fracassetti, D.; Romanini, E.; Colangelo, D.; Tirelli, A.; Lambri, M. Oxygen-Induced Faults in Bottled White Wine: A Review of Technological and Chemical Characteristics. *Food Chem.* **2021**, *348*, 128922. [CrossRef]
9. Fracassetti, D.; Di Canito, A.; Bodon, R.; Messina, N.; Vigentini, I.; Foschino, R.; Tirelli, A. Light-Struck Taste in White Wine: Reaction Mechanisms, Preventive Strategies and Future Perspectives to Preserve Wine Quality. *Trends Food Sci. Technol.* **2021**, *112*, 547–558. [CrossRef]
10. Clark, A.C. The Production of Yellow Pigments from (+)-Catechin and Dihydroxyfumaric Acid in a Model Wine System. *Eur. Food Res. Technol.* **2008**, *226*, 925–931. [CrossRef]
11. Nikolantonaki, M.; Coelho, C.; Noret, L.; Zerbib, M.; Vileno, B.; Champion, D.; Gougeon, R.D. Measurement of White Wines Resistance against Oxidation by Electron Paramagnetic Resonance Spectroscopy. *Food Chem.* **2019**, *270*, 156–161. [CrossRef]
12. Makhotkina, O.; Kilmartin, P.A. The Use of Cyclic Voltammetry for Wine Analysis: Determination of Polyphenols and Free Sulfur Dioxide. *Anal. Chim. Acta* **2010**, *668*, 155–165. [CrossRef]
13. Comuzzo, P.; Toniolo, R.; Battistutta, F.; Lizee, M.; Svigelj, R.; Zironi, R. Oxidative Behavior of (+)-Catechin in the Presence of Inactive Dry Yeasts: A Comparison with Sulfur Dioxide, Ascorbic Acid and Glutathione: Antioxidant Capacity of Yeast Derivatives Compared with Other Wine Additives. *J. Sci. Food Agric.* **2017**, *97*, 5158–5167. [CrossRef]
14. Gonzalez, A.; Vidal, S.; Ugliano, M. Untargeted Voltammetric Approaches for Characterization of Oxidation Patterns in White Wines. *Food Chem.* **2018**, *269*, 1–8. [CrossRef]
15. Zhang, Q.-A.; Shen, Y.; Fan, X.; Martín, J.F.G.; Wang, X.; Song, Y. Free Radical Generation Induced by Ultrasound in Red Wine and Model Wine: An EPR Spin-Trapping Study. *Ultrason. Sonochem.* **2015**, *27*, 96–101. [CrossRef] [PubMed]
16. Márquez, K.; Contreras, D.; Salgado, P.; Mardones, C. Production of Hydroxyl Radicals and Their Relationship with Phenolic Compounds in White Wines. *Food Chem.* **2019**, *271*, 80–86. [CrossRef] [PubMed]
17. Romanini, E.; Colangelo, D.; Lucini, L.; Lambri, M. Identifying Chemical Parameters and Discriminant Phenolic Compounds from Metabolomics to Gain Insight into the Oxidation Status of Bottled White Wines. *Food Chem.* **2019**, *288*, 78–85. [CrossRef] [PubMed]
18. Calò, A.; Scienza, A.; Costacurta, A. *Vitigni d'Italia*, 2nd ed.; Calderini Edagricole: Bologna, Italy, 2006; ISBN 978-88-506-5173-3.
19. Dias, D.A.; Smith, T.A.; Ghiggino, K.P.; Scollary, G.R. The Role of Light, Temperature and Wine Bottle Colour on Pigment Enhancement in White Wine. *Food Chem.* **2012**, *135*, 2934–2941. [CrossRef] [PubMed]
20. Cao, H.; Saroglu, O.; Karadag, A.; Diaconeasa, Z.; Zoccatelli, G.; Conte-Junior, C.A.; Gonzalez-Aguilar, G.A.; Ou, J.; Bai, W.; Zamarioli, C.M.; et al. Available Technologies on Improving the Stability of Polyphenols in Food Processing. *Food Front.* **2021**, *2*, 109–139. [CrossRef]
21. Hornedo-Ortega, R.; Reyes González-Centeno, M.; Chira, K.; Jourdes, M.; Teissedre, P.-L. Phenolic Compounds of Grapes and Wines: Key Compounds and Implications in Sensory Perception. In *Chemistry and Biochemistry of Winemaking, Wine Stabilization and Aging*; Cosme, F., Nunes, F.M., Filipe-Ribeiro, L., Eds.; IntechOpen: London, UK, 2021; ISBN 978-1-83962-575-6.

22. Brothersen, C.; McMahon, D.J.; Legako, J.; Martini, S. Comparison of Milk Oxidation by Exposure to LED and Fluorescent Light. *J. Dairy Sci.* **2016**, *99*, 2537–2544. [CrossRef] [PubMed]
23. Chemat, F.; Garrigues, S.; de la Guardia, M. Portability in Analytical Chemistry: A Green and Democratic Way for Sustainability. *Curr. Opin. Green Sustain. Chem.* **2019**, *19*, 94–98. [CrossRef]
24. Bradshaw, M.P.; Barril, C.; Clark, A.C.; Prenzler, P.D.; Scollary, G.R. Ascorbic Acid: A Review of Its Chemistry and Reactivity in Relation to a Wine Environment. *Crit. Rev. Food Sci. Nutr.* **2011**, *51*, 479–498. [CrossRef]
25. Barril, C.; Rutledge, D.N.; Scollary, G.R.; Clark, A.C. Ascorbic Acid and White Wine Production: A Review of Beneficial versus Detrimental Impacts: Ascorbic Acid and White Wine. *Aust. J. Grape Wine Res.* **2016**, *22*, 169–181. [CrossRef]
26. Fracassetti, D.; Limbo, S.; Pellegrino, L.; Tirelli, A. Light-Induced Reactions of Methionine and Riboflavin in Model Wine: Effects of Hydrolysable Tannins and Sulfur Dioxide. *Food Chem.* **2019**, *298*, 124952. [CrossRef] [PubMed]
27. Lagunes, I.; Vázquez-Ortega, F.; Trigos, Á. Singlet Oxygen Detection Using Red Wine Extracts as Photosensitizers: Singlet Oxygen Detection in Red Wine. *J. Food Sci.* **2017**, *82*, 2051–2055. [CrossRef] [PubMed]
28. Vignault, A.; González-Centeno, M.R.; Pascual, O.; Gombau, J.; Jourdes, M.; Moine, V.; Iturmendi, N.; Canals, J.M.; Zamora, F.; Teissedre, P.-L. Chemical Characterization, Antioxidant Properties and Oxygen Consumption Rate of 36 Commercial Oenological Tannins in a Model Wine Solution. *Food Chem.* **2018**, *268*, 210–219. [CrossRef] [PubMed]
29. Branca, E.G. Device and Method for the Analysis of Fluid Products. Patent PI 10UD2013A000091, 28 June 2013.
30. Zironi, R.; Buiatti, S.; Celotti, E. Evaluation of a New Colorimetric Method for the Determination of Catechins in Musts and Wines. *Wein Wiss.* **1992**, *47*, 1–7.
31. OIV (International Organization of Vine and Wine). *Compendium of International Methods of Wine and Must Analysis*; OIV: Paris, France, 2020; Volume 1, ISBN 978-2-85038-017-4.

Article

Light-Struck Taste in White Wine: Protective Role of Glutathione, Sulfur Dioxide and Hydrolysable Tannins

Daniela Fracassetti *, Sara Limbo, Natalia Messina, Luisa Pellegrino and Antonio Tirelli

Department of Food, Environmental and Nutritional Sciences, Università degli Studi di Milano, Via G. Celoria 2, 20133 Milan, Italy; sara.limbo@unimi.it (S.L.); nataliamessina.ita@gmail.com (N.M.); luisa.pellegrino@unimi.it (L.P.); antonio.tirelli@unimi.it (A.T.)
* Correspondence: daniela.fracassetti@unimi.it

Abstract: Light exposure of white wine can cause a light-struck taste (LST), a fault induced by riboflavin (RF) and methionine (Met) leading to the formation of volatile sulfur compounds (VSCs), including methanethiol (MeSH) and dimethyl disulfide (DMDS). The study aimed to investigate the impact of different antioxidants, i.e., sulfur dioxide (SO_2), glutathione (GSH) and chestnut tannins (CT), on preventing LST in model wine (MW) and white wine (WW), both containing RF and Met. Both MW and WW samples were added with the antioxidants, either individually or in different combinations, prior to 2-h light exposure and they were stored in the dark for 24 months. As expected, the light induced the degradation of RF in all the conditions assayed. Met also decreased depending on the antioxidants added. The presence of antioxidants limited the formation of LST as lower concentrations of VSCs were found in both MW and WW samples. In the latter matrix, neither MeSH nor DMDS were detected in the presence of CT, while only DMDS was found in WW+GSH, WW+SO_2+GSH and WW+CT+SO_2 samples at a concentration lower than the perception thresholds. Considering the antioxidants individually, the order of their effectiveness was CT ≥ GSH > SO_2 in WW under the adopted experimental conditions. The results indicate tannins as an effective enological tool for preventing LST in white wine and their use will be further investigated in different white wines under industrial scale.

Keywords: glutathione; sulfur dioxide; hydrolysable tannins; light-struck taste; storage; white wine

Citation: Fracassetti, D.; Limbo, S.; Messina, N.; Pellegrino, L.; Tirelli, A. Light-Struck Taste in White Wine: Protective Role of Glutathione, Sulfur Dioxide and Hydrolysable Tannins. *Molecules* **2021**, *26*, 5297. https://doi.org/10.3390/molecules26175297

Academic Editors: Fernando M. Nunes, Fernanda Cosme and Luís Filipe Ribeiro

Received: 30 July 2021
Accepted: 28 August 2021
Published: 31 August 2021

Publisher's Note: MDPI stays neutral with regard to jurisdictional claims in published maps and institutional affiliations.

Copyright: © 2021 by the authors. Licensee MDPI, Basel, Switzerland. This article is an open access article distributed under the terms and conditions of the Creative Commons Attribution (CC BY) license (https://creativecommons.org/licenses/by/4.0/).

1. Introduction

Light exposure of white wine, especially at wavelengths spanning from 370 to 450 nm, has a detrimental impact on its sensory characteristics. In particular, photo-induced chemical reactions can be responsible for the wine fault known as light-struck taste (LST) [1,2]. This defect actually arises from two opposite circumstances: the loss of floral and fruity notes [3], and the development of undesired flavors described as cooked cabbage, rotten eggs and onion [4]. The sulfur compounds related to this fault are methanethiol (MeSH) and dimethyl disulfide (DMDS) that are generated through the reaction between riboflavin (RF), a highly photo-sensitive vitamin, and methionine (Met), a sulfur-containing amino acid [5]. Two photo-oxidative mechanisms have been described, both involving RF. In the Type II mechanism, the excited RF transfers the excess of energy to molecular oxygen; as a consequence, singlet oxygen is generated, a very unstable, electrophilic species, capable of reacting with many compounds, including amino acids [6,7]. Once RF is reduced, it can reduce oxygen and return to its ground state [7,8]. In the Type I mechanism, when RF is exposed to light, it reaches the excited triplet state and reacts directly with electron donors, such as phenols and amino acids [9]. In particular, when Met acts as an electron donor, methional is generated. The latter compound is chemically unstable, photo-sensitive and easily decomposes to MeSH and acrolein through a retro-Michael reaction. Along with the early steps of photo-oxidation, MeSH can be generated by an alternative pathway that involves a direct cleavage of Met side chain [10]. Moreover, two molecules of MeSH can yield DMDS [11]. The olfactory perception thresholds for MeSH and DMDS in wine are 2–10 µg/L and 20–45 µg/L, respectively, the latter compound being less volatile [12]. Beside the photo-degradation of RF, other photo-induced reactions may involve tartaric acid and its complexes with iron ions [13]. The reactions result in glyoxylic acid formation, which in turn generates xanthylium ions, the species responsible for the browning of white wine [14].

Several varieties of white wine showed the tendency to develop LST. Previous studies showed that the risk of this fault decreases when RF concentration is lower than 50–80 µg/L [15–17]. The content of RF in grapes is too low for triggering LST [18], but it can increase during the fermentation process due to *Saccharomyces cerevisiae* metabolic activities. Interestingly, the level of RF in wine is strictly dependent on the *Saccharomyces* strain performing the alcoholic fermentation [19,20]. Consequently, beside protecting white wine from the light [2], oenological strategies suitable for limiting the risk of LST occurrence can be the use of low RF-producer yeast strains and RF removal prior to bottling [20]. The latter approach can be achieved by wine treatment with either bentonite (1 g/L) [15,18] or active charcoal (at relatively low concentrations, 0.1 g/L) [20]. These adjuvants are capable of removing up to about 70% of RF. However, the use of active charcoal as well as a high concentration of bentonite should be limited, as they may cause an aroma depletion of wine [18].

An additional oenological strategy for the prevention of LST can be the use of selected phenols. Maujean and Seguin [11] reported that the addition of flavan-3-ols can limit this wine fault, probably because of their light-shielding effect. Recently, the capability of hydrolysable tannins against LST occurrence was shown in model wine as they prevented the formation of sulfur compounds associated with LST [17]. More specifically, the protective effect of tannins can be ascribed to their competition with Met in donating electrons active in the reduction of RF [21]. In addition, Fracassetti and co-authors [17] hypothesized that singlet oxygen can oxidize tannins to quinones capable of binding MeSH; in this way, formation of DMDS is limited and LST resulted less perceived. Prevention of LST in white wine is of particular interest for the winemakers since this fault may cause the recall of bottled wine from the market [9]. As recently showed by Arapitsas and co-authors [22], LST is extremely persistent in wine and it is still perceived one year after the light exposure. However, the evolution of LST in white wine during its shelf life has not been investigated yet in the presence of antioxidant agents that can be commonly added. Among these, sulfur dioxide (SO_2) is the most widely used, as well as reduced glutathione (GSH) and hydrolysable tannins. GSH is able to reduce *o*-quinones back to cathecols [23]. GSH can also limit the loss of some aromas, prevent the atypical ageing of white wine and slow down the browning during ageing [24]. Among hydrolysable tannins, ellagitannins can protect phenols against oxidation more reactive to molecular oxygen than the native phenols of wine due to their large number of hydroxyl groups [25,26]. Even if tannins show a preventative effect against the appearance of LST [17], their addition in sparkling wine promoted the formation of sotolon, a marker of atypical ageing [27]. To the best of our knowledge, the effects of these antioxidants against LST have been not investigated. Major unanswered questions related to their use are (i) how a white wine susceptible to LST may change when exposed to light during the storage, and (ii) whether the developed LST persists over time.

Based on these questions, this study aimed to evaluate LST in a simple model wine solution (MW) and in a white wine (WW), both containing RF and Met, when initially exposed to light for a defined time and then stored in the dark for 24 months. The possible protective effect against the LST of the selected antioxidants, including SO_2, GSH and chestnut tannins (CT) was studied, by adding them, either individually or in different combinations, to both MW and WW.

2. Results and Discussion

The effects of selected antioxidant additives, namely SO_2, GSH and CT, added individually or in combination, were evaluated in both MW and WW after 24-month storage in the dark with and without a discreet exposure to light (2 h) prior the storage. The planned experiments would simulate the possible short-term light exposure of wine after bottling in a winery or on the shelf of a store, followed by the storage in the dark condition before commercialization or after purchase.

The amounts of GSH (average amount added: 50 ± 4 mg/L) and SO_2 (average amount added: 25 ± 3 mg/L) were chosen based on the results of a previous study [28]. The addition of GSH took into account the possible residual content of GSH in wine that can be up to 30 mg/L [29] and the supplementation allowed by the International Organization of Vine and Wine (OIV) (20 mg/L) [30]. The addition of RF (200 µg/L) approaches the amounts (150 µg/L [16] or even higher) that can occur in wine depending on the yeast strain performing the fermentation [20], while that of Met (4 mg/L) corresponds to the average amounts in wine (3–5 mg/L) [13,31,32]. Hydrolysable tannins showed the ability to prevent LST in model wine when added at 40 mg/L level [17]. A slightly higher concentration of 50 mg/L was adopted in the present study in order to further prevent the appearance of LST without promoting bitterness and astringency [33]. Since polyphenols can be involved in the oxidative pathways generating sotolon [27], a marker of atypical ageing of white wine, its presence was also considered.

2.1. Additives and Storage: Effects in Model Wine Solution

2.1.1. Storage in the Dark without Light Exposure

The effect of the tested antioxidants, added individually or in different combination, was firstly evaluated in MW samples stored without light exposure. As expected, RF was still present (193.5 ± 13.5 µg/L), which is not surprising since RF is relatively stable to heat-treatments, dehydration and usual food storage conditions [34,35]. In contrast, this compound is extremely sensitive to visible or UV light. The decrease of Met (concentration added: 4.63 ± 0.28 mg/L) was from small to negligible (-2%) and only occurred in the absence of additives (Table 1). Differently, the decrease of Met was dependent on the additives added and ranged from -24% in MW+SO_2 and MW+SO_2+GSH samples up to -100% in MW+GSH and MW+CT+GSH samples (Table 1). Among compounds expected to arise from the oxidation of Met [36], only Met sulfoxide was found, in accordance with the previous study carried out by NMR [21]. The absence of this compound in MW samples without additives allowed to exclude the possible oxidation of Met by the acidic environment or matrix components. In the presence of SO_2, Met sulfoxide could arise from aerobic oxidation of bisulfite, leading to several radical species [37]. Met was completely oxidized into Met sulfoxide in MW+GSH and MW+SO_2+GSH samples (Table 1). Under our experimental conditions, it seems that GSH behaved as a pro-oxidant instead of an antioxidant, possibly because of its efficiency in scavenging free radicals, thus generating thiyl radicals. The thiyl radical favors the formation of superoxide as well as singlet oxygen [38]. As a consequence, the oxidation of Met to Met sulfoxide could be promoted even because Met is one of the amino acids mainly targeted by singlet oxygen [39]. Consistently, Met sulfoxide was the main product explaining the loss of Met in the samples with added SO_2 and GSH, alone or in combination (Table 1). On the contrary, Met sulfoxide did not quantitatively correspond to Met lost in the presence of CT, with or without SO_2 and GSH, thus suggesting that compounds other than Met sulfoxide could be generated in these conditions [36].

GSH strongly decreased in all samples and it was not detected in MW+GSH treatment (Table 1). Cys was found in the samples where GSH was present, with the exception of the MW+GSH sample, and likely derived from the hydrolysis of this tripeptide [28] due to the long storage and the acidic environment adopted in this study.

Little differences were found in TPI that resulted slightly higher, though statistically significant, in the MW+CT+SO$_2$+GSH sample. Comparable absorbance values at 420 nm were found in the MW samples added with CT, with or without SO$_2$ and GSH.

The volatile sulfur compounds (VSCs) associated with LST were determined even for the trial without light exposure. None of VSCs, namely MeSH, DMDS and dimethyl trisulfide (DMTS), were detected in MW samples stored in the dark without light exposure. Similarly, no perception of the cooked cabbage note occurred in those samples (data not shown).

2.1.2. Storage in the Dark after Light Exposure

No residual RF was found in MW samples exposed to light prior to storage, irrespective of the antioxidant mixture added (data not shown). As observed for the samples stored without light exposure, the decrease of Met was related to the presence of additives, and it ranged between −41% in samples without any additives and −100% with MW+CT+GSH treatment (Table 1). In terms of Met loss, the impact of light exposure of MW samples was evident in most of the conditions tested. The exceptions were MW+CT and MW+CT+SO$_2$ samples, where the Met decrease was comparable in treatments with and without light exposure, and MW+CT+GSH samples, where Met was not revealed after storage (Table 1). Contrarily to what was observed in the absence of light exposure, both Met and GSH were still detected after storage in MW+GSH samples. This behavior is hard to explain; the efficient scavenging activity of GSH and the generated thiyl radicals [38] may participate in the photo-oxidative reactions and limit the oxidation of Met, while these radicals could oxidize Met in dark storage (Table 1). Residual GSH concentrations ranged from 2.39 ± 0.65 mg/L to 7.86 ± 0.63 mg/L in MW+GSH and MW+CT+GSH samples, respectively (Table 1). To a certain extent, under the adopted experimental conditions, the light exposure seems to limit the degradation/hydrolysis of GSH when added alone or in combination with CT. The scavenger and/or quencher activity of hydrolysable tannins may have a protective effect towards GSH since the residual level of GSH was higher in samples also added with CT. In any case, further investigations are needed in order to clarify GSH reactivity towards Met and its role in photo-degradative mechanisms. The lowest concentrations of Met sulfoxide were found in the CT-added samples, despite the little amounts of residual Met. These results suggest that the formation of Met sulfoxide is lower and Met can go through different oxidative fate [36] in the presence of hydrolysable tannins.

Light exposure showed a negligible effect on TPI in MW samples where CT was added (Table 1); the addition of the other antioxidants did not significantly influence the TPI. Differences were found in the absorbance values at 420 nm that were significantly higher in MW+CT sample, although to a small extent. In comparison with the samples stored in the dark without light exposure (Table 1), the absorbance values at 420 nm were nearly halved (Table 1). The photo-induced mechanisms did not lead to a browning phenomenon in the experimental conditions adopted, possibly because of the absence of the transition metals, catalyzers of oxidations [14,40].

Table 1. Concentrations of methionine, methionine sulfoxide, glutathione, cysteine, total phenol index and absorbance at 420 nm determined in model wine solution (MW) samples.

Treatment	Methionine mg/L (µmol/L)	Methionine Sulfoxide mg/L (µmol/L)	Glutathione mg/L	Cysteine mg/L	Total Phenol Index	Absorbance at 420 nm AU
Samples Stored in the Dark for 24 Months without Light Exposure						
MW	4.52 ± 0.32 [a] (30.30)	nc [e]	na	na	na	na
MW+SO$_2$	3.54 ± 0.25 [b] (23.74)	0.47 ± 0.03 [b] (2.87)	na	na	na	na
MW+GSH	nd [c]	5.18 ± 0.36 [a] (31.34)	Nd [c]	nd [c]	na	na
MW+SO$_2$+GSH	3.51 ± 0.25 [b] (23.54)	0.85 ± 0.06 [c] (5.14)	2.20 ± 0.11 [a]	0.46 ± 0.02 [a]	na	na
MW+CT	0.31 ± 0.02 [d] (2.07)	0.43 ± 0.03 [b] (2.63)	na	na	0.948 ± 0.006 [a]	0.039 ± 0.001 [a]
MW+CT+SO$_2$	0.49 ± 0.03 [d] (3.28)	0.27 ± 0.02 [b] (1.63)	na	na	0.962 ± 0.010 [a]	0.037 ± 0.002 [a]
MW+CT+GSH	nd [c]	0.50 ± 0.34 [b] (3.04)	0.28 ± 0.014 [b]	0.32 ± 0.02 [b]	0.946 ± 0.006 [a]	0.041 ± 0.002 [a]
MW+CT+SO$_2$+GSH	1.54 ± 0.11 [e] (10.29)	1.72 ± 0.12 [d] (10.39)	0.24 ± 0.012 [b]	0.30 ± 0.01 [b]	0.987 ± 0.007 [b]	0.043 ± 0.002 [a]
Samples Stored in the Dark for 24 Months after Light Exposure						
MW	2.71 ± 0.31 [a] (18.15)	0.88 ± 0.73 [b] (5.34)	na	na	na	na
MW+SO$_2$	2.50 ± 0.14 [a] (16.78)	2.19 ± 0.36 [a] (13.28)	na	na	na	na
MW+GSH	2.60 ± 0.58 [a] (17.42)	0.89 ± 0.47 [b] (5.37)	2.39 ± 0.65 [c]	0.31 ± 0.01 [c]	na	na
MW+SO$_2$+GSH	2.04 ± 0.08 [b] (13.69)	2.24 ± 0.06 [a] (13.58)	2.57 ± 0.83 [c]	0.33 ± 0.00 [c]	na	na
MW+CT	0.28 ± 0.07 [c] (1.85)	0.35 ± 0.03 [c] (2.13)	na	na	0.918 ± 0.006 [a]	0.028 ± 0.003 [a]
MW+CT+SO$_2$	0.27 ± 0.02 [c] (1.81)	0.28 ± 0.04 [c] (1.69)	na	na	0.922 ± 0.030 [a]	0.023 ± 0.001 [bc]
MW+CT+GSH	nd [d]	0.44 ± 0.06 [c] (2.67)	7.86 ± 0.63 [a]	0.37 ± 0.04 [b]	0.952 ± 0.007 [a]	0.027 ± 0.002 [ab]
MW+CT+SO$_2$+GSH	0.21 ± 0.04 [c] (1.39)	0.26 ± 0.04 [c] (1.58)	3.99 ± 1.04 [b]	1.07 ± 0.02 [a]	0.936 ± 0.050 [a]	0.020 ± 0.006 [c]

For samples coding see Table 6. The average added concentrations of RF and Met were 193.5 ± 13.5 µg/L and 4.63 ± 0.28 mg/L (31.1 µmol/L), respectively. Methionine sulfone was not detected in any of the samples analysed. Different letters along column mean significant differences ($p < 0.05$). Legend: nd, not detected; na, not analyzed.

The concentrations of MeSH and DMDS were influenced by the antioxidant added but were significantly higher in the MW sample without any antioxidants. The addition of all three antioxidants was most effective in MW, as only negligible amounts of MeSH and with no DMDS were found (Table 2).

Table 2. Concentrations (µg/L) of methanethiol (MeSH), dimethyl disulfide (DMDS) and dimethyl trisulfide (DMTS) in model wine (MW) samples stored in the dark after light exposure.

Treatment	MeSH		DMDS		DMTS		Ratio Sulfur Formed/Met Degraded
	µg/L	OAV	µg/L	OAV	µg/L	OAV	
MW	10.83 ± 0.99 [a]	36.1	47.65 ± 3.86 [a]	1.1–2.4	64.04 ± 7.17 [ab]	640	14.05
MW+SO_2	1.42 ± 0.13 [d]	4.7	nd	—	5.97 ± 0.67 [d]	59.7	0.82
MW+GSH	1.38 ± 0.13 [d]	4.6	0.67 ± 0.05 [b]	<0.03	64.94 ± 7.27 [ab]	649	7.79
MW+SO_2+GSH	2.93 ± 0.27 [c]	9.8	0.99 ± 0.08 [b]	<0.05	5.43 ± 0.61 [d]	54.3	0.87
MW+CT	1.14 ± 0.10 [de]	3.8	nd	—	73.52 ± 8.23 [a]	735	4.93
MW+CT+SO_2	6.07 ± 0.55 [b]	20.2	nd	—	27.16 ± 3.04 [c]	272	2.14
MW+CT+GSH	0.45 ± 0.04 [e]	1.5	nd	—	56.28 ± 6.30 [b]	563	3.56
MW+CT+SO_2+GSH	0.45 ± 0.04 [e]	1.5	nd	—	0.75 ± 0.08 [e]	7.5	0.07

The Odor Activity Values (OAVs) were calculated as the ratio between the amount found in the sample and the perception threshold for each volatile sulfur compound. The perception threshold concentrations considered are as follows: MeSH, 0.3 µg/L; DMDS, 20–45 µg/L; DMTS, 0.1 µg/L [12]. For samples coding see Table 6. Different letters mean significant differences ($p < 0.05$). Legend: nd, not detected.

This result was in accordance with the outcome of sensory analysis: The perception of the "cooked cabbage" note was negligible (2/9) for the MW+CT+SO_2+GSH sample, while the highest perception (7/9) was in the MW+CT+SO_2 sample (Figure 1). The formation of DMTS could be dependent on the long storage since it was absent in MW samples with added hydrolysable tannins just after the light exposure as previously observed [17]. DMTS could originate upon storage from the oxidation of methional and MeSH [41] and its formation could be prevented by SO_2, as lower levels of DMTS were found in the presence of SO_2 (Table 2). Interestingly, considering the sulfur conversion yield (sulfur formed/Met degraded), values lower than 1 were found in the treatments with SO_2 alone or in combination with GSH and CT. These findings suggest that in these samples, Met mainly acted as an electron-donor to reduce RF. In the other samples, additional chemical pathways were also involved leading to Met oxidation, i.e., reaction with singlet oxygen, forming Met sulfoxide and other oxidative products [36] as mentioned above.

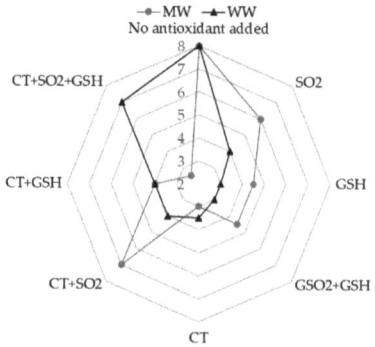

Figure 1. Sensory perception of the descriptor "cooked cabbage" related to the light-struck taste for both model wine solution (MW) and white wine (WW) both stored in the dark after light exposure. Data were obtained from medians of the scores indicated by the judges. For samples coding see to Table 6. Legend: SO_2, sulfur dioxide; GSH, glutathione; CT, chestnut tannins.

2.2. Additives and Storage: Effects in White Wine

2.2.1. Storage in the Dark without Light Exposure

In all WW samples, RF was still present (177.2 ± 4.2 µg/L) after 24-month storage in the dark without light exposure, while an overall decrease of Met was observed (Table 3) in the experimental conditions adopted. Such a decrease of Met was small (about −5%) in the WW_0 sample (not spiked white wine) whereas, in the additive-spiked WW samples, it ranged from −26% (WW+GSH) to −59% (WW+CT+SO_2+GSH). These data suggest that the degradation of Met could be due to its oxidative deamination [42] as it was limited by GSH and promoted by SO_2. The aerobic oxidation of bisulfite, leading to several radical species [37], might cause a higher loss of Met.

Cys content was comparable in all samples and was 3.10 ± 0.07 mg/L on average (Table 3). The strong decrease of GSH, observed in all the GSH-spiked samples, did not correspond to an increase of Cys. SO_2 did not prevent GSH oxidation since no significant differences were found between WW+GSH and WW+SO_2+GSH samples (Table 3). On the contrary, CT led to significantly higher concentrations of GSH that persisted in WW samples after 24-month storage in the dark. Such a difference in GSH levels can be ascribed to the ability of CT to consume oxygen [43] due to its galloyl- groups [44], thus protecting GSH against oxidation. The negligible effect of SO_2 against GSH oxidation was also revealed in the WW+CT+SO_2+GSH sample whose GSH concentration was not significantly different from that of the WW+CT+GSH sample.

Slight differences were found in both TPI and flavonoids depending on the different additives tested and their combinations (Table 3). The lowest levels of TPI and flavonoids were found in the presence of both GSH and SO_2. The absorbance values at 420 nm were significantly lower in the presence of SO_2, confirming the efficacy of this antioxidant in protecting the yellow color of white wine [27].

None of the VSCs, namely MeSH, DMDS and DMTS, were detected in this set of samples and, consistently, the perception of the cooked cabbage note was only negligible as the samples were scored 2/9 at maximum (data not shown). This finding indicates that an LST-susceptible wine, even if intentionally, does not develop this fault until it is protected against the light, e.g., by using dark bottles [22]. However, the light exposure of white wine in dark bottles can still have an indirect impact through the increase in temperature. Maury and co-authors [45] found major browning caused by the high level of xanthylium ions present in dark bottles and released due to high temperature. Proper oenological strategies and storage conditions are essential to preserve the wine quality after bottling.

2.2.2. Storage in the Dark after Light Exposure

Similar to MW samples, no RF was detected in WW samples independently of the antioxidants tested (data not shown).

Met content decreased in WW samples containing both GSH and SO_2 (−21%) or SO_2 only (−38%) (Table 3), suggesting the influence of the antioxidants on photo-degradative mechanisms and their competition with Met in both Type I and Type II pathways [13,17,21].

Changes in the profile of free amino acids in WW samples were found to be dependent on the antioxidants added (Figure A1 in Appendix A), with the exceptions of serine, aspartic acid, isoleucine, valine and lysine whose concentration decreased in all assayed conditions, and alanine, glutamine and phenylalanine showing negligible differences (data not shown). While tryptophan was not detected in any sample, possibly because of a concentration lower than the detection limit, Cys was detected only in WW samples added with GSH that, as already mentioned, can be its parent molecule [28]. For other amino acids, such as histidine and tyrosine, the addition of SO_2 and its combination with CT led to a small decrease (Figure A1 in Appendix A). Overall, the decrease of histidine, tyrosine, Met and Cys could be due to the reaction with singlet oxygen, indicating that amino acids other than Met can act as electron donors bringing RF back to its reduced state. Min and Boff [39] reported that singlet oxygen mainly reacts with five amino acids (tryptophan, histidine, tyrosine, Met and Cys). GSH could act as an electron donor in the reduction of

RF; in fact, even if it decreased up to 88% in WW samples stored in the dark, GSH contents halved in samples exposed to light in comparison to those stored in the dark (Table 3). Both CT and SO_2 did prevent GSH oxidation since significant differences were found in treatments with combined addition of the different additives (Table 3).

Changes in the profile of free amino acids in WW samples were found to be dependent on the antioxidants added (Figure A1 in Appendix A), with the exceptions of serine, aspartic acid, isoleucine, valine and lysine whose concentration decreased in all assayed conditions, and alanine, glutamine and phenylalanine showing negligible differences (data not shown). While tryptophan was not detected in any sample, possibly because of a concentration lower than the detection limit, Cys was detected only in WW samples added with GSH that, as already mentioned, can be its parent molecule [28]. For other amino acids, such as histidine and tyrosine, the addition of SO_2 and its combination with CT led to a small decrease (Figure A1 in Appendix A). Overall, the decrease of histidine, tyrosine, Met and Cys could be due to the reaction with singlet oxygen, indicating that amino acids other than Met can act as electron donors bringing RF back to its reduced state. Min and Boff [39] reported that singlet oxygen mainly reacts with five amino acids (tryptophan, histidine, tyrosine, Met and Cys). GSH could act as an electron donor in the reduction of RF; in fact, even if it decreased up to 88% in WW samples stored in the dark, GSH contents halved in samples exposed to light in comparison to those stored in the dark (Table 3). Both CT and SO_2 did prevent GSH oxidation since significant differences were found in treatments with combined addition of the different additives (Table 3).

The absorbance values at 420 nm were lower in WW samples that were exposed to light before the dark storage in comparison to those that were not. Furthermore, a major protective effect on yellow color was observed in the presence of SO_2 (0.061 ± 0.001 AU), GSH (0.073 ± 0.000 AU) or the combination of the two (0.071 ± 0.001 AU) (Table 3). In the presence of CT, the absorbance values at 420 nm were slightly higher (0.090 ± 0.006–0.103 ± 0.006 AU), but still halved compared to the same samples stored in the dark (Table 3). These findings differ from previous literature results since a browning increase was reported to be due to the light exposure [2,14,40]. Such a difference could depend on the wine tested in the study or the light source employed for the light exposure. We could expect the metal-mediated oxidative phenomena to occur since both iron and copper were present in WW although at low concentrations (1.95 mg/L and 0.24 mg/L for iron and copper, respectively). Further investigation is needed to better clarify this aspect.

The content of both MeSH and DMDS varied remarkably in WW samples, depending on the antioxidants added (Table 4), and both compounds were not detected in WW+CT sample. No DMTS was detected in all samples. No MeSH was found and the DMDS concentration was lower than the perception threshold in WW+GSH, WW+SO_2+GSH and WW+CT+SO_2 (Table 4). This result was also supported by the sensory analysis indicating no significant differences between the above-mentioned samples (Figure 1). The MeSH concentration was higher than the respective perception threshold in WW (Odor Activity Values (OAVs) 18.9–94.5), WW+SO_2 (OAVs 1.2–5.9) and WW+CT+GSH (OAVs 1.4–7.0) samples. DMDS led to an OAV up to 1.3 only in samples containing all three antioxidants investigated. We cannot exclude that the antioxidant activity of SO_2, when present at concentrations close to 100 mg/L, may limit the ability of hydrolysable tannins to work against LST formation, possibly because SO_2 can reduce the quinones back to phenols avoiding the thiol group of MeSH to perform this reduction [46]. Our results confirm LST to be an irreversible fault that can be perceived in wine stored in the dark for longer than one year [22].

Table 3. Concentrations of methionine, glutathione, cysteine, flavonoids, total phenol index and absorbance values at 420 nm determined in white wine (WW) samples.

Treatment	Methionine mg/L	Glutathione mg/L	Cysteine mg/L	Flavonoids mg Catechin/L	Total Phenol Index	Absorbance at 420 nm AU
Samples Stored in the Dark for 24 Months without Light Exposure						
WW_0	5.62 ± 0.45 [b]	nd	3.10 ± 0.15 [a]	623.4 ± 26.9 [ac]	36.3 ± 0.1 [a]	0.218 ± 0.006 [a]
WW	6.23 ± 0.50 [b]	nd	3.21 ± 0.16 [a]	607.3 ± 26.2 [ac]	36.0 ± 0.1 [a]	0.131 ± 0.005 [b]
$WW+SO_2$	5.41 ± 0.43 [b]	nd	3.15 ± 0.16 [a]	599.5 ± 10.8 [a]	36.3 ± 0.0 [a]	0.127 ± 0.006 [b]
WW+GSH	7.13 ± 0.57 [a]	5.88 ± 0.27 [b]	3.08 ± 0.15 [a]	589.6 ± 22.6 [a]	37.2 ± 0.3 [b]	0.179 ± 0.007 [c]
$WW+SO_2+GSH$	6.03 ± 0.48 [b]	6.13 ± 0.31 [b]	3.08 ± 0.15 [a]	575.6 ± 6.6 [b]	34.6 ± 0.5 [c]	0.155 ± 0.008 [d]
WW+CT	6.21 ± 0.50 [b]	nd	3.01 ± 0.15 [a]	630.8 ± 18.9 [c]	36.9 ± 0.3 [a]	0.185 ± 0.006 [c]
$WW+CT+SO_2$	5.41 ± 0.43 [b]	nd	2.99 ± 0.15 [a]	644.0 ± 19.2 [c]	39.7 ± 0.5 [d]	0.166 ± 0.009 [d]
WW+CT+GSH	5.94 ± 0.47 [b]	8.90 ± 0.44 [a]	3.16 ± 0.16 [a]	639.0 ± 44.1 [c]	41.8 ± 0.3 [e]	0.213 ± 0.010 [a]
$WW+CT+SO_2+GSH$	3.91 ± 0.31 [c]	9.47 ± 0.42 [a]	3.11 ± 0.15 [a]	646.4 ± 12.1 [c]	38.4 ± 0.5 [d]	0.198 ± 0.009 [e]
Samples Stored in the Dark for 24 Months after Light Exposure						
WW_0	2.64 ± 0.21 [b]	nd	nd	630.4 ± 27.2 [ab]	35.9 ± 0.1 [a]	0.070 ± 0.001 [a]
WW	6.14 ± 2.29 [a]	nd	nd	620.5 ± 26.8 [ab]	35.1 ± 0.1 [ab]	0.075 ± 0.006 [a]
$WW+SO_2$	5.96 ± 1.74 [a]	nd	nd	600.1 ± 10.8 [ab]	34.2 ± 0.0 [b]	0.061 ± 0.001 [b]
WW+GSH	6.60 ± 0.53 [a]	3.74 ± 0.15 [a]	5.21 ± 0.07 [c]	600.5 ± 23.0 [ab]	34.5 ± 0.3 [ab]	0.073 ± 0.000 [a]
$WW+SO_2+GSH$	7.64 ± 0.42 [a]	3.82 ± 0.35 [a]	7.07 ± 2.04 [b]	608.1 ± 7.0 [ab]	34.1 ± 0.5 [b]	0.071 ± 0.001 [a]
WW+CT	6.81 ± 0.12 [a]	nd	nd	672.6 ± 20.1 [a]	39.0 ± 1.0 [c]	0.103 ± 0.006 [c]
$WW+CT+SO_2$	6.58 ± 0.23 [a]	nd	nd	572.9 ± 129.1 [b]	39.3 ± 0.4 [c]	0.090 ± 0.006 [d]
WW+CT+GSH	7.45 ± 0.08 [a]	4.88 ± 0.35 [a]	6.69 ± 0.41 [b]	574.7 ± 39.6 [b]	38.1 ± 2.3 [c]	0.101 ± 0.000 [cd]
$WW+CT+SO_2+GSH$	6.74 ± 0.38 [a]	5.61 ± 2.11 [a]	7.66 ± 1.25 [a]	653.4 ± 12.2 [ab]	38.7 ± 0.7 [c]	0.096 ± 0.009 [cd]

For samples coding see Table 6. The average added concentrations of RF was 177.2 ± 4.2 µg/L; the average concentration of spiked Met was 9.63 ± 0.38 mg/L. Different letters along column mean significant differences ($p < 0.05$). Legend: nd, not detected; WW_0: glutathione- and riboflavin-free white wine, Met concentration was 5.90 ± 0.35 mg/L.

Table 4. Concentrations (µg/L) of methanethiol (MeSH) and dimethyl disulfide (DMDS) determined in white wine (WW) samples.

Treatment	MeSH		DMDS		Molar Ratio Sulfur Formed/ Met Degraded
	µg/L	OAV	µg/L	OAV	
WW	189.09 ± 17.21 [a]	18.9–94.5	2.83 ± 0.23 [d]	<0.14	17.09
WW+SO$_2$	11.80 ± 1.07 [b]	1.2–5.9	2.48 ± 0.20 [d]	<0.12	1.21
WW+GSH	nd	—	1.34 ± 0.11 [d]	<0.07	0.14
WW+SO$_2$+GSH	nd	—	4.82 ± 0.39 [c]	0.11–0.24	0.77
WW+CT	nd	—	nd	—	0.00
WW+CT+SO$_2$	nd	—	18.86 ± 1.53 [b]	0.4–0.9	1.96
WW+CT+GSH	14.03 ± 1.28 [b]	1.4–7.0	16.69 ± 1.35 [b]	0.4–0.8	4.43
WW+CT+SO$_2$+GSH	nd	—	25.18 ± 2.04 [a]	0.6–1.3	2.76

No volatile sulfur compounds were revealed in the WW$_0$ sample (glutathione- and riboflavin-free white wine, no Met added). The Odor Activity Value (OAV) was calculated as the ratio between the amount found in the sample and the perception threshold of the volatile sulfur compound. The perception threshold concentrations considered are as follows: MeSH, 2–10 µg/L; DMDS, 20–45 µg/L [12]. For samples coding see Table 6. Different letters mean significant differences ($p < 0.05$). Legend: nd, not detected.

Sotolon is a compound mainly associated with atypical (or oxidative) white wine ageing [47]. A previous study showed that the use of phenol-based preparations to replace SO$_2$ could cause an increase of sotolon content [27]. In the experimental conditions adopted here, negligible amounts of sotolon were detected in all tested samples (Table 5). The highest concentrations of sotolon were observed in the WW sample (3.96 ± 0.72 µg/L) followed by WW+CT+SO$_2$+GSH (2.53 ± 0.53 µg/L). In any case, the concentration of sotolon in all WW samples was lower than its olfactory perception threshold (7–8 µg/L) in white wine [48] indicating that none of the tested antioxidants, singularly or in combination, were responsible for atypical ageing.

Table 5. Concentrations (µg/L) of sotolon determined in white wine (WW) samples.

Treatment	Sotolon
WW$_0$	3.96 ± 0.72 [a]
WW	1.06 ± 0.07 [c]
WW+SO$_2$	nd
WW+GSH	nd
WW+SO$_2$+GSH	nd
WW+CT	trace
WW+CT+SO$_2$	0.59 ± 0.02 [d]
WW+CT+GSH	0.85 ± 0.07 [c]
WW+CT+SO$_2$+GSH	2.13 ± 0.36 [b]

For samples coding see Table 6. Different letters mean significant differences ($p < 0.05$). Legend: nd, not detected; WW$_0$: glutathione- and riboflavin-free white wine, no Met added.

The overall profile of volatile compounds (VOCs) was considered in WW samples exposed to light before storage. Thirty VOCs were detected corresponding to 3 acids, 8 alcohols and 19 esters (Figure A2 in Appendix A). Differences were found in relation to the antioxidants added. The significant increase occurring in the presence of antioxidants were related to nonanoic acid ethyl ester, 3-henex-1-ol and 2,4-hexadienoic ethyl ester in particular where CT was added. These compounds are associated with green and fat, grass and apple and peach notes, respectively. The two esters, isopropyl 3,4 hexadionate and decanoic acid ethyl ester, both responsible for fruity notes, mostly decreased in the presence of antioxidants. These findings indicate the loss of fruity aromas due to the light exposure [3,49,50], although the white wine used in this study was not characterized by evident floral and fruity notes. Further research will be carried out to clarify this aspect using a more aromatic wine.

2.3. Comparison

When added individually, the three antioxidants had different effectiveness in preventing the development of LST. The relative order was $SO_2 > CT > GSH$ in MW and $CT \geq GSH > SO_2$ in WW. Therefore, the attempt made to understand the role of the different antioxidants when used in combination by Principal Component Analysis (PCA) was carried out for the two systems (MW and WW) separately. All the parameters investigated in this study were included.

In case of MW, PC1 and PC2 together explained 68% of variance and the samples were clustered as (i) MW, MW+GSH, (ii) MW+SO$_2$, MW+SO$_2$+GSH, (iii) MW+CT, MW+CT+SO$_2$, MW+CT+GSH and (iv) MW+CT+SO$_2$+GSH (Figure 2). The use of GSH alone led to a small difference in comparison to MW, while CT alone seemed to play its protective role in a manner similar to that achieved when combined with SO$_2$ or GSH.

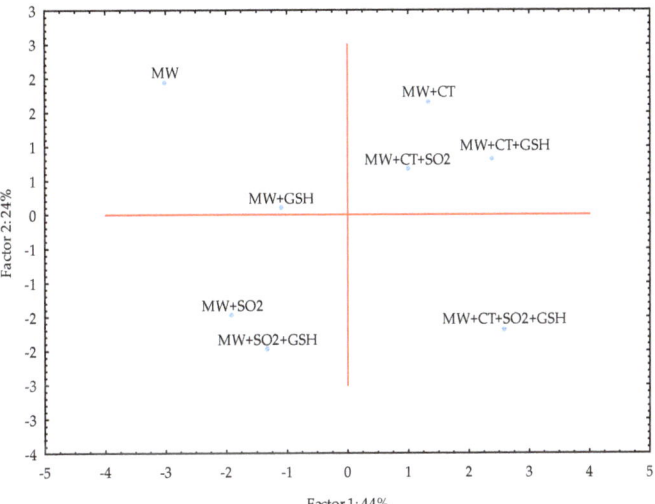

Figure 2. Principal Component Analysis (PCA) for the model wine solution (MW) stored in the dark after the light exposure. For samples coding refer to Table 6. Legend: SO$_2$, sulfur dioxide (20 mg/L), GSH, glutathione (50 mg/L), CT, chestnut tannins (50 mg/L).

PC1 and PC2 together explained 64% of the variance for WW samples that resulted clustered as follows: (i) WW, (ii) WW+SO$_2$, (iii) WW+CT, WW+CT+SO$_2$, WW+GSH, (iv) WW+SO$_2$+GSH, WW+CT+GSH and (v) WW+CT+SO$_2$+GSH (Figure 3). It appears evident that the addition of all the three antioxidants made the WW sample clearly distinguishable from the other samples, as it was found for MW samples. Both CT and GSH alone led to similar evolution of WW; moreover, when GSH was used with either SO$_2$ or CT, the evolution of LST in white wine could occur in a similar way, as it was observed for MW.

The study was carried out in both model wine and white wine due to the complexity of the latter. A very simple model wine was thus designed to avoid interferences and accurately follow the light-induced reactions of RF and Met in the presence of selected antioxidants. With the exception of the addition of SO$_2$ and CT+GSH, the treatments led to comparable results in both MW and WW as showed by the respective PCA (Figures 2 and 3). Even if in WW the intensity of LST differed in comparison to MW, the effectiveness of CT alone and in combination with SO$_2$ and SO$_2$+GSH was evidenced for the white wine used in the study under our experimental conditions.

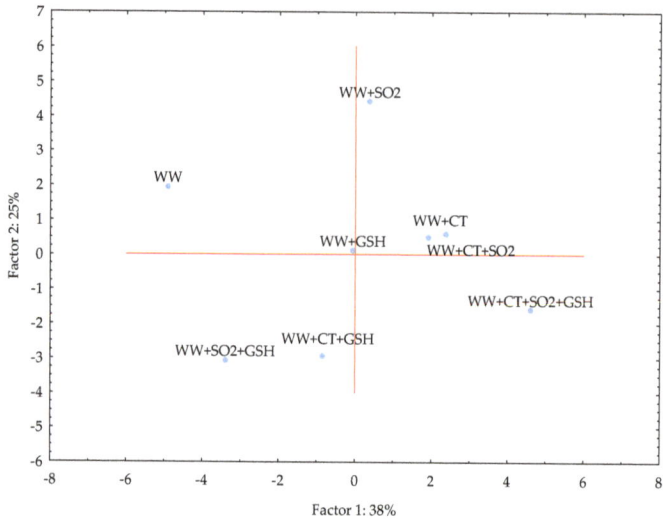

Figure 3. Principal Component Analysis (PCA) for the white wine (WW) stored in the dark after the light exposure. For samples coding refer to Table 6. Legend: SO_2, sulphur dioxide (20 mg/L), GSH, glutathione (50 mg/L), CT, chestnut tannins (50 mg/L).

3. Materials and Methods

3.1. Chemicals and Reagents

Methanol, ethanol, acetonitrile, dichloromethane, riboflavin, citric acid, tartaric acid, boric acid, mercaptoethanol, *o*-phtaldehyde (OPA), amino acid multi standard (containing acidic, neutral, and basic amino acids), riboflavin (RF), d_6-dimethyl sulphide (d_6-DMS), isopropyl disulphide, dimethyl disulphide (DMDS), dimethyl trisulphide (DMTS), *p*-benzoquinone (pBQ), 3-mercaptopropanoic acid (3MPA), glutathione, trifluoroacetic acid and hydrochloric acid were purchased from Sigma-Aldrich (St. Louis, MO, USA). Sodium metabisulfite was purchased from J.T. Baker (Deventer, The Netherlands). All the chemicals were of analytical grade, at least. HPLC grade water was obtained by a Milli-Q system (Millipore Filter Corp., Bedford, MA, USA).

Commercial hydrolysable tannins from chestnut wood intended for oenological use were provided by Dal Cin (Concorezzo, Italy).

The model wine solution (MW) was made of 5.0 g/L tartaric acid and 12% ethanol (*v/v*), adjusted to pH 3.2 with sodium hydroxide (Merck, Darmstadt, Germany).

The white wine (WW_0) produced with Trebbiano grape in vintage 2016 was collected at a local winery just after bottling and analyzed. The concentration of Met in WW_0 was 5.90 ± 0.35 mg/L and that of total SO_2 was 80 ± 2 mg/L, while no RF nor GSH were detected.

3.2. Experimental Plan

The experimental plan consisted of assessing the effect of different additives on the evolution of wine (i) stored in the dark or (ii) exposed to light for 2 h and then stored in the dark. Both MW and WW, added with RF (200 µg/L) and Met (4 mg/L), were considered. The three tested additives were SO_2 (20 mg/L), GSH (50 mg/L) and CT (50 mg/L), added individually or in different combinations for a total of 8 trials for MW and 8 trials for WW. WW_0 without the addition of RF and Met was also considered (Table 6).

Table 6. Experimental plan and sample coding according to treatment.

Sample Coding		Antioxidant(s) Added
Model Wine (MW)	White Wine (WW)	
—	WW_0	No addition
MW	WW	No antioxidant added
$MW+SO_2$	$WW+SO_2$	Sulfur dioxide
MW+GSH	WW+GSH	Glutathione
$MW+SO_2+GSH$	$WW+SO_2+GSH$	Sulfur dioxide/Glutathione
MW+CT	WW+CT	Chestnut tannins
$MW+CT+SO_2$	$WW+CT+SO_2$	Chestnut tannins/Sulfur dioxide
MW+CT+GSH	WW+CT+GSH	Chestnut tannins/Glutathione
$MW+CT+SO_2+GSH$	$WW+CT+SO_2+GSH$	Chestnut tannins/Sulfur dioxide/Glutathione

The codes "MW" and "WW" indicate the model wine solution and white wine, respectively, added with riboflavin and methionine.

In order to perform exposure to light under standardized conditions, MW and WW were placed in clear glass bottles (100 mL) that were hermetically sealed without headspace and exposed for 2 h to fluorescent light bulbs emitting 3172 Lumen at 6500 K, with high emission in the absorption wavelengths of RF (370 and 440 nm). A laboratory-made lightning device was used, consisting of three fluorescence light bulbs, placed 40 cm from each other. Each bottle was positioned between two light bulbs, i.e., at a 20 cm distance [17]. The light-exposed bottles were then stored at $18 \pm 2\,°C$ in the dark for 24 months. The same sample sets of both MW and WW, not light-exposed, were kept in the dark at identical conditions, as a control.

The concentration of RF, GSH, Met, volatile sulphur compounds (VSCs), i.e., methanethiol (MeSH), DMDS and DMTS, were determined. Two oxidation compounds from Met, namely methionine sulfoxide (Met sulfoxide) and methionine sulfone (Met sulfone), were quantified in MW samples. The total polyphenol index (TPI) and absorbance at 420 nm were assessed only in MW samples containing CT and in all the WW samples. Additionally, the flavonoids, free amino acid profile, sotolon and the overall volatile profile were analyzed in WW samples. The sensory analysis was carried out for all the trials in both MW and WW for the samples light exposed and kept protected against the light.

3.3. Determination of Riboflavin

The method reported by Fracassetti et al. [51] was applied for the measurement of RF content with some modifications [17]. Briefly, sample solutions were passed through a 0.22-µm PVDF filter (Millipore, Billerica, MA, USA) and 50 µL aliquot was injected in an Acquity HClass UPLC (Waters, Milford, MA, USA) system equipped with a photo diode array detector 2996 (Waters). The detection wavelength was 440 nm. The separation was carried out with (solvent A) 90% 50 mmol citrate buffer at pH 2.5 and 10% methanol (v/v) and (solvent B) 10% 50 mmol citrate buffer at pH 2.5 and 90% methanol (v/v) in gradient mode (70% B in 8 min) at a flow rate of 0.6 mL/min. Calibration curves were obtained for RF concentrations in the range 10–500 µg/L. Data acquisition and processing were performed by Empower 2 software (Waters).

3.4. Determination of Glutathione and Cysteine

Glutathione and cysteine (Cys) were determined by derivatization with p-benzoquinone (pBQ) [29]. Briefly, MW and WW samples (2 mL) were derivatized with pBQ (100 µL, 8 mM) followed by the addition of 3MPA (1 mL, 1.5 M). The reaction mix was filtered through a 0.22 µm pore-size PVDF membrane (Millipore, Billerica, MA, USA) and analyzed by an Acquity HClass UPLC (Waters) system equipped with a photo diode array detector 2996 (Waters) using a phenyl-hexyl column (250 × 4.6 mm, 5 µm, 110 Å, Phenomenex, Torrance, CA, USA). The separation was carried out with (solvent A) water/trifluoroacetic acid 0.05% (v/v) and (solvent B) methanol in gradient mode (from 10% B to 35% B in

18 min) at a flow rate of 1 mL/min [29]. The detection wavelength was 303 nm; data acquisition and processing were performed by Empower 2 software (Waters).

3.5. Determination of Volatile Sulphur Compounds and Other Volatile Compounds

The analysis of volatile sulfur compounds (VSCs) was performed by Solid Phase Micro Extraction (SPME)-GC/MS following the method described by Fracassetti et al. [17]. Duplicate injections were carried out for each sample. Results are expressed as the relative concentration (µg/L) for MeSH referred to as d_6-DMS; DMDS and DMTS amounts were determined by the external standard method (0.5–100 µg/L). The Odor Activity Values (OAVs) were determined as the ratio between the amount of the VSC found in the sample and the respective perception threshold. The perception threshold concentrations considered were as follows: MeSH, 0.3 µg/L in MW and 2–10 µg/L in WW; DMDS, 20–45 µg/L; DMTS, 0.1 ug/L [12]. The ratio between the moles of sulfur compounds formed, obtained by summing MeSH, DMDS and DMTS concentrations, and the moles of sulfur lost as degraded Met was calculated.

For the WW samples, the overall profiles of volatile compounds (VOCs) were further evaluated. VOCs were identified according to the NIST library and for an R match higher than 95% [52]. Data are expressed as the ratio between the area value found in WW_0, set equal to 1, and in samples submitted to the different treatments as labelled in Table 6.

3.6. Determination of Methionine, Methionine Sulfoxide and Methionine Sulfone

Methionine, Met sulfoxide and Met sulfone concentrations were quantified in MW samples by UPLC as o-phthalaldehyde (OPA) derivatives under the conditions described by Fracassetti et al. [17] with some modifications. The derivatization solution was prepared in a 10 mL volumetric flask by dissolving 250 mg of OPA in 1.5 mL of ethanol, adding 200 µL of 2-mercaptoethanol, and making up to the volume with borate buffer 0.4 M at pH 10.5. The pre-column derivatization was performed as follows: 500 µL of borate buffer 0.4 M at pH 10.5 were added with 200 µL of sample and 100 µL of OPA solution; the reaction mixture was vortexed for 2 min and 640 µL of phosphoric acid 1.5% (v/v) were added [36]. The reaction mixture was filtered with 0.22 µm PVDF filers (Millipore) and injected. The chromatographic separation of OPA derivatives was carried out using an Acquity HClass UPLC (Waters) system equipped with a photo diode array detector 2996 (Waters). The column was a Nova-Pak C18 (150 mm × 3.9 mm column, 4 µm particle size stationary phase) (Waters) maintained at 40°C. The solvents were (solvent A) citrate buffer 10 mM at pH 7.5 and (solvent B) acetonitrile/methanol/water in proportion 45/45/10 $(v/v/v)$. The separation was carried out at 1 mL/min in gradient mode in which B was from 5% to 47% in 22 min. The detection wavelength was 338 nm. The concentrations of Met, Met sulfoxide and Met sulfone were determined by the external standard method (0.1–5 mg/L). Data acquisition and processing were performed by Empower 2 software (Waters).

3.7. Determination of the Free Amino Acidic Profile

Free amino acids were quantified in WW samples according to the method of Fracassetti et al. [20] with some modifications by using an Acquity HClass UPLC (Waters) system equipped with a photo diode array detector 2996 (Waters). The pre-column derivatization procedure was performed as follows: 750 µL of borate buffer 0.4 M at pH 10.5 were added with 300 µL of sample and 150 µL of OPA solution. The reaction mixture was vortexed for 2 min, filtered through a 0.22 µm PVDF filer (Millipore) and injected. The OPA-derivatized amino acids were separated in a Kinetex Phenyl-Hexyl, 150 mm × 4.6 mm column, with 2.6 µm particle size (Phenomenex) maintained at 50 °C. Eluting solvents were (solvent A) citrate buffer 10 mM at pH 7.5 and (solvent B) acetonitrile/methanol/citrate buffer 10 mM at pH 7.5 in proportion 45/45/10 $(v/v/v)$. The separation was carried out at 1 mL/min in gradient mode operating as follows: 5% B for 3 min; from 5% to 15% B at 6.5 min; from 15% to 20% B at 9 min; from 20% to 30% B at 12 min; from 30% to 40% at 15.5 min; from 40% to 80% at 23 min. The detection wavelength was 338 nm. Amino acids, namely Met, aspartic acid, glutamic acid, asparagine, serine, glutamine, histidine, threonine, arginine, alanine, tyrosine, valine, phenylalanine, isoleucine, leucine, ornithine and lysine, were identified and determined by the external standard method (0.1–20 mg/L). Data acquisition and processing were performed by Empower 2 software (Waters).

3.8. Determination of Total Flavonoids, Total Phenol Index and Absorbance at 420 nm

Total flavonoids, total phenol index and absorbance at 420 nm were determined in all WW samples and in MW samples where CT was added.

For the assessment of total flavonoid content, the samples were properly diluted with a hydrochloric ethanol solution (ethanol/water/hydrochloric acid 37%, 70/30/1 $v/v/v$) in order to obtain an absorption value lower than 1 ± 0.05 AU at 280 nm. The absorption spectra of the sample were recorded in the wavelength range 700–230 nm and the quantification of flavonoids was carried out according to Corona et al. [53]. The results are expressed as mg catechin/L, taking into account the derivative of the peak registered at 280 nm and the molar extinction coefficient of catechin in hydrochloric ethanol.

Total phenol index (TPI) was measured based on the absorption value at 280 nm after proper dilution of the sample with water in order to obtain an absorption value lower than 1 ± 0.05 AU at 280 nm. TPI was calculated by multiplying the absorbance value at 280 for the dilution factor [54,55].

The absorption values at 420 nm were considered in order to estimate the impact of the tested additives on yellow color/browning [56].

3.9. Determination of Sotolon

Sotolon was measured in WW samples following the preparation described by Gabrielli et al. [57]. Briefly, 3 g of NaCl were dissolved in 30 mL wine in a 100-mL bottle, then 40 mL of dichloromethane (DCM) were added. The bottle was hermetically closed and shaken for 10 min with a wrist action stirrer (Griffin Flask Shaker). The mixture was centrifuged 5 min at $5000 \times g$ and the DCM was separated by a separatory funnel and recovered. The solvent extraction procedure was carried out three times. Eventually, the three organic solvent fractions were jointly collected and added with 2 g of anhydrous sodium sulfate. DCM was evaporated under vacuum, then the dry material was dissolved into 2 mL of methanol 5%, which was purified by a PVPP 50 mg SPE cartridge and the eluted solution was recovered. The quantification of sotolon was carried out by UPLC-UV [57].

3.10. Sensory Analysis

A panel constituted by nine expert judges (5 males, 4 females, aged 25–55) carried out the olfactory scoring for the "cooked cabbage" descriptor. The score ranged from 1 (not perceived) to 9 (extremely perceived). The panelists were firstly trained using MW samples spiked with Met (4 mg/L) and two different levels of RF (200 µg/L or 400 µg/L) and exposed to light for two hours using the above-described illuminating device (Section 3.2) in order to make the judges confident about the perception of cooked cabbage note. Sniffing sessions were then carried out using WW samples (Met 4 mg/L, RF 200 µg/L and 400 µg/L, light exposure for two hours). The judges were calibrated by sniffing MW solutions spiked with Met (4 mg/L) and RF (200 µg/L) exposed to light for increasing time up to two hours. Each MW and WW sample was evaluated just after the bottle opening and served at temperature $18 \pm 2\,°C$.

3.11. Statistical Analysis

The statistical analysis was performed with SPSS Win 12.0 program (SPSS Inc., Chicago, IL, USA). One-way ANOVA was carried out to determine the significant differences related to chemical parameters and sensory analysis. Significant differences were judged by a post-hoc Fischer LSD ($p < 0.05$). The principal component analysis (PCA) was performed with Statistica 12 software (Statsoft Inc., Tulsa, OK, USA) on auto-scaled data for an overall overview of the effect due to the different additives added and their combination considering the chemical parameters and the sensory data.

4. Conclusions

The use of additives against the appearance of LST in white wine is a crucial aspect in wine technology since a variety of oenological strategies exists potentially counteracting the sensory modifications after bottling. Therefore, understanding the mechanisms behind each of these is of utmost interest. For this reason, the photo-induced mechanisms were investigated in a model solution. This approach allows the easier interpretation of chemical pathways taking place in wine since interfering reactions could be avoided.

The hydrolysable tannins showed to have a protective effect against the formation of LST in the white wine adopted in this study. Nonetheless, the intensity of LST differed in the tested white wine in comparison to model wine. The prevention of LST by means of hydrolysable tannins, alone and in combination with SO_2, was found in both the matrices investigated, supporting the capability of tannins to counteract the formation of LST. The simultaneous addition of tannins and SO_2 produced a different effect than CT alone. These results suggest that a higher addition of SO_2 could not prevent LST in white wine, but, on the contrary, it could favor the VSC-dependent spoilage. Differently, the use of hydrolysable tannins prior to bottling could be an effective oenological approach to limit the occurrence of LST. The combined use of other antioxidants (i.e., SO_2+GSH) can be also effective.

Future perspectives will be to evaluate LST formation in white wines produced under an industrial scale with hydrolysable tannins added at bottling. Their addition will be investigated in other white wines, both still and sparkling, and rosé wines to further evidence their capability against LST.

Author Contributions: Conceptualization, D.F. and A.T.; methodology, D.F. and N.M.; software, D.F.; formal analysis, N.M.; investigation, D.F. and N.M.; resources, D.F. and A.T.; data curation, D.F.; writing—original draft preparation, D.F. and N.M.; writing—review and editing, S.L., L.P. and A.T.; funding acquisition, D.F., S.L. and A.T. All authors have read and agreed to the published version of the manuscript.

Funding: The study was supported by European Agricultural Fund for Rural Development (Enofotoshield project; D.d.s. 1 luglio 2019—n. 9551, B.U. R.L. Serie Ordinaria n. 27—04 luglio 2019) and Piano di Sostegno alla Ricerca 2017–2018—Linea 2—Università degli Studi di Milano.

Institutional Review Board Statement: Not applicable.

Informed Consent Statement: Not applicable.

Data Availability Statement: Not applicable.

Acknowledgments: The authors are grateful to Andrea Baratti for his technical support.

Conflicts of Interest: The authors declare no conflict of interest.

Sample Availability: Not available.

Appendix A

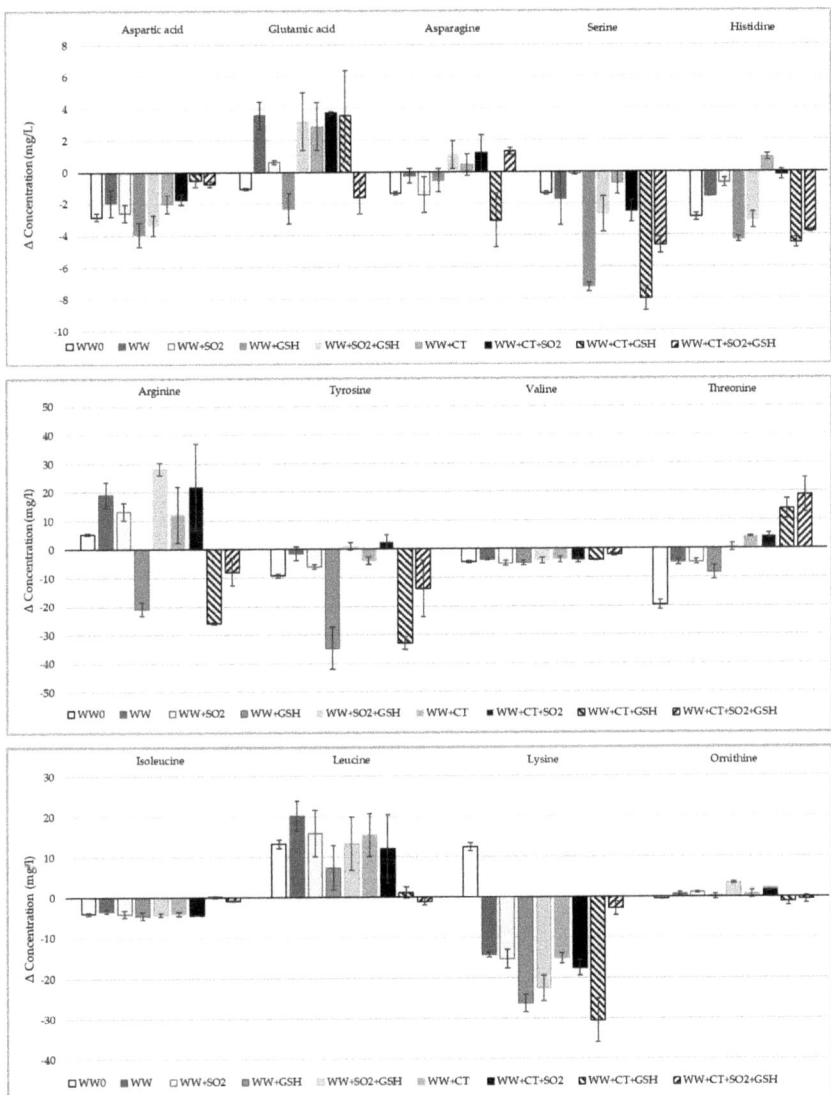

Figure A1. Differences of the amino acid concentrations between the white wine (WW) samples without light exposure and the corresponding WW samples with light exposure. For samples coding, see Table 6. Data are not shown for alanine, glutamine and phenylalanine as negligible differences were found.

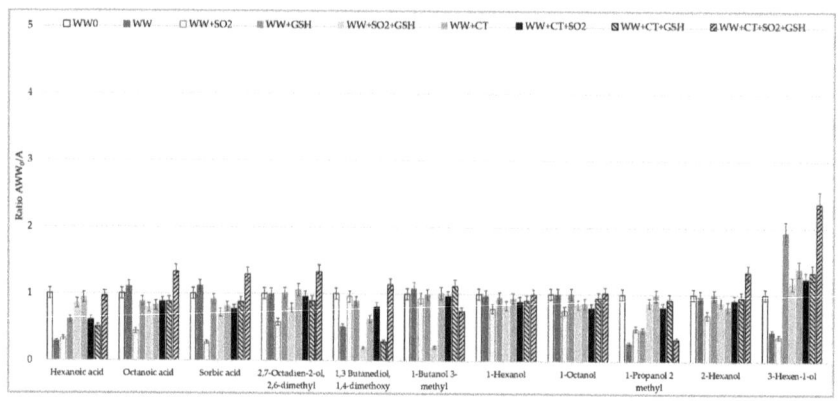
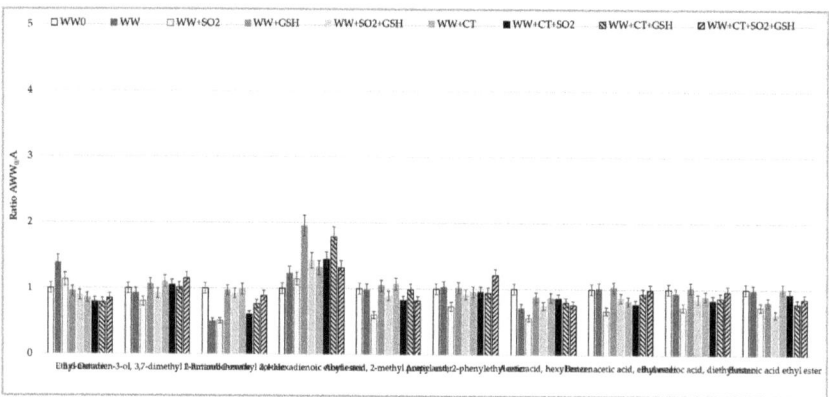
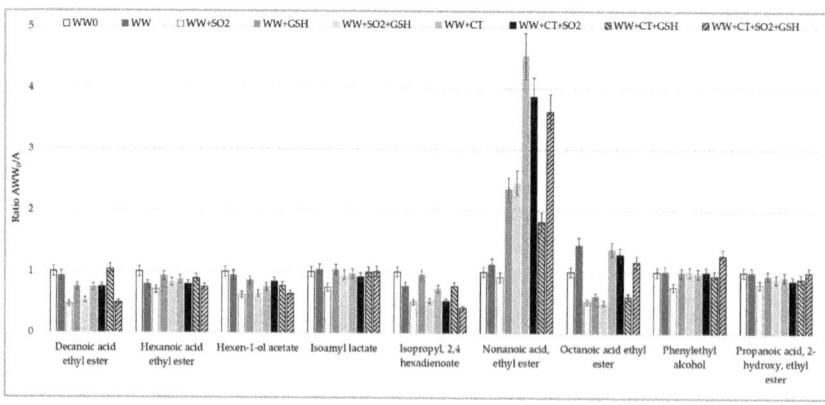

Figure A2. Volatile compound profiles determined in white wine (WW) samples exposed to light prior the storage in the dark. WW_0 sample (glutathione- and riboflavin-free white wine, no Met added) was fixed as 1 and the ratio among it and WW samples with the tested additives were calculated. For samples coding see Table 6.

References

1. Bekbölet, M. Light effects on food. *J. Food Prot.* **1990**, *53*, 430–440. [CrossRef]
2. Clark, A.C.; Dias, D.A.; Smith, T.A.; Ghiggino, K.P.; Scollary, G.R. Iron III tartrate as a potential precursor of light-induced oxidative degradation of white wine: Studies in a model wine system. *J. Agric. Food Chem.* **2011**, *59*, 3575–3581. [CrossRef] [PubMed]

3. D'Auria, M.; Emanuele, L.; Racioppi, R. The effect of heat and light on the composition of some volatile compounds in wine. *Food Chem.* **2009**, *117*, 9–14. [CrossRef]
4. Maujean, A.; Haye, M.; Feuillat, M. Contribution à l'étude des "goûts de lumière" dans le vin de Champagne. II. Influence de la lumière sur le potentiel d'oxydoreduction. Correlation avec la teneur en thiols du vin. *OENO One* **1978**, *12*, 277–290. [CrossRef]
5. Maujean, A.; Seguin, N. Contribution à l'étude des goûts de lumière dans les vins de Champagne. 3. Les réactions photochimiques responsables des goûts de lumière dans le vin de Champagne. *Sci. Aliment.* **1983**, *3*, 589–601.
6. Foote, C.S. Photosensitized oxidation and singlet oxygen: Consequences in biological systems. In *Free Radicals in Biology*; Pryor, W.A., Ed.; Academic Press: New York, NY, USA, 1976; Volume 2, pp. 85–133.
7. Cardoso, D.R.; Libardi, S.H.; Skibsted, L.H. Riboflavin as a photosensitizer. Effects on human health and food quality. *Food Funct.* **2012**, *3*, 487–502. [CrossRef]
8. Sheraz, M.A.; Kazi, S.H.; Ahmed, S.; Anwar, Z.; Ahmad, I. Photo, thermal and chemical degradation of riboflavin. *Beilstein J. Org. Chem.* **2014**, *10*, 1999–2012. [CrossRef] [PubMed]
9. Fracassetti, D.; Di Canito, A.; Bodon, R.; Messina, N.; Vigentini, I.; Foschino, R.; Tirelli, A. Light-struck taste in white wine: Reaction mechanisms, preventive strategies and future perspectives to preserve wine quality. *Trends Food Sci. Technol.* **2021**, *112*, 547–558. [CrossRef]
10. Asaduzzaman, M.; Scampicchio, M.; Biasioli, F.; Bremer, P.J.; Silcock, P. Methanethiol formation during the photochemical oxidation of methionine-riboflavin system. *Flavour Fragr. J.* **2020**, *35*, 34–41. [CrossRef]
11. Maujean, A.; Seguin, N. Contribution à l'étude des goûts de lumière dans les vins de Champagne. 4. Approches a une solution œnologique des moyens de prévention des goûts de lumière. *Sci. Aliment.* **1983**, *3*, 603–613.
12. Fracassetti, D.; Vigentini, I. Occurrence and analysis of sulfur compounds in wine. In *Grapes and Wines–Advances in Production, Processing, Analysis and Valorization*; InTechOpen: London, UK, 2017; pp. 225–251. Available online: https://www.intechopen.com/chapters/58638 (accessed on 30 August 2021).
13. Grant-Preece, P.; Barril, C.; Schmidtke, L.M.; Scollary, G.R.; Clark, A.C. Light-induced changes in bottled white wine and underlying photochemical mechanisms. *Crit. Rev. Food Sci. Nutr.* **2017**, *57*, 743–754. [CrossRef]
14. Grant-Preece, P.; Barril, C.; Leigh, M.; Schmidtke, L.M.; Clark, A.C. Impact of fluorescent lighting on the browning potential of model wine solutions containing organic acids and iron. *Food Chem.* **2018**, *243*, 239–248. [CrossRef]
15. Pichler, U. Analisi della riboflavina nei vini bianchi e influenza della sua concentrazione. *L'Enotecnico* **1996**, *32*, 57–62.
16. Mattivi, F.; Monetti, A.; Vrhovsek, U.; Tonon, D.; Andrés-Lacueva, C. High-performance liquid chromatographic determination of the riboflavin concentration in white wines for predicting their resistance to light. *J. Chromatogr. A* **2000**, *888*, 121–127. [CrossRef]
17. Fracassetti, D.; Limbo, S.; Pellegrino, L.; Tirelli, A. Light-induced reactions of methionine and riboflavin in model wine: Effects of hydrolysable tannins and sulfur dioxide. *Food Chem.* **2019**, *298*, 124952. [CrossRef]
18. Riberau-Gayon, P.; Glories, Y.; Maujean, A.; Dubourdieu, D. *Handbook of Enology*; John Wiley & Sons Ltd: Chichester, UK, 2006.
19. Santos, M.A.; García-Ramírez, J.J.; Revuelta, J.L. Riboflavin Biosynthesis in *Saccharomyces cerevisiae*. *J. Biol. Chem.* **1995**, *270*, 437–444. [CrossRef]
20. Fracassetti, D.; Gabrielli, M.; Encinas, J.; Manara, M.; Pellegrino, L.; Tirelli, A. Approaches to prevent the light-struck taste in white wine. *Aust. J. Grape Wine Res.* **2017**, *23*, 329–333. [CrossRef]
21. Fracassetti, D.; Tirelli, A.; Limbo, S.; Mastro, M.; Pellegrino, L.; Ragg, E.M. Investigating the role of antioxidant compounds in riboflavin-mediated photo-oxidation of methionine: A 1H-NMR approach. *ACS Omega* **2020**, *5*, 26220–26229. [CrossRef] [PubMed]
22. Arapitsas, P.; Dalledonne, S.; Scholz, M.; Catapano, A.; Carlin, S.; Mattivi, F. White wine light-strike fault: A comparison between flint and green glass bottles under typical supermarket conditions. *Food Packag. Shelf Life* **2020**, *24*, 100492. [CrossRef]
23. Makhotkina, O.; Kilmartin, P.A. Uncovering the influence of antioxidants on polyphenol oxidation in wines using an electrochemical method: Cyclic voltammetry. *J. Electroanal. Chem.* **2009**, *633*, 165–174. [CrossRef]
24. Lavigne, V.; Dubourdieu, D. Affinamento sulle fecce e freschezza dei vini bianchi. *Vigne Vini* **2004**, *31*, 58–66.
25. Vignault, A.; González-Centeno, M.R.; Pascual, O.; Gombau, J.; Jourdes, M.; Moine, V.; Teissedre, P.-L. Chemical characterization, antioxidant properties and oxygen consumption rate of 36 commercial oenological tannins in a model wine solution. *Food Chem.* **2018**, *268*, 210–219. [CrossRef] [PubMed]
26. Ricci, A.; Parpinello, G.P.; Teslić, N.; Kilmartin, P.A.; Versari, A. Suitability of the cyclic voltammetry measurements and DPPH• spectrophotometric assay to determine the antioxidant capacity of food-grade oenological tannins. *Molecules* **2019**, *24*, 2925. [CrossRef]
27. Fracassetti, D.; Gabrielli, M.; Costa, C.; Tomás-Barberán, F.A.; Tirelli, A. Characterization and suitability of polyphenols-based formulas to replace sulfur dioxide for storage of sparkling white wine. *Food Control.* **2016**, *60*, 606–614. [CrossRef]
28. Fracassetti, D.; Coetzee, C.; Vanzo, A.; Ballabio, D.; du Toit, W.J. Oxygen consumption in South African Sauvignon blanc wines: Role of glutathione, sulphur dioxide and certain phenolics. *S. Afr. J. Enol. Vitic.* **2013**, *34*, 156–169. [CrossRef]
29. Fracassetti, D.; Tirelli, A. Monitoring of glutathione concentration during winemaking by a reliable high-performance liquid chromatography analytical method. *Aus. J. Grape Wine Res.* **2015**, *21*, 389–395. [CrossRef]
30. Resolution Oeno 446–2015. Treatment of Wine with Glutathione. Available online: https://www.oiv.int/public/medias/1687/oiv-oeno-446-2015-en.pdf (accessed on 30 July 2021).

31. Amerine, M.A.; Ough, C.S. Alcohols. In *Methods for Analysis of Musts and Wines*; Amerine, M.A., Ough, C.S., Eds.; John Wiley and Sons: New York, NY, USA, 1980.
32. Sartor, S.; Burin, V.M.; Caliari, V.; Bordignon-Luiz, M.T. Profiling of free amino acids in sparkling wines during over-lees aging and evaluation of sensory properties. *LWT-Food Sci. Technol.* **2021**, *140*, 110847. [CrossRef]
33. Robichaud, J.L.; Noble, A.C. Astringency and bitterness of selected phenolics in wine. *J. Sci. Food Agric.* **1990**, *53*, 343–353. [CrossRef]
34. Bitsch, R.; Bitsch, I. HPLC determination of riboflavin in fortified foods. In *Fortified foods with Vitamins: Analytical Concepts to Assure Better and Safer Products*; Rychlik, M., Ed.; John Wiley and Sons: New York, NY, USA, 2011.
35. Golbach, J.L.; Ricke, S.C.; O'Bryan, C.A.; Crandall, P.G. Riboflavin in nutrition, food processing and analysis—a review. *J Food Res.* **2014**, *3*, 23–35. [CrossRef]
36. Barata-Vallejo, S.; Ferreri, C.; Postigo, A.; Chatgilialoglu, C. Radiation chemical studies of methionine in aqueous solution: Understanding the role of molecular oxygen. *Chem. Res. in Toxicol.* **2010**, *23*, 258–263. [CrossRef] [PubMed]
37. Inoue, M.; Hayatsu, H. The interactions between bisulfite and amino acids. The formation of methionine sulfoxide from methionine in the presence of oxygen. *Chem. Pharm. Bull.* **1971**, *19*, 1286–1289. [CrossRef]
38. Winterbourn, C.C. Revisiting the reactions of superoxide with glutathione and other thiols. *Arch. Biochem. Biophys.* **2016**, *595*, 68–71. [CrossRef] [PubMed]
39. Min, D.B.; Boff, J.M. Chemistry and reaction of singlet oxygen in foods. *Compr. Rev. Food Sci. Food Saf.* **2002**, *1*, 58–72. [CrossRef]
40. Clark, A.C.; Prenzler, P.D.; Scollary, G.R. Impact of the condition of storage of tartaric acid solution on the production and stability of glyoxylic acid. *Food Chem.* **2007**, *102*, 905–916. [CrossRef]
41. Gijs, L.; Perpète, P.; Timmermans, A.; Collin, S. 3-Methylthiopropionaldehyde as precursor of dimethyl trisulfide in aged beers. *J. Agric. Food Chem.* **2000**, *48*, 6196–6199. [CrossRef] [PubMed]
42. Câmara, J.S.; Alves, M.A.; Marques, J.C. Changes in volatile composition of Madeira wines during their oxidative ageing. *Anal. Chim. Acta* **2006**, *563*, 188–197. [CrossRef]
43. Fracassetti, D.; Tirelli, A. Effetti della composizione del vino rosso sulla cinetica di consumo dell'ossigeno in presenza di tannini enologici. In Proceedings of the 11° Enoforum 2019, Vicenza, Italy, 21–23 May 2019.
44. Danilewicz, J.C. Mechanism of autoxidation of polyphenols and participation of sulfite in wine: Key role of iron. *Am. J. Enol Vitic.* **2011**, *62*, 319–328. [CrossRef]
45. Maury, C.; Clark, A.C.; Scollary, G.R. Determination of the impact of bottle colour and phenolic concentration on pigment development in white wine stored under external conditions. *Anal. Chim. Acta* **2010**, *660*, 81–86. [CrossRef] [PubMed]
46. Cilliers, J.J.L.; Singleton, V.L. Caffeic acid autoxidation and the effects of thiols. *J. Agric. Food Chem.* **1990**, *38*, 1789–1796. [CrossRef]
47. Lavigne, V.; Pons, A.; Darriet, P.; Dubourdieu, D. Changes in the sotolon content of dry white wines during barrel and bottle aging. *J. Agric. Food Chem.* **2008**, *56*, 2688–2693. [CrossRef] [PubMed]
48. Guichard, E.; Pham, T.T.; Etiévant, P. Quantitative determination of sotolon in wines by high-performance liquid chromatography. *Chromatographia* **1993**, *37*, 539–542. [CrossRef]
49. Benítez, P.; Castro, R.; Natera, R.; García Barroso, C. Changes in the polyphenolic and volatile content of "fino" Sherry wine exposed to high temperature and ultraviolet and visible radiation. *Eur. Food Res. Technol.* **2006**, *222*, 302–309. [CrossRef]
50. Díaz, I.; Castro, R.I.; Ubeda, C.; Loyola, R.; Felipe Laurie, V. Combined effects of sulfur dioxide, glutathione and light exposure on the conservation of bottled Sauvignon blanc. *Food Chem.* **2021**, *356*, 129689. [CrossRef] [PubMed]
51. Fracassetti, D.; Limbo, S.; D'Incecco, P.; Tirelli, A.; Pellegrino, L. Development of a HPLC method for the simultaneous analysis of riboflavin and other flavin compounds in liquid milk and milk products. *Eur. Food Res. Technol.* **2018**, *244*, 1545–1554. [CrossRef]
52. Fracassetti, D.; Camoni, D.; Montresor, L.; Bodon, R.; Limbo, S. Chemical characterization and volatile profile of Trebbiano di Lugana wine: A case study. *Foods* **2020**, *9*, 956. [CrossRef]
53. Corona, O.; Squadrito, M.; Vento, G.; Tirelli, A.; Di Stefano, R. Over-evaluation of total flavonoids in grape skin extracts containing sulphur dioxide. *Food Chem.* **2015**, *172*, 537–542. [CrossRef]
54. Di Stefano, R.; Cravero, M.C.; Gentilini, N. Metodi per lo studio dei polifenoli dei vini. *L'Enotecnico* **1989**, *5*, 83–89.
55. Fracassetti, D.; Gabrielli, M.; Corona, O.; Tirelli, A. Characterisation of Vernaccia Nera (*Vitis vinifera* L.) grapes and wine. *S. Afr. J. Enol. Vitic.* **2017**, *38*, 72–81. [CrossRef]
56. Li, H.; Guo, A.; Wang, H. Mechanisms of oxidative browning of wine. *Food Chem.* **2008**, *108*, 1–13. [CrossRef]
57. Gabrielli, M.; Fracassetti, D.; Tirelli, A. UHPLC quantification of sotolon in white wine. *J. Agric. Food Chem.* **2014**, *62*, 4878–4883. [CrossRef]

Review

Non-*Saccharomyces* as Biotools to Control the Production of Off-Flavors in Wines

Antonio Morata *, Iris Loira, Carmen González and Carlos Escott

enotecUPM, Departamento de Química y Tecnología de Alimentos, Universidad Politécnica de Madrid, Avenida Puerta de Hierro 2, 28040 Madrid, Spain; iris.loira@upm.es (I.L.); carmen.gchamorro@upm.es (C.G.); carlos.escott@gmail.com (C.E.)
* Correspondence: antonio.morata@upm.es

Abstract: Off-flavors produced by undesirable microbial spoilage are a major concern in wineries, as they affect wine quality. This situation is worse in warm areas affected by global warming because of the resulting higher pHs in wines. Natural biotechnologies can aid in effectively controlling these processes, while reducing the use of chemical preservatives such as SO_2. Bioacidification reduces the development of spoilage yeasts and bacteria, but also increases the amount of molecular SO_2, which allows for lower total levels. The use of non-*Saccharomyces* yeasts, such as *Lachancea thermotolerans*, results in effective acidification through the production of lactic acid from sugars. Furthermore, high lactic acid contents (>4 g/L) inhibit lactic acid bacteria and have some effect on *Brettanomyces*. Additionally, the use of yeasts with hydroxycinnamate decarboxylase (HCDC) activity can be useful to promote the fermentative formation of stable vinylphenolic pyranoanthocyanins, reducing the amount of ethylphenol precursors. This biotechnology increases the amount of stable pigments and simultaneously prevents the formation of high contents of ethylphenols, even when the wine is contaminated by *Brettanomyces*.

Keywords: wine; yeasts; non-*Saccharomyces*; off-smells; volatile acidity; ethylphenols; pyranoanthocyanins; pH control; bioprotection

1. Introduction

Wine quality is strongly and negatively affected by some microbial metabolites with low sensory thresholds and negative olfactory impact, including reduced sulfur compounds [1,2], volatile acidity [3], ethylphenols [4,5], and acetaldehyde [6]. The sensory impact is quite variable because sensory thresholds can range from very low values (H_2S, 1.6 µg/L) to very high concentrations (volatile acidity 0.3–0.6 g/L), so the range is about 1 million times (Table 1). This makes the analytical approach very specific and makes the use of sensitive and reproducible techniques based mainly on gas chromatography–mass spectrometry (GC-MS or GC-MS/MS) instruments essential for their determination [7]. These analytical methods often require specific sample preparations and concentrations: headspace (HS), dynamic headspace (DHS), solid-phase microextraction (SPME), and Twister [7].

Table 1. Wine off-flavors produced by microbial metabolites.

Compound	Sensory Threshold	Off-Flavor Concentration [1]	Descriptor	Reference
H_2S	1.6 µg/L	>1.6 µg/L	Rotten eggs/putrefaction	[2]
Volatile acidity	0.3–0.6 g/L	>0.8 g/L legal limit 1.2 g/L	Vinegar	[7–9]
Ethyl acetate	12 mg/L	>150 mg/L	Glue, solvent	[7]
4-Ethylphenol	230 µg/L	>425 µg/L	Phenolic, stable, leather, horse sweat	[5,10,11]
Acetaldehyde	100–125 mg/L	>125 mg/L	Fruity, rotten apples, nut-like, sherry	[6]

[1] The sensory threshold and off-flavor perception can be variable depending on the structure, composition, and sensory buffering effect of the wine.

The control of fermentative purity as well as the development of wild spoilage microorganisms in wines are related to pH and sulfur dioxide contents. In warm areas affected by global warming, pH values have been increasing in recent years, which are associated with higher alcoholic strength and intense phenolic ripening [12–14], especially in varieties that accumulate high potassium contents in berries [15,16]. A high pH produces wines that are more chemically and microbiologically unstable, and therefore are more susceptible to microbial spoilage, including off-flavor formation. Wine pH can range from 2.8 to 4.5, although most wines are in the 3 to 4 range. However, wines below 3.5 are very stable and usually less affected by microbial developments, while wines with pHs close to 4 are very risky as many spoilage bacteria and yeasts can easily develop in them during processing and especially during aging and storage.

Yeast species can help in the biocontrol of off-flavor formation via bioprotection as a result of competition with or the elimination of wild spoilage microorganisms [17–19], by acidity production [20–23], by nutrient competition and depletion [24,25], by the depletion of off-flavor precursors [26], or by the adsorption of defective molecules on cell walls [27,28] (Figure 1).

This review is focused on the elimination of off-flavors by using non-*Saccharomyces* yeasts that are able to control pH by bioacidification or to decrease the concentration of precursors of molecules responsible for sensory defects.

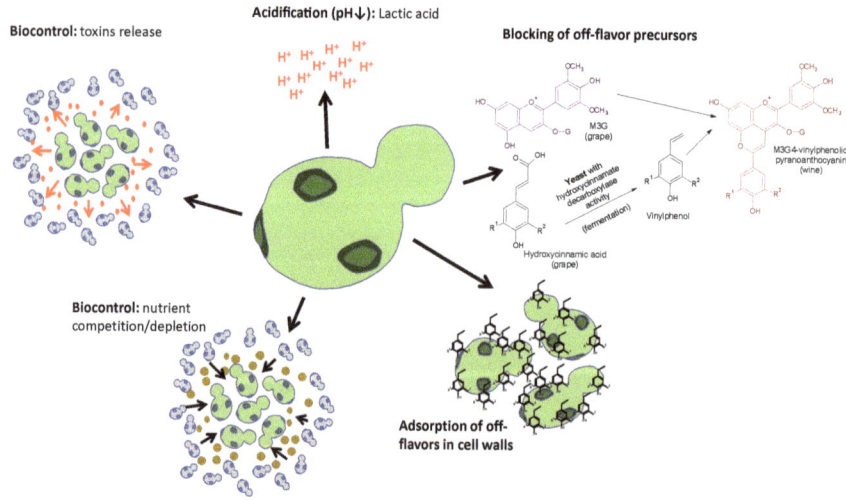

Figure 1. Summary of the mechanisms of biocontrol of off-flavor formation by yeasts. M3G: malvidin-3-O-glucoside; M3G4-vinylphenolic: malvidin-3-O-glucoside-4-vinylphenolic.

2. Bioprotection

Bioprotection is a current concept, so its definition is still under discussion. However, it can be considered the active or passive use of some microorganisms to preserve foods and beverages and to exclude other spoilage microorganisms, thus avoiding the production of off-flavors, sensory alterations, or even the formation of toxic molecules. Bioprotection is a hot topic in enology and foods; several reviews have recently been published [29–31]. Bioprotection can be achieved by the production of molecules or metabolites with antimicrobial effects such as organic acids [32]; toxins such as killer factors [33–35]; deleterious chelates such as pulcherrimin [25,36]; glucanases [30,37]; ethanol produced in the fermentation of sugars; as well as by nutrient depletion [38] (Figure 1).

The application of bioprotective microorganisms can be scheduled at several stages of the winemaking process [31]. During the prefermentative stage or during prefermentative maceration, they can be applied directly on the grapes (harvesting machine) to control wild yeast [39,40], mold, and bacteria populations [19,41]. During fermentation, they can be applied to control the development of spoilage yeasts and bacteria and the oxidative processes, and frequently also to improve the sensory profile of wines by producing flavor compounds [18,42]. Finally, they can be applied after fermentation to protect and stabilize wines during barrel and bottle aging.

Bioprotection has been proposed as an effective biotool to reduce SO_2 levels in wines [19,41]. The non-*Saccharomyces Torulaspora delbrueckii* (Td) and *Metschnikowia pulcherrima* (Mp) have been used to produce industrial fermentations without added sulfites. These bioprotective non-*Saccharomyces* may control some spoilage microorganisms and prevent chemical and enzymatic oxidation [41,43].

Regarding off-flavor formation, *Metschnikowia fructicola* has been successfully used to reduce the production of ethyl acetate by apiculate yeasts such as *Hanseniaspora uvarum* during cold soak [44].

3. Bioacidification by *Lachancea thermotolerans* (Lt)

Acidification and pH control are key tools in enology to preserve wine stability and prevent microbial spoilage. Tartaric acid, the strongest acid in grapes, is systematically used in many wines, particularly in warm areas, to improve chemical stability, enhance color and stabilize anthocyanins, increase the levels of active molecular SO_2, and improve wine freshness [45]. Additionally, other acids such as malic, lactic, and citric acid can be used in enology, as can alternative physicochemical processes such as exchange resins and electrodialysis [45]. Acidification with up to 4 g/L tartaric acid is allowed in wines [9].

Bioacidification with *Saccharomyces cerevisiae* (Sc) through malic acid production has been previously studied. Some Sc strains can produce up to 1 g/L [46] when acidification occurs at the beginning of fermentation (days 2–6). However, the production at the highest level takes place in musts with low malic acid content and low acidity. Moreover, the effect on pH from increasing malic acid by 1 g/L is low and can be degraded by lactic acid bacteria, and thus malic acidification by Sc is not an effective biotechnology in winemaking. Other acids such as lactic, fumaric, and citric acids are also produced by Sc, but at low concentrations and with little impact on wine pH.

Lactic acid is also used for wine acidification, and the sensory effect is better in postfermentation acidifications. Even when often associated with dairy products, the sensory profile of lactic acid is fresh and citric [47]. Lactic acid is also authorized by the OIV for wine acidification. However, the use of lactic acid bioproduction by *Lachancea thermotolerans* is a natural and powerful biotool to control wine pH [20–23,48,49]. Some strains are capable of producing more than 16 g/L [21]. This amount is likely to be excessive in enological applications, and thus the use of strains with yields ranging from 5 to 8 g/L may be more appropriate [50–52]. With these conditions, it is easy to achieve pH reductions of 0.4–0.5 units [50–52], which is more effective than the usual acidification with tartaric acid. The production of L(+)-lactic acid [53] is done by the metabolization of sugars so that some reduction in alcoholic strength can be obtained, ranging from

0.2 [50] to 0.9% vol. [52]. Additionally, this strong reduction in pH favors higher levels of molecular SO_2 under enological conditions (Table 2), which is very effective in controlling undesirable spoilage microorganisms and off-flavor production. The usual total SO_2 contents (40–60 mg/L) at the typical high pH of grape juice in warm areas (3.7–4.0) can produce ineffective molecular SO_2 contents (<0.5 mg/L) (Table 2). These conditions are suitable for the development of spoilage microorganisms that may increase the content of off-flavors or even allergenic or toxic molecules in wines, such as biogenic amines or ethyl carbamate, which can increase during wine aging. The same levels of total SO_2 at pH 3.4–3.5 that can be obtained by Lt acidification during fermentation can easily produce molecular SO_2 levels above 0.8 mg/L, resulting in a more protective effect and a safer situation (Table 2).

Table 2. Bioacidification by *Lachancea thermotolerans* strain L31 and effect on pH and molecular SO_2 in several fermentations with red and white varieties in different Spanish regions. Colors indicate the effectiveness of molecular SO_2 from unideal (red) or less than optimal (yellow) to optimal (green) depending on the increase in acidity.

Variety (Region)	Inoculation	Lactic Acid (g/L) and Initial→Final pH	Effect of Acidity on the Molecular SO_2 (mg/L) *	Reference
Tempranillo (Ribera del Duero)	Sequential with *S. cerevisiae*	0.91→6.60 g/L 3.90→3.63	0.42→0.77	[50]
Tempranillo (Ribera del Duero)	Mixed with *O. oeni* and sequential with *S. cerevisiae*	0.91→7.50 g/L 3.90→3.31	0.42→1.56	[50]
Tempranillo (Mancha)	Sequential with *S. cerevisiae*	3.8→3.4	0.50→1.22	Unpublished
Albariño (Rias Baixas-O Rosal)	Sequential with *S. cerevisiae*	0.05→2.7 g/L 3.12→2.85	2.07→3.63	[50]
Airén (La Mancha)	Sequential with *S. cerevisiae*	0.05→4.20 g/L 3.75→3.35	0.51→1.25	[51]

* Comparison for a total content of SO_2 of 50 mg/L.

Therefore, Lt fermentations are a potent biotool to promote wine stability by reducing pH and increasing molecular SO_2 levels. The use of Lt in sequential fermentations produces a significant reduction in pH and a concomitant effect on molecular SO_2 (Figure 2). To produce complete fermentations without residual sugars, it is necessary to inoculate some non-*Saccharomyces* yeasts because most Lt strains have fermentative powers ranging from 7% to 9% v/v [20,23].

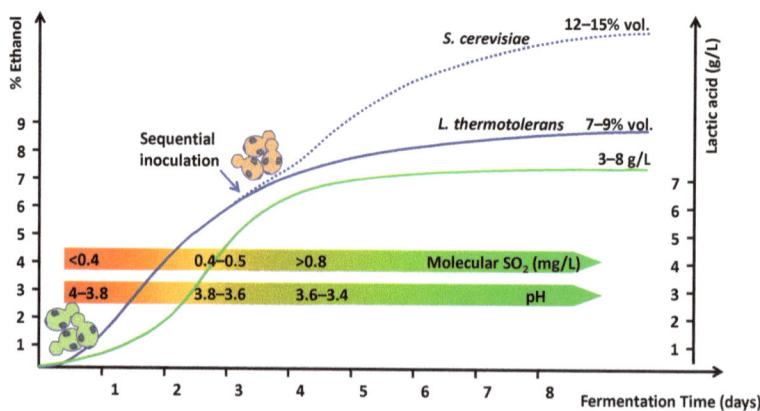

Figure 2. Sequential fermentations with *Lachancea thermotolerans* (green yeasts—bottom left) and *Saccharomyces cerevisiae* (pink yeasts—center). Effect on lactic acid production (green line), alcohol content (blue line), pH, and molecular SO_2.

A typical stage in the stabilization of red wines is the need to carry out malolactic fermentation (MLF) and improve the sensory profile. However, MLF results in a reduction of pH, and in warm areas the sensory perception can be flat with a less crispy acidity sensation in the mouth. In these winemaking regions, it could be interesting to preserve malic acidity by inhibiting MLF and simultaneously lower the pH by producing lactic acid with Lt. It should be noted that lactic acid is a strong inhibitor of MLF at high doses, which occurs in many enzymatic processes and is known as product inhibition. It has been observed that lactic acid concentrations above 4 g/L produce a strong inhibition of MLF [32] and significantly decrease lactic acid bacteria populations. At lower values (2 g/L), a significant delay of MLF is observed [32]. Therefore, in addition to pH control by Lt acidification, effective inhibition of MLF can be obtained when lactic acid production is higher than 4 g/L. Furthermore, other malic-acid-preserving additives such as fumaric acid or chitosan can be used to control MLF [54]; the former is in the final stages of evaluation at the OIV [55] and the latter is also authorized for organic wines. Additionally, it has been observed that fumaric acid production can be increased by engineered Sc to more than 5 g/L [56]. Overproduction is done by overexpression of the RoPYC gene, so perhaps in some countries where the use of engineered yeasts is allowed, it can aid in inhibiting MLF along with Lt.

Another interesting application of Lt is the control of volatile acidity levels and most likely of ethyl acetate contents as well [57]. In addition, several authors have reported low volatile acidity contents (<0.5 g/L) in sequential fermentations with Lt [20,22,50,58], even in ternary fermentations with Lt and other non-*Saccharomyces*, such as *Metschnikowia pulcherrima*, *Torulaspora delbrueckii* or *Hanseniaspora vineae* [59], and ethyl acetate contents similar to Sc controls [50].

Lactic acid production and pH reduction by Lt also have concomitant effects on color due to the increased amounts of pyrylium cation in the wine, resulting in a hyperchromic effect and color protection [60,61].

Furthermore, Lt fermentations have shown preliminary positive effects on *Brettanomyces* control, likely due to acidification and the high contents of lactate [52].

4. Apiculate Yeasts and Volatile Acidity

Traditionally, apiculate yeasts (*Hanseniaspora/Kloeckera* species), usually involved in the early fermentation phases, have been considered overproducers of volatile acidity and ethyl acetate [62]. Pure culture fermentations of *Hanseniaspora uvarum* and *Kloeckera apiculata* have been reported to produce up to 0.98 and 1.5 g/L acetic acid, and 408 and 225 mg/L acetoin, respectively. In fact, they usually release high contents of acetate esters during fermentation; such is the case for the accumulation of ethyl acetate with concentrations between 450 and 760 mg/L. However, not all species behave in the same way and some of them, such as *Hanseniaspora vineae* (Hv), have shown a high ability to decrease volatile acidity in sequential fermentations with *Saccharomyces cerevisiae* compared to single Sc fermentations [63,64]. In triplicate fermentations of white wines, the Sc control produced 0.45 g/L acetic acid, but the sequential fermentation with Hv/Sc produced 0.36 g/L [65]. Additionally, Hv can produce significant amounts of floral and fruity acetate esters, benzenoids, and terpenes, improving the aroma profile of flat neutral varieties [63,64,66,67]. Furthermore, Hv is better adapted to the fermentation process and it is possible to select strains capable of reaching 10% ethanol [68]. The use of other *Hanseniaspora* species such as *H. opuntiae* fermenting Cabernet Sauvignon red grapes has also shown low volatile acidity values together with positive fruity and floral profiles [69,70].

5. Biocompatibility

The use of non-*Saccharomyces* in ternary cultures (two non-*Saccharomyces* species and one *Saccharomyces* species) in sequential or mixed fermentations has several advantages in terms of aroma improvement, control of spoilage microorganisms, and depletion of off-flavors; however, it is very important to ensure the biocompatibility of the strains used.

When Lt has been used in co-inoculation with Hv, Td, and Mp, the latter has shown very good compatibility with Lt, reaching even higher levels of acidification than using only Lt in sequential fermentation with Sc [59]. However, the simultaneous use of Lt and Td decreased acidification, and the pH was higher compared to Lt alone, but lower than in the Sc control. The Lt and Hv strains showed the worst effectiveness on pH reduction, despite our high expectations of the complementary effect of both yeasts on acidity and aroma. This may be caused by the extra consumption of thiamine and pantothenate by Hv and the potential depletion of these important micronutrients, particularly of thiamine. The genes for thiamine biosynthesis in Hv and other *Hanseniaspora* species have not yet been identified, and this may explain the increased requirements of this vitamin in Hv fermentations [68,71,72]. Thiamine consumption and depletion may affect the development of other non-*Saccharomyces* species when used in co-fermentation.

6. Depletion of Off-Flavor Precursors

The production of some off-flavors that are extremely deleterious to wine quality, such as ethylphenols (EPs) [4,5], is highly dependent on precursor content. EPs are formed from hydroxycinnamic acids (HCAs) or their tartaric esters (TE-HCAs) by the sequential activities of hydroxycinnamate decarboxylase (HCDC) and vinylphenol reductase (VPR) from *Brettanomyces/Dekkera* [4,10]. Several technologies have been proposed to control *Brettanomyces* in wines, including emerging non-thermal technologies, some additives, and biotechnologies [4,73]. Many Sc strains express HCDC activity, but VPR activity has not been described in this species. Some Sc strains express HCDC activity with high intensity with the ability to transform most hydroxycinnamic acids into vinylphenols (VPs) (Figure 3). Moreover, it has been observed that these VPs can spontaneously react with grape anthocyanins to form vinylphenolic pyranoanthocyanins (VPAs) [74,75], which are stable pigments under enological conditions, as they are less affected by pH, oxidations, and sulfur dioxide bleaching than grape anthocyanins [61,76–79]. The use of Sc with an appropriate expression of HCDC activity is a powerful and natural biotool to favor the enzymatic metabolization of HCAs to VPs and the subsequent reaction with grape anthocyanins to form VPAs. This biological process blocks the EP precursors into stable VPAs, which are positive in terms of color stability, but also preserves the wines from the effect of *Brettanomyces/Dekkera*. When 10 commercial yeasts with verified HCDC activity were used to ferment red musts and subsequently contaminated with *Brettanomyces*, the 4EP content ranged from 22 to 498 µg/L, which is below or close to the sensory threshold of 4EP in wines [26]. However, in the control yeast (without HCDC activity), the 4EP content was 1150 µg/L, more than twice the sensory threshold [26]. Furthermore, most of the HCAs in grapes are found as tartaric esters (TE-HCAs); that is, caftaric, coutaric, and fertaric acids are reservoirs of HCAs that can be released by acid hydrolysis during aging. The use of cinnamyl esterase enzymes during fermentation can release the HCAs which, using Sc with HCDC activity, can be transformed into VPs and subsequently into VPAs by condensation with grape anthocyanins [26]. In addition, the use of some non-*Saccharomyces* yeast strains such as *Torulaspora delbrueckii* or *Metschnikowia pulcherrima* can enhance the formation of VPAs [60,80].

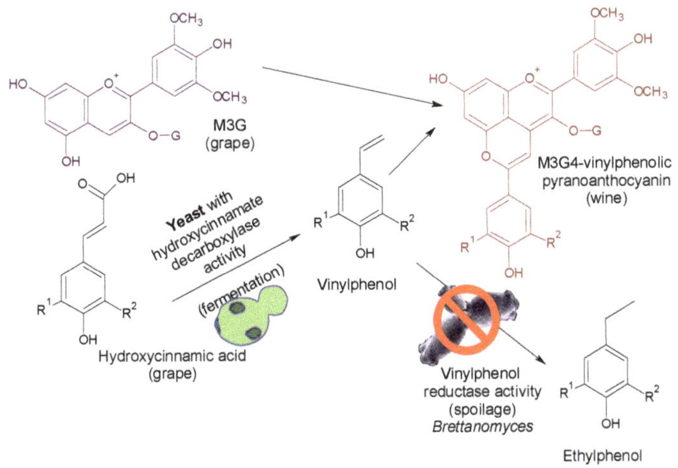

Figure 3. The depletion of ethylphenol precursors by the metabolization of hydroxycinnamic acids to vinylphenols and the blocking of this by reaction with grape anthocyanins to form vinylphenolic pyranoanthocyanins.

7. Increasing the Implantation of Non-*Saccharomyces* as Bioprotective Tools Using Emerging Non-Thermal Technologies

To achieve a good effectiveness with non-*Saccharomyces* yeasts in off-flavor control through bioprotection, acidification, and improved sulfur dioxide efficiency or precursor depletion, it is necessary to reach a good implantation of the intended species. One of the main drawbacks of non-*Saccharomyces* yeasts is the low fermentative power (<10% vol. and in many species <4% vol.) and the low fermentative yield, which generally results in poor implantation compared to Sc. To improve this, the use of non-thermal emerging technologies is compelling and effective because they can effectively eliminate wild yeasts, but also have little impact on the sensory components of the musts, that is, the aroma, pigments, and flavors [73,81–83].

The emerging non-thermal technologies include pressurization technologies (high hydrostatic pressure, HHP [84], and ultra-high-pressure homogenization, UHPH [85]), pulsed electric fields (PEFs) [86], pulsed light (PL) [87], irradiation (βI) [88], cold plasma (CD) [89], and ultrasound (US) [90]. All of them except US have demonstrated a good capacity to inactivate wild yeasts and even bacteria in grapes and musts, and preserve sensory and nutritional quality. HHP can produce reductions in wild yeast populations of more than 4-log [91,92], but residual bacterial counts can remain. UHPH is capable of producing sterilization with the elimination of yeast, bacteria, and even spores, depending on the in-valve temperature [85,93]. Pulsed technologies (i.e., PEFs and PL) have shown an inactivation capacity around or above 2-log for wild yeasts in grapes [94–96]. The antimicrobial performance of PEFs can be greatly enhanced in combination with mild temperatures (50 °C) [97].

The inactivation of wild yeasts by emerging non-thermal technologies in grapes or grape must is a useful technology to facilitate the implantation of non-*Saccharomyces* starters that can be used to control off-flavor formation. Several non-thermal technologies have shown high efficiency in increasing the implantation of non-*Saccharomyces* yeasts, such as HHP [92] and PEFs [95]. The high effectiveness of UHPH also makes it a leading technology not only for improving yeast implantation, but also for reducing SO_2 levels due to its ability to inactivate oxidative enzymes [85].

In addition, when used on grapes, several of these non-thermal technologies are able to increase the extraction of phenolic compounds, thus improving the tannin and anthocyanin content of the wine. An increase in anthocyanin extraction ranging from

23% to 63% by HHP [98,99], 21% to 29% by PEFs [100], and the same contents but with reductions of more than 50% in maceration time by US [101] have been published.

8. Conclusions

The use of non-*Saccharomyces* in wine fermentation is a verified biotechnology to improve the sensory profile, and is also a powerful biotool to control off-flavor formation by the biocontrol of spoilage microorganisms, by pH control and the improvement of molecular SO_2 contents by acidification, and by the depletion of precursors, among many other potential future possibilities.

Author Contributions: Conceptualization, A.M.; writing—original draft preparation, A.M. and I.L.; writing—review and editing, A.M., I.L., C.G. and C.E.; funding acquisition, A.M. All authors have read and agreed to the published version of the manuscript.

Funding: This study was funded by the Ministerio de Ciencia, Innovación y Universidades, project: RTI2018-096626-B-I00.

Institutional Review Board Statement: Not applicable.

Data Availability Statement: The review study did not report any data.

Conflicts of Interest: The authors declare no conflict of interest.

Sample Availability: Samples of the compounds ... are available from the authors.

References

1. Swiegers, J.H.; Pretorius, I.S. Modulation of volatile sulfur compounds by wine yeast. *Appl. Microbiol. Biotechnol.* **2007**, *74*, 954–960. [CrossRef]
2. Ugliano, M.; Kolouchova, R.; Henschke, P.A. Occurrence of hydrogen sulfide in wine and in fermentation: Influence of yeast strain and supplementation of yeast available nitrogen. *J. Ind. Microbiol. Biotechnol.* **2011**, *38*, 423–429. [CrossRef]
3. Bely, M.; Rinaldi, A.; Dubourdieu, D. Influence of assimilable nitrogen on volatile acidity production by *Saccharomyces cerevisiae* during high sugar fermentation. *J. Biosci. Bioeng.* **2003**, *96*, 507–512. [CrossRef]
4. Suárez, R.; Suárez-Lepe, J.A.; Morata, A.; Calderón, F. The production of ethylphenols in wine by yeasts of the genera *Brettanomyces* and *Dekkera*: A review. *Food Chem.* **2007**, *102*, 10–21. [CrossRef]
5. Malfeito-Ferreira, M. Two decades of "Horse Sweat" taint and *Brettanomyces* yeasts in wine: Where do we stand now? *Beverages* **2018**, *4*, 32. [CrossRef]
6. Liu, S.Q.; Pilone, G.J. An overview of formation and roles of acetaldehyde in winemaking with emphasis on microbiological implications. *Int. J. Food Sci. Technol.* **2000**, *35*, 49–61. [CrossRef]
7. Culleré, L.; López, R.; Ferreira, V. Chapter 20—The instrumental analysis of aroma-active compounds for explaining the flavor of red wines. In *Red Wine Technology*; Morata, A., Ed.; Elsevier, Academic Press: Amsterdam, The Netherlands, 2018; pp. 283–307. ISBN 9780128144008.
8. Ribéreau-Gayon, P.; Glories, Y.; Maujean, A.; Dubourdieu, D. Alcohols and other volatile compounds. In *Handbook of Enology: The Chemistry of Wine Stabilization and Treatments (Volume 2)*; John Wiley & Sons: Chichester, UK, 2006; pp. 51–64.
9. OIV International Code of Oenological Practice. Available online: https://www.oiv.int/public/medias/3741/e-code-annex-maximum-acceptable-limits.pdf (accessed on 15 June 2021).
10. Chatonnet, P.; Dubourdie, D.; Boidron, J.-n.; Pons, M. The origin of ethylphenols in wines. *J. Sci. Food Agric.* **1992**, *60*, 165–178. [CrossRef]
11. Curtin, C.; Bramley, B.; Cowey, G.; Holdstock, M.; Kennedy, E.; Lattey, K.; Coulter, A.; Henschke, P.; Francis, L.; Godden, P. Sensory perceptions of 'Brett' and relationship to consumer preference. In Proceedings of the Thirteenth Australian Wine Industry Technical Conference, Adelaide, Australia, 29 July–2 August 2007; Blair, R.J., Williams, P.J., Pretorius, I.S., Eds.; Australian Wine Industry Technical Conference Inc.: Adelaide, Australia, 2007; pp. 207–211.
12. Mira de Orduña, R. Climate change associated effects on grape and wine quality and production. *Food Res. Int.* **2010**, *43*, 1844–1855. [CrossRef]
13. Sadras, V.O.; Petrie, P.R.; Moran, M.A. Effects of elevated temperature in grapevine. II juice pH, titratable acidity and wine sensory attributes. *Aust. J. Grape Wine Res.* **2013**, *19*, 107–115. [CrossRef]
14. Mozell, M.R.; Thachn, L. The impact of climate change on the global wine industry: Challenges & solutions. *Wine Econ. Policy* **2014**, *3*, 81–89. [CrossRef]
15. Mpelasoka, B.S.; Schachtman, D.P.; Treeby, M.T.; Thomas, M.R. A review of potassium nutrition in grapevines with special emphasis on berry accumulation. *Aust. J. Grape Wine Res.* **2003**, *9*, 154–168. [CrossRef]
16. Rogiers, S.Y.; Coetzee, Z.A.; Walker, R.R.; Deloire, A.; Tyerman, S.D. Potassium in the grape (*Vitis vinifera* L.) berry: Transport and function. *Front. Plant Sci.* **2017**, *8*, 1629. [CrossRef] [PubMed]

17. Berbegal, C.; Spano, G.; Fragasso, M.; Grieco, F.; Russo, P.; Capozzi, V. Starter cultures as biocontrol strategy to prevent *Brettanomyces bruxellensis* proliferation in wine. *Appl. Microbiol. Biotechnol.* **2018**, *102*, 569–576. [CrossRef] [PubMed]
18. Roudil, L.; Russo, P.; Berbegal, C.; Albertin, W.; Spano, G.; Capozzi, V. Non-*Saccharomyces* commercial starter cultures: Scientific trends, recent patents and innovation in the wine sector. *Recent Pat. Food. Nutr. Agric.* **2020**, *11*, 27–39. [CrossRef]
19. Simonin, S.; Roullier-Gall, C.; Ballester, J.; Schmitt-Kopplin, P.; Quintanilla-Casas, B.; Vichi, S.; Peyron, D.; Alexandre, H.; Tourdot-Maréchal, R. Bio-Protection as an alternative to sulphites: Impact on chemical and microbial characteristics of red wines. *Front. Microbiol.* **2020**, *11*, 1308. [CrossRef] [PubMed]
20. Comitini, F.; Gobbi, M.; Domizio, P.; Romani, C.; Lencioni, L.; Mannazzu, I.; Ciani, M. Selected non-*Saccharomyces* wine yeasts in controlled multistarter fermentations with *Saccharomyces cerevisiae*. *Food Microbiol.* **2011**, *28*, 873–882. [CrossRef] [PubMed]
21. Banilas, G.; Sgouros, G.; Nisiotou, A. Development of microsatellite markers for *Lachancea thermotolerans* typing and population structure of wine-associated isolates. *Microbiol. Res.* **2016**, *193*, 1–10. [CrossRef] [PubMed]
22. Hranilovic, A.; Gambetta, J.M.; Schmidtke, L.; Boss, P.K.; Grbin, P.R.; Masneuf-Pomarede, I.; Bely, M.; Albertin, W.; Jiranek, V. Oenological traits of *Lachancea thermotolerans* show signs of domestication and allopatric differentiation. *Sci. Rep.* **2018**, *8*, 1–13. [CrossRef]
23. Morata, A.; Loira, I.; Tesfaye, W.; Bañuelos, M.A.; González, C.; Suárez Lepe, J.A. *Lachancea thermotolerans* applications in wine technology. *Fermentation* **2018**, *4*, 53. [CrossRef]
24. Zhang, X.; Li, B.; Zhang, Z.; Chen, Y.; Tian, S. Antagonistic yeasts: A promising alternative to chemical fungicides for controlling postharvest decay of fruit. *J. Fungi* **2020**, *6*, 158. [CrossRef]
25. Sipiczki, M. *Metschnikowia pulcherrima* and related pulcherrimin-producing yeasts: Fuzzy species boundaries and complex antimicrobial antagonism. *Microorganisms* **2020**, *8*, 1029. [CrossRef]
26. Morata, A.; Vejarano, R.; Ridolfi, G.; Benito, S.; Palomero, F.; Uthurry, C.; Tesfaye, W.; González, C.; Suárez-Lepe, J.A. Reduction of 4-ethylphenol production in red wines using HCDC+ yeasts and cinnamyl esterases. *Enzym. Microb. Technol.* **2013**, *52*, 99–104. [CrossRef] [PubMed]
27. Chassagne, D.; Guilloux-Benatier, M.; Alexandre, H.; Voilley, A. Sorption of wine volatile phenols by yeast lees. *Food Chem.* **2005**, *91*, 39–44. [CrossRef]
28. Palomero, F.; Ntanos, K.; Morata, A.; Benito, S.; Suárez-Lepe, J.A. Reduction of wine 4-ethylphenol concentration using lyophilised yeast as a bioadsorbent: Influence on anthocyanin content and chromatic variables. *Eur. Food Res. Technol.* **2011**, *232*, 971–977. [CrossRef]
29. Muccilli, S.; Restuccia, C. Bioprotective role of yeasts. *Microorganisms* **2015**, *3*, 588–611. [CrossRef]
30. Escott, C.; Loira, I.; Morata, A.; Bañuelos, M.A.; Suárez-Lepe, J.A. Wine spoilage yeasts: Control strategy. In *Yeast—Industrial Applications*; Morata, A., Loira, I., Eds.; InTechOpen: London, UK, 2017. [CrossRef]
31. Nardi, T. Microbial resources as a tool for enhancing sustainability in winemaking. *Microorganisms* **2020**, *8*, 507. [CrossRef] [PubMed]
32. Morata, A.; Bañuelos, M.A.; López, C.; Song, C.; Vejarano, R.; Loira, I.; Palomero, F.; Suarez Lepe, J.A. Use of fumaric acid to control pH and inhibit malolactic fermentation in wines. *Food Addit. Contam.—Part A Chem. Anal. Control. Expo. Risk Assess.* **2020**, *37*, 228–238. [CrossRef]
33. Woods, D.R.; Bevan, E.A. Studies on the nature of the killer factor produced by *Saccharomyces cerevisiae*. *J. Gen. Microbiol.* **1968**, *51*, 115–126. [CrossRef]
34. Young, T.W.; Yagiu, M. A comparison of the killer character in different yeasts and its classification. *Antonie Leeuwenhoek* **1978**, *44*, 59–77. [CrossRef]
35. Schmitt, M.J.; Breinig, F. Yeast viral killer toxins: Lethality and self-protection. *Nat. Rev. Microbiol.* **2006**, *4*, 212–221. [CrossRef] [PubMed]
36. Morata, A.; Loira, I.; Escott, C.; del Fresno, J.M.; Bañuelos, M.A.; Suárez-Lepe, J.A. Applications of *Metschnikowia pulcherrima* in wine biotechnology. *Fermentation* **2019**, *5*, 63. [CrossRef]
37. Comitini, F.; Ciani, M. The zymocidial activity of *Tetrapisispora phaffii* in the control of *Hanseniaspora uvarum* during the early stages of winemaking. *Lett. Appl. Microbiol.* **2010**, *50*, 50–56. [CrossRef]
38. Gobert, A.; Tourdot-Maréchal, R.; Sparrow, C.; Morge, C.; Alexandre, H. Influence of nitrogen status in wine alcoholic fermentation. *Food Microbiol.* **2019**, *83*, 71–85. [CrossRef] [PubMed]
39. Chacon-Rodriguez, L.; Joseph, C.M.L.; Nazaris, B.; Coulon, J.; Richardson, S.; Dycus, D.A. Innovative use of non-*Saccharomyces* in bio-protection: *T. delbrueckii* and *M. pulcherrima* applied to a machine harvester. *Catal. Discov. Pract.* **2020**, *4*, 82–90. [CrossRef]
40. Pelonnier-Magimel, E.; Windholtz, S.; Masneuf-Pomarède, I.; Barbe, J.C. Sensory characterisation of wines without added sulfites via specific and adapted sensory profile. *Oeno One* **2020**, *54*, 671–685. [CrossRef]
41. Simonin, S.; Alexandre, H.; Nikolantonaki, M.; Coelho, C.; Tourdot-Maréchal, R. Inoculation of *Torulaspora delbrueckii* as a bio-protection agent in winemaking. *Food Res. Int.* **2018**, *107*, 451–461. [CrossRef]
42. Rubio-Bretón, P.; Gonzalo-Diago, A.; Iribarren, M.; Garde-Cerdán, T.; Pérez-Álvarez, E.P. Bioprotection as a tool to free additives winemaking: Effect on sensorial, anthocyanic and aromatic profile of young red wines. *LWT* **2018**, *98*, 458–464. [CrossRef]
43. Windholtz, S.; Redon, P.; Lacampagne, S.; Farris, L.; Lytra, G.; Cameleyre, M.; Barbe, J.C.; Coulon, J.; Thibon, J.; Masneuf-Pomarède, I. Non-*Saccharomyces* yeasts as bioprotection in the composition of red wine and in the reduction of sulfur dioxide. *LWT* **2021**, *149*, 111781. [CrossRef]

44. Johnson, J.; Fu, M.; Qian, M.; Curtin, C.; Osborne, J.P. Influence of select non-*Saccharomyces* yeast on *Hanseniaspora uvarum* growth during prefermentation cold maceration. *Am. J. Enol. Vitic.* **2020**, *71*, 278–287. [CrossRef]
45. Comuzzo, P.; Battistutta, F. Chapter 2—Acidification and pH Control in Red Wines. In *Red Wine Technology*; Morata, A., Ed.; Elsevier, Academic Press: Amsterdam, The Netherlands, 2018; pp. 17–34. ISBN 9780128144008.
46. Yéramian, N.; Chaya, C.; Suárez Lepe, J.A. L-(-)-malic acid production by *Saccharomyces* spp. during the alcoholic fermentation of wine (1). *J. Agric. Food Chem.* **2007**, *55*, 912–919. [CrossRef]
47. Morata, A.; Escott, C.; Bañuelos, M.A.; Loira, I.; Del Fresno, J.M.; González, C.; Suárez-lepe, J.A. Contribution of non-*Saccharomyces* yeasts to wine freshness. A review. *Biomolecules* **2020**, *10*, 34. [CrossRef]
48. Gobbi, M.; Comitini, F.; Domizio, P.; Romani, C.; Lencioni, L.; Mannazzu, I.; Ciani, M. *Lachancea thermotolerans* and *Saccharomyces cerevisiae* in simultaneous and sequential co-fermentation: A strategy to enhance acidity and improve the overall quality of wine. *Food Microbiol.* **2013**, *33*, 271–281. [CrossRef]
49. Balikci, E.K.; Tanguler, H.; Jolly, N.P.; Erten, H. Influence of *Lachancea thermotolerans* on cv. Emir wine fermentation. *Yeast* **2016**, *33*, 313–321. [CrossRef]
50. Morata, A.; Bañuelos, M.A.; Vaquero, C.; Loira, I.; Cuerda, R.; Palomero, F.; González, C.; Suárez-Lepe, J.A.; Wang, J.; Han, S.; et al. *Lachancea thermotolerans* as a tool to improve pH in red wines from warm regions. *Eur. Food Res. Technol.* **2019**, *245*, 885–894. [CrossRef]
51. Vaquero, C.; Loira, I.; Bañuelos, M.A.; Heras, J.M.; Cuerda, R.; Morata, A. Industrial performance of several *Lachancea thermotolerans* strains for ph control in white wines from warm areas. *Microorganisms* **2020**, *8*, 830. [CrossRef]
52. Hranilovic, A.; Albertin, W.; Capone, D.L.; Gallo, A.; Grbin, P.R.; Danner, L.; Bastian, S.E.P.; Masneuf-Pomarede, I.; Coulon, J.; Bely, M.; et al. Impact of *Lachancea thermotolerans* on chemical composition and sensory profiles of Merlot wines. *Food Chem.* **2021**, *349*, 129015. [CrossRef] [PubMed]
53. Gatto, V.; Binati, R.L.; Lemos Junior, W.J.F.; Basile, A.; Treu, L.; de Almeida, O.G.G.; Innocente, G.; Campanaro, S.; Torriani, S. New insights into the variability of lactic acid production in *Lachancea thermotolerans* at the phenotypic and genomic level. *Microbiol. Res.* **2020**, *238*, 126525. [CrossRef] [PubMed]
54. Castro Marín, A.; Colangelo, D.; Lambri, M.; Riponi, C.; Chinnici, F. Relevance and perspectives of the use of chitosan in winemaking: A review. *Crit. Rev. Food Sci. Nutr.* **2020**, 1–15. [CrossRef] [PubMed]
55. OIV. Treatment with Fumaric Acid in Wine to Inhibit MLF. *Resolution OENO-TECHNO 15-581*. 2020. Available online: https://www.oiv.int/en/technical-standards-and-documents/resolutions-of-the-oiv/oenology-resolutions (accessed on 28 July 2021).
56. Xu, G.; Chen, X.; Liu, L.; Jiang, L. Fumaric acid production in *Saccharomyces cerevisiae* by simultaneous use of oxidative and reductive routes. *Bioresour. Technol.* **2013**, *148*, 91–96. [CrossRef]
57. Vilela, A. *Lachancea thermotolerans*, the non-*Saccharomyces* yeast that reduces the volatile acidity of wines. *Fermentation* **2018**, *4*, 56. [CrossRef]
58. Escott, C.; Morata, A.; Ricardo-Da-Silva, J.M.; Callejo, M.J.; Del Carmen González, M.; Suarez-Lepe, J.A. Effect of *Lachancea thermotolerans* on the formation of polymeric pigments during sequential fermentation with *Schizosaccharosmyces pombe* and *Saccharomyces cerevisiae*. *Molecules* **2018**, *23*, 2353. [CrossRef] [PubMed]
59. Vaquero, C.; Loira, I.; Heras, J.M.; Carrau, F.; González, C.; Morata, A. Biocompatibility in ternary fermentations with *Lachancea thermotolerans*, other non-*Saccharomyces* and *Saccharomyces cerevisiae* to control pH and improve the sensory profile of wines from warm areas. *Front. Microbiol.* **2021**, *12*, 656262. [CrossRef] [PubMed]
60. Escott, C.; Del Fresno, J.M.; Loira, I.; Morata, A.; Tesfaye, W.; del Carmen González, M.; Suárez-Lepe, J.A. Formation of polymeric pigments in red wines through sequential fermentation of flavanol-enriched musts with non-*Saccharomyces* yeasts. *Food Chem.* **2018**, *239*, 975–983. [CrossRef] [PubMed]
61. Morata, A.; Escott, C.; Loira, I.; Manuel Del Fresno, J.; González, C.; Suárez-Lepe, J.A. Influence of *Saccharomyces* and non-*Saccharomyces* yeasts in the formation of pyranoanthocyanins and polymeric pigments during red wine making. *Molecules* **2019**, *24*, 4490. [CrossRef]
62. Ciani, M.; Maccarelli, F. Oenological properties of non-*Saccharomyces* yeasts associated with wine-making. *World J. Microbiol. Biotechnol.* **1997**, *14*, 199–203. [CrossRef]
63. Medina, K.; Boido, E.; Fariña, L.; Gioia, O.; Gomez, M.E.; Barquet, M.; Gaggero, C.; Dellacassa, E.; Carrau, F. Increased flavour diversity of Chardonnay wines by spontaneous fermentation and co-fermentation with *Hanseniaspora vineae*. *Food Chem.* **2013**, *141*, 2513–2521. [CrossRef] [PubMed]
64. Martin, V.; Valera, M.J.; Medina, K.; Boido, E.; Carrau, F. Oenological impact of the *Hanseniaspora/Kloeckera* yeast genus on wines—A review. *Fermentation* **2018**, *4*, 76. [CrossRef]
65. Del Fresno, J.M.; Escott, C.; Loira, I.; Herbert-Pucheta, J.E.; Schneider, R.; Carrau, F.; Cuerda, R.; Morata, A. Impact of *Hanseniaspora vineae* in alcoholic fermentation and ageing on lees of high-quality white wine. *Fermentation* **2020**, *6*, 66. [CrossRef]
66. Zhang, B.; Xu, D.; Duan, C.; Yan, G. Synergistic effect enhances production in the mixed fermentation of 2-phenylethyl acetate *Hanseniaspora vineae* and *Saccharomyces cerevisiae*. *Process Biochem.* **2020**, *90*, 44–49. [CrossRef]
67. Del Fresno, J.M.; Escott, C.; Loira, I.; Carrau, F.; Cuerda, R.; Schneider, R.; Bañuelos, M.A.; González, C.; Suárez-Lepe, J.A.; Morata, A. The impact of *Hanseniaspora vineae* fermentation and ageing on lees on the terpenic aromatic profile of white wines of the albillo variety. *Int. J. Mol. Sci.* **2021**, *22*, 2195. [CrossRef]

68. Valera, M.J.; Boido, E.; Dellacassa, E.; Carrau, F. Comparison of the glycolytic and alcoholic fermentation pathways of *Hanseniaspora vineae* with *Saccharomyces cerevisiae* wine yeasts. *Fermentation* **2020**, *6*, 78. [CrossRef]
69. Luan, Y.; Zhang, B.Q.; Duan, C.Q.; Yan, G.L. Effects of different pre-fermentation cold maceration time on aroma compounds of *Saccharomyces cerevisiae* co-fermentation with *Hanseniaspora opuntiae* or *Pichia kudriavzevii*. *LWT Food Sci. Technol.* **2018**, *92*, 177–186. [CrossRef]
70. Feng, C.T.; Du, X.; Wee, J. Microbial and chemical analysis of non-*Saccharomyces* yeasts from Chambourcin hybrid grapes for potential use in winemaking. *Fermentation* **2021**, *7*, 15. [CrossRef]
71. Seixas, I.; Barbosa, C.; Mendes-Faia, A.; Güldener, U.; Tenreiro, R.; Mendes-Ferreira, A.; Mira, N.P. Genome sequence of the non-conventional wine yeast *Hanseniaspora guilliermondii* UTAD222 unveils relevant traits of this species and of the *Hanseniaspora* genus in the context of wine fermentation. *DNA Res.* **2019**, *26*, 67–83. [CrossRef]
72. Steenwyk, J.L.; Opulente, D.A.; Kominek, J.; Shen, X.X.; Zhou, X.; Labella, A.L.; Bradley, N.P.; Eichman, B.F.; Čadež, N.; Libkind, D.; et al. Extensive loss of cell-cycle and DNA repair genes in an ancient lineage of bipolar budding yeasts. *PLoS Biol.* **2019**, *17*, e3000255. [CrossRef]
73. Pinto, L.; Baruzzi, F.; Cocolin, L.; Malfeito-Ferreira, M. Emerging technologies to control *Brettanomyces* spp. in wine: Recent advances and future trends. *Trends Food Sci. Technol.* **2020**, *99*, 88–100. [CrossRef]
74. Morata, A.; Gómez-Cordovés, M.C.; Calderón, F.; Suárez, J.A. Effects of pH, temperature and SO_2 on the formation of pyranoanthocyanins during red wine fermentation with two species of *Saccharomyces*. *Int. J. Food Microbiol.* **2006**, *106*, 123–129. [CrossRef] [PubMed]
75. Morata, A.; González, C.; Suárez-Lepe, J.A. Formation of vinylphenolic pyranoanthocyanins by selected yeasts fermenting red grape musts supplemented with hydroxycinnamic acids. *Int. J. Food Microbiol.* **2007**, *116*, 144–152. [CrossRef] [PubMed]
76. Bakker, J.; Timberlake, C.F. Isolation, identification and characterization of new color-stable anthocyanins occurring in some red wines. *J. Agric. Food Chem.* **1997**, *45*, 35–43. [CrossRef]
77. De Freitas, V.; Mateus, N. Formation of pyranoanthocyanins in red wines: A new and diverse class of anthocyanin derivatives. *Anal. Bioanal. Chem.* **2011**, *401*, 1467–1477. [CrossRef] [PubMed]
78. Morata, A.; Loira, I.; Suárez Lepe, J.A. Influence of yeasts in wine colour. In *Grape and Wine Biotechnology*; InTech: Rijeka, Croatia, 2016. [CrossRef]
79. Escribano-Bailón, M.T.; Rivas-Gonzalo, J.C.; García-Estévez, I. Wine color evolution and stability. In *Red Wine Technology*; Elsevier, Academic Press: Amsterdam, The Netherlands, 2018; pp. 195–205. ISBN 9780128144008. [CrossRef]
80. Escott, C.; Morata, A.; Loira, I.; Tesfaye, W.; Suarez-Lepe, J.A. Characterization of polymeric pigments and pyranoanthocyanins formed in microfermentations of non-*Saccharomyces* yeasts. *J. Appl. Microbiol.* **2016**, *121*, 1346–1356. [CrossRef]
81. Morata, A.; Loira, I.; Vejarano, R.; González, C.; Callejo, M.J.; Suárez-Lepe, J.A. Emerging preservation technologies in grapes for winemaking. *Trends Food Sci. Technol.* **2017**, *67*, 36–43. [CrossRef]
82. Paniagua-Martínez, I.; Ramírez-Martínez, A.; Serment-Moreno, V.; Rodrigues, S.; Ozuna, C. Non-thermal technologies as alternative methods for *Saccharomyces cerevisiae* inactivation in liquid media: A review. *Food Bioprocess Technol.* **2018**, *11*, 487–510. [CrossRef]
83. Gómez-López, V.M.; Pataro, G.; Tiwari, B.; Gozzi, M.; Meireles, M.Á.A.; Wang, S.; Guamis, B.; Pan, Z.; Ramaswamy, H.; Sastry, S.; et al. Guidelines on reporting treatment conditions for emerging technologies in food processing. *Crit. Rev. Food Sci. Nutr.* **2021**, 1–25. [CrossRef] [PubMed]
84. San Martín, M.F.; Barbosa-Cánovas, G.V.; Swanson, B.G. Food processing by high hydrostatic pressure. *Crit. Rev. Food Sci. Nutr.* **2002**, *42*, 627–645. [CrossRef]
85. Morata, A.; Guamis, B. Use of UHPH to obtain juices with better nutritional quality and healthier wines with low levels of SO_2. *Front. Nutr.* **2020**, *7*, 598286. [CrossRef] [PubMed]
86. Raso, J.; Calderón, M.L.; Góngora, M.; Barbosa-Cánovas, G.V.; Swanson, B.G. Inactivation of *Zygosaccharomyces bailii* in fruit juices by heat, high hydrostatic pressure and pulsed electric fields. *J. Food Sci.* **1998**, *63*, 1042–1044. [CrossRef]
87. Santamera, A.; Escott, C.; Loira, I.; Del Fresno, J.M.; González, C.; Morata, A. Pulsed light: Challenges of a non-thermal sanitation technology in the winemaking industry. *Beverages* **2020**, *6*, 45. [CrossRef]
88. Morata, A.; Bañuelos, M.A.; Tesfaye, W.; Loira, I.; Palomero, F.; Benito, S.; Callejo, M.J.; Villa, A.; González, M.C.; Suárez-Lepe, J.A. Electron beam irradiation of wine grapes: Effect on microbial populations, phenol extraction and wine quality. *Food Bioprocess Technol.* **2015**, *8*, 1845–1853. [CrossRef]
89. Sainz-García, E.; López-Alfaro, I.; Múgica-Vidal, R.; López, R.; Escribano-Viana, R.; Portu, J.; Alba-Elías, F.; González-Arenzana, L. Effect of the atmospheric pressure cold plasma treatment on Tempranillo red wine quality in batch and flow systems. *Beverages* **2019**, *5*, 50. [CrossRef]
90. García Martín, J.F.; Sun, D.W. Ultrasound and electric fields as novel techniques for assisting the wine ageing process: The state-of-the-art research. *Trends Food Sci. Technol.* **2013**, *33*, 40–53. [CrossRef]
91. Delfini, C.; Conterno, L.; Carpi, G.; Rovere, P.; Tabusso, A.; Cocito, C.; Amati, A. Microbiological stabilisation of grape musts and wines by High Hydrostatic Pressures. *J. Wine Res.* **1995**, *6*, 143–151. [CrossRef]
92. Bañuelos, M.A.; Loira, I.; Escott, C.; Del Fresno, J.M.; Morata, A.; Sanz, P.D.; Otero, L.; Suárez-Lepe, J.A. Grape processing by High Hydrostatic Pressure: Effect on use of non-*Saccharomyces* in must fermentation. *Food Bioprocess Technol.* **2016**, *9*, 1769–1778. [CrossRef]

93. Bañuelos, M.A.; Loira, I.; Guamis, B.; Escott, C.; Del Fresno, J.M.; Codina-Torrella, I.; Quevedo, J.M.; Gervilla, R.; Chavarría, J.M.R.; de Lamo, S.; et al. White wine processing by UHPH without SO_2. Elimination of microbial populations and effect in oxidative enzymes, colloidal stability and sensory quality. *Food Chem.* **2020**, *332*, 127417. [CrossRef] [PubMed]
94. Garde-Cerdán, T.; Arias-Gil, M.; Marsellés-Fontanet, A.R.; Ancín-Azpilicueta, C.; Martín-Belloso, O. Effects of thermal and non-thermal processing treatments on fatty acids and free amino acids of grape juice. *Food Control* **2007**, *18*, 473–479. [CrossRef]
95. Vaquero, C.; Loira, I.; Raso, J.; Álvarez, I.; Delso, C.; Morata, A. Pulsed Electric Fields to improve the use of non-*Saccharomyces* starters in red wines. *Foods* **2021**, *10*, 1472. [CrossRef] [PubMed]
96. Escott, C.; López, C.; Loira, I.; González, C.; Bañuelos, M.A.; Tesfaye, W.; Suárez-Lepe, J.A.; Morata, A. Improvement of must fermentation from late harvest cv. Tempranillo grapes treated with pulsed light. *Foods* **2021**, *10*, 1416. [CrossRef] [PubMed]
97. Wu, Y.; Mittal, G.S.; Griffiths, M.W. Effect of pulsed electric field on the inactivation of microorganisms in grape juices with and without antimicrobials. *Biosyst. Eng.* **2005**, *90*, 1–7. [CrossRef]
98. Corrales, M.; García, A.F.; Butz, P.; Tauscher, B. Extraction of anthocyanins from grape skins assisted by high hydrostatic pressure. *J. Food Eng.* **2009**, *90*, 415–421. [CrossRef]
99. Morata, A.; Loira, I.; Vejarano, R.; Bañuelos, M.A.; Sanz, P.D.; Otero, L.; Suárez-Lepe, J.A. Grape processing by High Hydrostatic Pressure: Effect on microbial populations, phenol extraction and wine quality. *Food Bioprocess Technol.* **2015**, *8*, 277–286. [CrossRef]
100. López, N.; Puértolas, E.; Condón, S.; Álvarez, I.; Raso, J. Effects of pulsed electric fields on the extraction of phenolic compounds during the fermentation of must of Tempranillo grapes. *Innov. Food Sci. Emerg. Technol.* **2008**, *9*, 477–482. [CrossRef]
101. Pérez-Porras, P.; Bautista-Ortín, A.B.; Jurado, R.; Gómez-Plaza, E. Using high-power ultrasounds in red winemaking: Effect of operating conditions on wine physico-chemical and chromatic characteristics. *LWT* **2021**, *138*, 110645. [CrossRef]

Review

A Review of Ladybug Taint in Wine: Origins, Prevention, and Remediation

Gary J. Pickering [1,2,3,4,*] **and Andreea Botezatu** [5]

[1] Departments of Biological Sciences and Psychology, Brock University, St. Catharines, ON L2S 3A1, Canada
[2] Cool Climate Oenology and Viticulture Institute, Brock University, St. Catharines, ON L2S 3A1, Canada
[3] National Wine and Grape Industry Centre, Charles Sturt University, Wagga Wagga, NSW 2678, Australia
[4] Sustainability Research Centre, University of the Sunshine Coast, Sippy Downs, QLD 4556, Australia
[5] Department of Horticultural Sciences, Texas A&M University, College Station, TX 77843-2133, USA; abotezatu@tamu.edu
* Correspondence: gpickering@brocku.ca

Abstract: Ladybug taint (also known as ladybird taint) is a relatively recently recognized fault that has been identified in wines from a wide range of terroirs. Alkyl-methoxypyrazines—particularly 2-isopropyl-3-methoxypyrazine—have been determined as the causal compounds, and these are introduced into grape must during processing, when specific species of vineyard-dwelling Coccinellidae are incorporated into the harvested fruit. *Coccinella septempunctata*, and especially the invasive *Harmonia axyridis*, are the beetles implicated, and climate change is facilitating wider dispersal and survivability of *H. axyridis* in viticultural regions worldwide. Affected wines are typically characterized as possessing excessively green, bell pepper-, and peanut-like aroma and flavor. In this paper, we review a range of vineyard practices that seek to reduce Coccinellidae densities, as well as both "standard" and novel wine treatments aimed at reducing alkyl-methoxypyrazine load. We conclude that while prevention of ladybug taint is preferable, there are several winery interventions that can remediate the quality of wine affected by this taint, although they vary in their relative efficacy and specificity.

Keywords: ladybird taint; methoxypyrazines; wine quality; wine faults; grape quality

1. What Is Ladybug Taint?

Ladybug taint (LBT, also known as ladybird taint) was first associated with the presence of large numbers of *Harmonia axyridis* (Pallas) (Coleoptera: Coccinellidae) beetles in vineyards at harvest, and corresponding off-odors in the subsequent wines. *H. axyridis*, more colloquially known as the multicolored Asian ladybeetle (MALB), originates in northeastern Asia, and was first introduced to North America in 1916 as a form of biocontrol against aphids and some small, soft-bodied pests [1,2]. It was introduced to France in 1982 [3], while in Canada it was first reported in southern Québec between 1992 and 1994 [4]. When introduced, the beetle typically extends its range into non-target regions and crops, including grape vineyards, and is now present in many winegrowing countries and regions around the globe. In the vineyard, MALB adults aggregate on grape clusters during autumn, and are often picked with the grapes and transported into the winery [5] (Figure 1). Their presence during crushing and winemaking operations can lead to unpleasant aromas and flavors in the subsequent wine, known collectively as LBT [6–8].

Sporadic anecdotal evidence that MALBs have been negatively impacting wine quality has been circulating since the 20th century, but the first widely reported link was noted in 2001 in some North American wine regions, where winemakers described an unpleasant aroma and taste in wines from that vintage similar to "burnt peanut butter" and reminiscent of crushed lady beetles [6]. This observation led to an investigation by Pickering et al. [6], who showed that LBT affects both the aroma and flavor of wines. The

authors added MALBs to red and white juice and must, and used descriptive analysis and a trained sensory panel to characterize the wines thus produced. White wines displayed higher intensities of bell pepper, asparagus, and peanut aroma and flavor compared with control wines, while red wines showed higher intensities of peanut, asparagus/bell pepper, and earthy/herbaceous aroma and flavor. At the same time, bitterness (more intense), sourness (more intense), and sweetness (less intense) were also affected in MALB-treated red wines, while in whites the intensity of fruit and floral descriptors was reduced compared with control wines. These effects increased with the number of beetles added to the juice/must [6].

Figure 1. A *Harmonia axyridis* (MALB) beetle on a grape berry prior to harvest.

Similar sensory profiles were subsequently reported by Pickering et al. [8,9] and Ross and Weller [10] in other wines produced in the presence of MALBs. The taint appears to be stable over time; affected wines were described similarly at bottling, and again after 10 months of aging [9]. Interestingly, dead beetles are also capable of producing LBT, as demonstrated by the addition of MALBs to red must at various stages postmortem, and subsequent sensory analysis of the wines [7]. At one day postmortem, the beetles negatively influenced the sensory profiles of the wines; however, at three days postmortem and beyond, these effects disappeared. These findings are relevant to tolerance levels for MALB in the winery, as dead beetles can be inadvertently incorporated in with the grapes at harvest—particularly after insecticides having been applied in the vineyard.

1.1. Alkyl-Methoxypyrazines Are the Molecules Responsible

Coccinellidae emit a mixture of odor-active compounds that most likely serve several behavioral functions, including defense, aggregation, and mate-attraction. One particular group of compounds—alkyl-methoxypyrazines (MPs)—have been closely scrutinized for their role in LBT (Figure 2). Cai et al. [11] used multidimensional gas chromatography–olfactometry–mass spectrometry to determine the odorants emitted by MALBs, and found that 2-*iso*propyl-3-methoxypyrazine (IPMP) was the most abundant and potent of the pyrazines released. IPMP has been described sensorially as displaying aromas of "peanuts", "potatoes", "peas", and "earthy" [11]. Three more MPs were also identified (2-*iso*butyl-3-methoxypyrazine (IBMP), 2-*sec*-butyl-3-methoxypyrazine (SBMP), and 2,5-dimethyl-3-methoxypyrazine (DMMP)), and their corresponding aromas reported as "peanut, potato, earthy, spicy", "nutty, potato, peanut", and "moldy, earthy", respectively [11], consistent with terms used for describing LBT-affected wines [6,8–10]. A more recent study reported a different DMMP isomer—3,5-dimethyl-2-methoxypyrazine—as a component of Coccinellidae hemolymph [12]. IPMP, IBMP, and SBMP have all been detected in the headspace above MALB, with IPMP being the most prevalent [13,14].

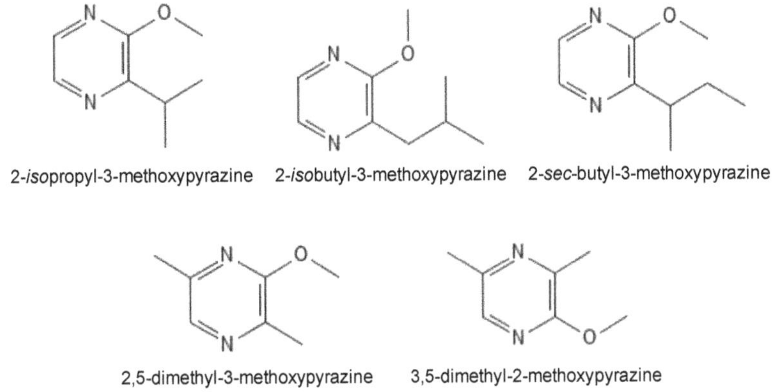

2-isopropyl-3-methoxypyrazine   2-isobutyl-3-methoxypyrazine   2-sec-butyl-3-methoxypyrazine

2,5-dimethyl-3-methoxypyrazine   3,5-dimethyl-2-methoxypyrazine

Figure 2. Alkyl-methoxypyrazines identified in Coccinellidae.

Pickering et al. [9] provided the first direct evidence that LBT was caused by MPs. They showed that IPMP levels were higher in white (38 ng/L) and red (30 ng/L) wines fermented with 10 MALB beetles/liter compared with control wines containing no beetles (8 ng/L). Several subsequent studies have confirmed that IPMP concentration in wine increases in the presence of MALBs [7,15,16]. Further implicating the role of IPMP in LBT, Pickering et al. [9] showed that the intensity of earthy/herbaceous descriptors strongly correlates with IPMP concentrations in red wines. Furthermore, wine produced with the addition of MALB and wine spiked with 15 ng/L IPMP display very similar sensory profiles [17].

One study reported no association between IPMP and MALB in wine. Galvan et al. [18] analyzed wines produced from artificially infested Frontenac grapes and from Leon Millot grapes harvested from vineyards with different degrees of "natural" MALB infestation (20% and 50% of the clusters infested with one or more *H. axyridis* adults). They found no differences in concentrations of IPMP between wines with different infestation levels, but they noted that the addition of MALBs did affect the wines' sensory profiles. The difference between this and previous findings in relation to IPMP may stem from the different analytical techniques employed. SBMP, rather than IPMP, has alternately been suggested as the causal compound for LBT in grape juice [19]; however, the method used to form this conclusion (frequency of detection) is not as robust as other approaches that have used compound concentrations. While Botezatu et al. [20] also reported numerically higher levels of SBMP compared to IPMP in Vidal and Cabernet Sauvignon wines produced with MALB additions, their aroma extract dilution analysis—considered a gold standard for identifying fault compounds in complex matrices—confirmed the dominant role of IPMP. Thus, while other MPs may certainly play a minor part in LBT, the balance of the literature shows that IPMP is the main contributor to the characteristic sensory profile of the taint.

It is important to note that MPs are endogenously produced by certain wine grape varieties; therefore, the presence of MPs in wine is not in and of itself indicative of LBT. In varietals such as Sauvignon Blanc, Cabernet Sauvignon, Cabernet Franc, and Carmenère, these endogenous MPs can confer typicality to their wines at low concentrations, although at elevated levels they negatively impact quality by contributing unbalanced "greenness" and "earthiness" [21–26]. In the case of endogenously occurring MPs, IBMP is normally the predominant pyrazine, both in concentration and odor-impact [27,28] It has therefore been proposed that the relative ratio of the different MP species in wine might serve as a "diagnostic" test for Coccinellidae influence. Specifically, the results of Botezatu et al. [28] suggest that an IPMP:IBMP ratio > 1 is indicative of juice/wine contaminated with Coccinellidae beetles.

1.2. How Much Is Too Much?

For grape growers and winemakers, it is important to know "how much is too much?". That is, what densities of MALB in the vineyard should be of concern and necessitate intervention, and what concentrations of MPs are required to elicit LBT in wine?

MPs are extremely odor-active aroma compounds. Most aromatic and volatile components in wine are present and active in the µg/L to mg/L range [29,30], while MP concentrations are measured in parts per trillion or ng/L [22], and contribute to aroma at these trace concentrations [31–34]. The detection threshold for IPMP in red wine is 1–2 ng/L [32,34] and 0.3–1.6 ng/L in white wine [34], with similar thresholds reported in grape juices [35]. Factors influencing the IPMP detection threshold in wine—and therefore, how LBT will be experienced—include wine style, mode of evaluation (ortho- vs. retro-nasal assessment), and familiarity with the odorant [34]. For instance, the researchers reported a significantly lower detection threshold for IPMP in a neutral-tasting Chardonnay compared with both a Gewürztraminer and a red blend. Additionally, a trend of greater sensitivity to LBT (i.e., lower detection thresholds) with increasingly familiarity with LBT was observed. The authors suggest that learning of the characteristic sensory cues associated with IPMP occurs in individuals with greater experience of LBT. A corollary, they speculate, is that as awareness of LBT increases in the marketplace, individual thresholds and perhaps acceptance of the taint may decrease [34]. Finally, wine style, mode of evaluation, and familiarity with the taint aside, consumers vary in their sensitivity to IPMP and, thus, how they will respond to wine affected by Coccinellidae; several-hundred-fold differences in detection thresholds have been noted between some individuals [34].

Cudjoe et al. [13] calculated the concentrations of MPs in individual MALBs as ≈27.5 µg/beetle for IPMP, 3.2 µg/beetle for IBMP, and 2.6 µg/beetle for SBMP. What do these values mean for tolerance levels of Coccinellidae in the vineyard? Pickering et al. [8] calculated that 1530 beetles/ton of white grapes or 1260 beetles per ton of red grapes would be needed before LBT could be detected in the subsequent wines. That translates to 1.3–1.5 beetles/kg of grapes; however the authors note that these limits could vary based on grape variety and the processing methods used, and thus recommend a more conservative limit of 200–400 beetles/ton of grapes. Galvan et al. [36], using a different approach, derived similar values, estimating the threshold at which 10% of the population can detect LBT as 1.9 beetles/kg grapes (equivalent to 1900 beetles/t), or 0.27 beetle/grape cluster for Frontenac grapes. Ross et al. [37] similarly estimated the threshold to be 1.8 MALB/kg (1800 beetles/t) in Concord grapes.

The density of MALBs in vineyards at or close to harvest certainly exceeds these "threshold" limits in some years, and this occurs on a periodic basis in some regions, possibly reflecting variations in the population of their preferred prey species. Vineyards in North America located near soybean or grain crops appear to be especially vulnerable. Likely, the harvest of these crops removes principal prey species—such as the aphid Hemiptera: Aphididae—and the beetles migrate to adjacent grapevines in search of further food sources and/or to make use of the vines as potential overwintering sites [3]. The position of "zero tolerance" for ladybugs advocated by some wineries and wine agencies is not supported by the scientific evidence. Coccinellidae do not taint the grapes directly; they must be physically incorporated into postharvest operations. They have long been part of the fauna found in grape vineyards, and only become a threat to wine quality in years when densities are excessive. As detailed above, acceptable levels for these beetles have been determined, and can be used to inform what, if any, interventions are needed by winegrowers.

1.3. Coccinellidae Species Responsible

Much of the focus to this point has been on MALBs; are all Coccinellidae equal with respect to their capacity to cause LBT in juice and wine? Cudjoe et al. [13] investigated the relative MP composition of three Coccinellidae species common to North America: *Coccinella septempunctata* (seven-spot), *H. axyridis* (MALB) and *Hippodamia convergens*. IPMP was reported in all three species, with the highest concentration (0.81 µg/mg) in *H. convergens* and the lowest in *C. septempunctata* (0.008 µg/mg), while SBMP and IBMP were above the instrumental limits of detection only in MALBs and *H. convergens*. Kögel et al. [14] also reported on the presence of MPs in *C. septempunctata* and *H. axyridis*, with IPMP found to be the most abundant in both species, followed by SBMP (in *H. axyridis*) and IBMP (in *C. septempunctata*).

Botezatu et al. [20] investigated the relative effects of contamination with seven-spot ladybugs and MALBs on wine quality. They added 0 or 10 beetles of each Coccinellidae species/kg of Vidal and Cabernet Sauvignon grapes at crush, and compared MP concentrations and sensory profiles in the finished wines. The addition of beetles led to similar effects on the sensory profiles of the wines, consistent with ladybug taint, regardless of the species. IPMP concentrations were not significantly different between species for Vidal wines, while in Cabernet sauvignon IPMP was significantly higher in wines made with the addition of seven-spot ladybugs. The difference between this result and that of Cudjoe et al. [13] may be attributed to methodological variations (live beetles in the Botezatu study versus frozen and thawed beetles in the Cudjoe study) and/or the different analytical techniques employed. Kögel et al. [16] also reported increases in IPMP concentrations in wines processed with either MALBs or seven-spot ladybugs.

Given that MALBs and seven-spot ladybugs are probably the most common Coccinellidae species in the majority of the world's wine regions, these results should underline the importance of monitoring for the presence of ladybeetles in vineyards at harvest. Both species show similar capacity to cause LBT in resulting wines and, thus, do not need to be differentiated in the vineyard when deciding on intervention thresholds, although it is unlikely that seven-spot ladybugs reach the densities required to affect wine quality as often as do MALBs.

2. Preharvest Prevention in the Vineyard

Coccinellidae beetles tend to appear in vineyards in the fall, after undertaking dispersal flights from their original feeding habitats both in response to temperature changes and as a result of the crops that initially hosted them being harvested [38]. Once in the vineyard, they do not damage healthy fruit, but they do feed on previously damaged grapes [39]. This underlines the importance of good vineyard management practices in order to prevent or reduce damaged fruit, such as bird-displacement measures, good canopy management, and the proper use of antifungal products. Since aphids are the main food source for MALBs, good weed management late in the season should be observed in order to reduce weed populations that can host aphids attractive to MALBs.

One key element of an integrated pest management strategy against *H. axyridis* in vineyards is effective surveying for beetle densities before harvest. Galvan et al. [36] assessed the usefulness of various sampling plans, and found binomial sampling to be a more accurate method to determine beetle densities than enumerative plans.

In addition to good vineyard management practices, other methods can be applied to reduce Coccinellidae populations, when detected, to below the tolerance levels outlined in Section 1.2. These fall into three general strategies: insecticides, which kill the beetles; semiochemical push–pull approaches, which combine both repellants and attractants to affect the spatial distribution of beetles; and repellent sprays that drive the beetles away from the vineyard (Figure 3).

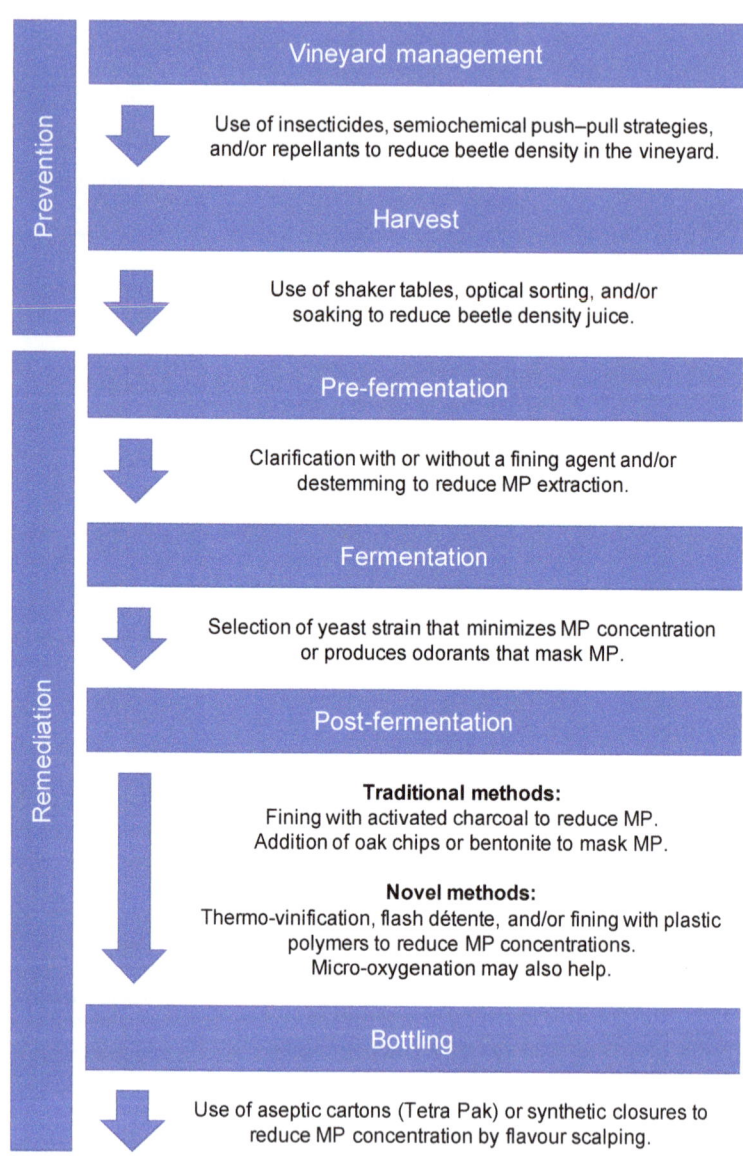

Figure 3. Summary of methods used to prevent or remediate ladybug taint during wine production.

2.1. Insecticides

Application of insecticides is the most common intervention used in North America to control Coccinellidae in vineyards. In Canada, Malathion 85 E (malathion) and Ripcord™ 400 EC (cypermethrin) can be used with a preharvest interval of seven and three days, respectively. Ripcord™ 400 EC has an extended repellency effect, and both products have good knockdown success [40]. In the United States, accepted insecticides include Venom® 70SG (dinotefuran), Clutch® 50WDG (clothianidin), and Mustang® Max EC (permethrin), with preharvest intervals of 0–1 days [40]. As noted in Section 1, MALBs can still elicit LBT in subsequent wines up to three days after they have been killed [7]; therefore, care should be taken during harvesting operations to minimize their incorporation with the fruit.

An issue with insecticides—particularly those with longer preharvest intervals—is that re-infestation can occur rapidly, since ladybeetles are very mobile and their densities can vary greatly from day to day [41]. Additionally, improper use of these products can increase the potential for elevated or unsafe levels of pesticide residues on grapes, or in juice and wine [42]. These potential shortcomings point to the need for alternative or complementary approaches for controlling MALBs in vineyards.

2.2. Semiochemical-Based Push–Pull Strategies

Semiochemical-based push–pull strategies use a combination of highly repellant and highly attractive stimuli in order to control the distribution and abundance of beetles in the vineyard. Ideally, the beetles are "pushed" away from the vineyard by the presence of the repellent compounds, while at the same time are attracted towards other areas–such as trapping zones or protected areas—by the "pull" compounds. Leroy et al. [43] evaluated semiochemicals from aphids (Z,E-nepetalactone, [E]-β-farnesene, α-pinene, and β-pinene), coccinellids ([-]-β-caryophyllene), and the nettle *Urtica dioica* L. as potential attractants for MALB—first in a wind tunnel, and later in a potato field. They reported Z,E-nepetalactone as the most efficient attractant in the wind tunnel experiments, and that MALBs were also responsive to it in the potato field. This latter result suggests that Z,E-nepetalactone may be effective under ecologically valid conditions as an efficient approach for controlling MALBs.

Following the observation that MALB beetles prefer feeding on damaged and overripe grapes [5], Glemser et al. [40] tested grape-derived, nontoxic compounds at two concentrations (high/low) for their potential efficacy as components of a push–pull strategy. They also included MPs in their study, as IBMP, IPMP, and SBMP had previously been identified as attractants to seven-spot beetles [44]. Results showed that MALBs were attracted to acetic acid (both concentrations), acetaldehyde (low concentration), acetic acid plus ethanol, acetic acid plus isobutanol, and IPMP (low concentration, 0.1 ng/L). As part of the same study, but reported elsewhere [42], the beetles were repelled by both ethyl acetate and a mixture of acetic acid plus acetaldehyde. Taken overall, these results indicate the potential for specific compounds and mixtures associated with grape spoilage/fermentation processes to form part of a semiochemical-based push–pull strategy for MALBs, although considerable work remains to operationalize these initial findings.

2.3. Repellent Sprays

Glemser et al. [40] looked at the potential of sulfur dioxide (SO_2) in the form of potassium metabisulfite (KMS) as a repellent compound against MALBs in vineyards. SO_2 is widely used in the winemaking industry as an antimicrobial agent, an antioxidant, and a general preservative for wine quality. Therefore, using SO_2 as a spray could be a simple and inexpensive solution, and concerns about residual presence on the grapes would be minimal, given its use in winemaking. The authors found that when sprayed in the vineyard, KMS significantly reduced the number of MALB beetles on grape vines (≈30% fewer with a 5 g/L solution, and 50–60% fewer on vines treated with 10 g/L KMS). However, a later study showed no repellant effects in the field for a 10 g/L KMS solution, with the authors suggesting that stronger winds around the application date may have dissipated the SO_2 from the treated vines [38]. Encouragingly, KMS (5 g/L solution) sprayed in the vineyards at 14, 7, 3, or 1 day before harvest did not affect free SO_2 concentrations in the resultant grape juice [40]. Additionally, the authors report no visible phytotoxicity effects on vines post KMS treatment, but do caution against potential environmental pollution issues (e.g., acid rain).

A more recent report from Glemser et al. [38] examined a series of products that were either already registered for use on grapes or had been previously reported as repellant to other insects. In lab-based repellency trials, carvacrol (a monoterpenoid found in essential oils), Timorex Gold (tea tree oil), pine oil, and granite dust were effective at reducing the number of MALB beetles by more than 80%, with pine oil being the most effective, even 72 h after application. However, in the field, only Biobenton (bentonite; up to 39% reduction), Buran (garlic powder + KMS), and Solfobenton (KMS and pine oil) were effective.

Other repellent compounds have also been identified, such as terpenoids from catnip oil (*Nepeta cataria* L.), grapefruit [45], camphor, menthol [46], and even the mosquito repellent DEET (N,N-diethyl-3-methylbenzamide) [47]. However, while suitable for spraying on buildings, these products have not been tested or approved for use in vineyards. Additionally, the potential for residual effects on juice and wine flavor need to be considered for any repellency spray and, thus, sensory analyses should be included in future studies before recommendations on their use can be made.

3. Postharvest Prevention and Remediation

When grapes infested with Coccinellidae arrive at the winery, or wine is made and subsequently identified as being affected by LBT, what options are available for fixing the problem? A significant body of research exists around this question, in part because solutions to LBT in many cases will also be transferable for improving wine made from grapes of suboptimal ripeness, given that elevated MP levels are the common cause. Both conventional treatments and newer approaches under development are reviewed below and summarized in Figure 3.

3.1. Removing Beetles

Perhaps the most intuitive and simple action is to remove beetles from the grapes at or after harvest prior to further processing. Specialized shaker tables have been designed to facilitate this, and are very effective, although they are generally restricted to hand-harvested fruit, and throughput can be limited [48]. Anecdotally, optical sorters have also been reported to be very effective. These systems use high-speed cameras and image-processing software to separate grapes from beetles (and other non-grape material), and can be incorporated into the grape harvester. Alternatively, some wineries report soaking the grapes in water, allowing for the beetles to rise to the surface, where they can be removed, but this can potentially lead to dilution of the sugar concentration and affect wine quality [48]. Several interventions that are part of traditional winemaking practice have been evaluated in order to determine their effectiveness at removing MPs from must/wine or reducing the severity of LBT.

3.2. Traditional Approaches to Treating Faulted Wine

In considering interventions for musts/wines affected by Coccinellidae, the winemaker should also be cognizant of minimizing the contribution from the endogenous MP load of the grapes, especially in high-MP varieties such as Cabernet Sauvignon, Cabernet Franc, and Carmenère, or when the grapes have not reached full flavor maturity (Section 1.1). This includes de-stemming, as over half of grape-derived MPs come from the stems [24,49], and consideration of maceration time, with most MP extraction occurring during the first 24 h [50,51].

Clarification of white wine juice is effective in reducing MP levels by up to approximately 50% of their initial levels [52]. The authors tested Chardonnay juice that had been clarified either through the use of bentonite (1 g/L) or natural settling (24 h and 48 h), and found that regardless of the clarification method, wine produced from clarified juice had significantly lower IPMP concentrations compared with wines produced from unclarified juice. They also showed that the longer the settling period, the greater the decrease in IPMP for both naturally clarified and bentonite-clarified juices.

The use of selected yeast strains to decrease MP levels in wines or mediate their sensorial expression has also been investigated. The abstract of the presentation by Treloar and Howell [53] suggests the ability of some yeast strains to reduce IBMP concentrations in red wine. However, limited data were provided, including no information on the analytical method(s) employed, making it difficult to evaluate these claims. Sorptive processes involving yeast cell walls are known to occur with other classes of volatile compounds (e.g., [54]), so this possibility cannot be discounted for MPs. However, the findings of Pickering et al. [55] provide a cautionary note; they used Cabernet Sauvignon juice spiked with 30 ng/L IPMP, and fermented it with various commercial *Saccharomyces* yeast strains (Lalvin BM45, Lalvin EC1118, Lalvin ICV-D21, and Lalvin ICV-D80). The effects were mixed, with some yeasts unexpectedly increasing IPMP levels (BM45, 29% increase), while others had no effects on IPMP concentration. From a sensorial perspective, the D80 strain produced wines with the highest intensity ratings for LBT-related attributes, while D21 had the lowest ratings. Since MP levels were not affected by these two strains, the effects are most likely attributable to matrix composition and the masking effects of other aroma compounds. The authors recommend D21 as a suitable choice for wines with high IPMP levels, regardless of their source [55]. Sala et al. [50] examined the effects of malolactic fermentation after alcoholic fermentation on MPs, and concluded that malolactic fermentation had no effect on MP levels in the wine.

Fining is a common way of dealing with various wine faults and instabilities, so several fining agents were examined by Pickering et al. [56] as possible LBT remediation options. The authors assessed the application of activated charcoal, bentonite, oak chips, deodorized oak chips, white light, and ultraviolet light in both red and white wines produced with the addition of MALB beetles. The relative efficacy of the treatments varied between red and white wines, with red wines generally more responsive, particularly in terms of flavor improvement. In reds, oak chips did not affect IPMP concentrations, but were the most effective addition in reducing the intensity of LBT-related attributes. Additionally, asparagus/bell pepper flavor was significantly lower in red wines treated with bentonite, charcoal, and deodorized oak. In white wine, activated charcoal reduced IPMP concentrations by 34%; however, LBT-related characteristics were unchanged. Oak chips also led to a significant decrease in asparagus flavor in whites, as well as a trend for lower intensities for all LBT attributes. The positive impact of oak chips on both white and red wines was attributed by the authors to perceptual masking of LBT by other aromatic constituents in the wines, since the addition of deodorized chips did not reduce the intensity of the taint-related attributes. Light had no effect on IPMP concentrations nor sensory profiles in either type of wine.

The sorptive capacity of packaging leading to direct removal of volatile compounds (termed flavor scalping) has been previously established in the food literature, as well as exploited commercially—particularly with polymer packaging and nonpolar flavor compounds [57]. Thus, the possibility that the choice of wine packaging and closure type might reduce MP levels in finished wine has also been investigated as a potentially "noninvasive" approach to remediating LBT.

The effects of both closure and packaging on MPs and other odorants in Riesling and Cabernet Franc wines were investigated by Blake et al. [58]. They spiked each wine with 30 ng/L of IBMP, SBMP, and IPMP, and then bottled and stoppered the wines using five cork-type closures—a natural cork, an agglomerate cork, a roll-on tamper-evident (ROTE) screw cap, an extruded synthetic cork and a molded synthetic closure. A portion of the MP-enriched wines was alternatively stored in aseptic cartons (Tetra Pak). All three MPs were affected by the closure or packaging type to some extent. Wines stored in Tetra Pak cartons had the lowest IPMP concentrations, with a 23% and 41% reduction for Riesling and Cabernet Franc, respectively, possibly due to the migration of IPMP to the aluminum surface layer of the carton, with subsequent adsorption on the oxide layer [58]. Concentrations of SBMP were similarly lower than initial levels in Tetra Pak-stored wines (average decrease of 27%). The authors also reported a 10–21% decrease in IPMP concentration after 18 months of bottle-aging in wines closed with synthetic molded closures. IBMP decreased significantly with 18-month aging in all conditions, with the greatest decrease reported with Tetra Pak and synthetic molded closures, and the smallest change with natural cork [58].

Subsequently, the adsorptive capacity of synthetic corks for MPs has been further demonstrated by Pickering et al. [59]. A Chardonnay wine spiked with 40 ng/L each of

IBMP, SBMP, and IPMP was soaked for 140 h with either natural corks, agglomerate corks, molded synthetic closures, or extruded synthetic closures at two different closure addition levels (5 and 10 units/L). All closures significantly reduced MP concentrations in the wine, with the greatest efficacy observed with synthetic closures (70–89% reduction). SBMP was most affected by closure treatments. Unfortunately, sensory analysis of the wines was not performed due to volume restrictions; thus, the sensorial impact of these "treatments" on LBT and overall quality remains to be determined. Finally, these trials show that some MPs do decrease simply as a function of aging. However, as other wine odorants also change and new ones are formed during aging/maturation [60], a strategy of just letting the wines age to remove/reduce LBT before releasing them may not be advisable—indeed, the sensory results (after 10 months aging) of Pickering et al. [9] suggest that it is not, although careful barrel aging in some styles should reduce LBT, as supported by some anecdotal reports to author G.J.P.

3.3. Novel Interventions

In addition to the traditional methods for processing wine or treating tainted wine reviewed above, several novel approaches or novel applications of existing technologies, have been evaluated with regard to their capacity to remediate LBT or reduce MP concentrations in juice and wine.

3.3.1. Heat

Thermovinification involves the heating of red musts for a short time to 60–80 °C, which can reduce IBMP content in red wine by 29–67% [61]. More recently, Kögel et al. [16] reported a moderate decrease in all MPs examined (i.e., IPMP, IBMP, SBMP, and DMMP) subsequent to thermovinification of Pinot Noir. However, care should be taken, as this technique may lead to undesirable sensory modifications in wines [62].

Thermoflash, known widely in Europe as flash détente or flash release, is a version of thermovinification used to reduce the fermentation time of red wine and improve quality through increased extraction of tannins, color, and aroma compounds [63]. It works by heating the must of crushed grapes to a set temperature (usually around 185 °F/85 °C), then sending the heated grapes to a high-vacuum chamber, where the temperature decreases sharply and causes the water in the skins' cells to evaporate rapidly, and the water in the berries to turn to steam. The fast expansion of the steam causes the cell walls of the vacuoles in the skins to explode, leading to immediate color and tannin extraction while also releasing aromatic compounds. The evaporated water (6–10% of initial must volume) is then put through a condenser, and remains as a separate byproduct, or is added back to the must later for fermentation [63]. Non-peer-reviewed data indicate that the IBMP content of Cabernet Sauvignon processed with this technology dropped from 19.2 ng/L IBMP pre-treatment to < 1 ng/L after, while the flash water contained 112.4 ng/L IBMP [64]. Reduction of vegetal notes was also reported in Cabernet Sauvignon wine processed through flash détente in France [65], while winemakers in California reported the presence of easily identifiable MP-related aromas in the water condensate [64]. Similar results from Australia showed that pre-flashed Zinfandel berries had an IBMP content of 8.4 ng/kg, the juice contained less than 3.0 ng/L, and the flash water had 27.8 ng/L. Unfortunately, without peer-reviewed data, the validity of these results remains to be determined, although the purported reproducibility of the findings in multiple countries and varieties suggests that it is an approach with considerable potential for reducing MP loads. Sensory impacts on wines thus treated also need to be further elucidated.

3.3.2. Oxygen

Micro-oxygenation (MOX) is the process of intentionally introducing very small, measured doses of oxygen into wines, in order to bring about desirable changes in color, aroma, and mouthfeel. It requires the use of specialized equipment to regulate the amount of oxygen that is administered [66].

Anecdotal reports from wineries suggest that MOX of wines may be effective at reducing MP levels and/or perceptually masking the effects of LBT in wines. Limited peer-reviewed studies add to these observations: Saenz-Navajas et al. [67] investigated the impact of micro-oxygenation on the sensory properties of Merlot wines, while monitoring viable yeasts and SO_2 levels. While they did not measure MP levels in the wine, they found that MOX led to a reduction in the green vegetable aromas usually associated with them. Similar results were demonstrated with Cencibel wines by Cejudo-Basante et al. [68] and Cejudo-Basante et al. [69]. Contrary to these findings, Oberhostler et al. [70] reported an increase in some vegetative aromas following MOX in a red wine blend.

3.3.3. The Panacea of Plastics?

Ryona et al. [71] reported silicone to be effective at reducing MP levels in grape juice. Their hypothesis was first confirmed on a model juice, and then demonstrated on four grape juices/musts (a Chardonnay white, a Riesling white, a Cabernet Franc rosé, and a Cabernet Franc red). Reductions in MPs ranged from 53% to 93%, with significant decreases observed immediately after the addition of silicone to wine produced without skin contact. As a caveat, decreases in other volatile aroma compounds were also noted, indicating a need for the sensory evaluation of wines treated with silicone.

Botezatu and Pickering [72] further investigated the capacity of plastic polymers to reduce MP levels in wine. In an initial phase, they treated wines with high levels of IPMP, SBMP, and IBMP (20 ng/L each) with 13 different plastic polymers, allowing for the identification of three potential candidates (silicone, ethylene and vinyl acetate, and a polylactic-acid-based biodegradable polymer) based both on their capacity to reduce MP concentrations and their overall influence on the wines' sensory profiles. The three polymers reduced overall MP concentrations by 18% (ethylene and vinyl acetate), 28% (polylactic acid), and 30% (silicone). The capacity of these selected polymers to reduce MP levels in red wine was then tested as a function of contact time. Significant decreases in all target MPs were observed after 24 h treatment: polylactic acid reduced IPMP and IBMP concentrations by 52% and 36%, respectively, while silicone reduced IPMP and IBMP by 96% and 100%, respectively. Ethylene and vinyl acetate was less effective in lowering MP levels (7% IPMP and 23% IBMP after 24 h).

Botezatu et al. [73] subsequently investigated the capacity of two plastic polymers—one silicone-based, the other polylactic-acid-based—applied with varying surface areas to reduce concentrations of IPMP, IBMP, and SBMP in a Merlot wine. All surface areas tested (50 cm^2/L, 200 cm^2/L, and 600 cm^2/L) showed reductions in all three MPs, and the polylactic acid polymer was the more effective of the two, with reductions in IPMP concentration of up to 75% at a surface area of 600 cm^2/L [73]. Analysis of non-target volatile aroma compounds indicated minimal changes; however, results from the sensory evaluation of the treated wines were less clear, often not showing the expected reductions in the intensity of LBT-related descriptors. At the same time, the polymers did not contribute negatively to the aroma or flavor profiles of the treated wines.

3.3.4. Just Nuke It

A more unusual approach to remediating LBT has been reported on by Wilson [74] and Wilson et al. [75], with the application of irradiation to tainted wine. In their pilot study, a dose of 100 Gy was reported to decrease the perceived intensity of LBT; however, the authors did not report on how other sensory attributes of the wines were affected by the treatment, especially with respect to the potential oxidative damage to flavor compounds from such a treatment. We are not aware of any further studies that have investigated this approach.

3.4. The Challenge of Specificity

A fundamental limitation of most of the interventions to date aimed at remediating LBT is that they lack specificity for the causal compounds—alkyl-methoxypyrazines. That is, while they may lead to a decrease in MP concentrations through both direct and indirect mechanisms, other—typically desirable—aroma and flavor compounds are also removed from the juice/wine or adversely affected. Two approaches that seek to address this challenge are the use of an odorant-binding protein, and imprinted polymers.

Odorant-binding proteins and major urinary proteins (MUPs) are small extracellular proteins belonging to the lipocalin family, and can bind MPs with micromolar affinities [76,77]. Inglis et al. [78] describe the overexpression of piglet odorant-binding protein (pIOBP) and murine major urinary protein (mMUP2) in the yeast *Pichia pastoris*, their subsequent purification and concentration, and the very high affinity of these proteins for MPs in a grape juice matrix. Subsequent reports have focused on mMUP2, and this protein, when applied to Chardonnay juice, binds IBMP and IPMP, and the resulting complexes can be successfully fined out with bentonite [79]. The effectiveness of mMUP2 is excellent: when applied to juice that is subsequently fined with bentonite and filtered with a 10 kDa polyethersulfone membrane, the system removed > 99% of IPMP and IBMP [78]. However, the performance of mMUP2 within the more challenging matrix of wine, where ethanol and phenolic compounds may reduce its efficacy, remains to be fully elucidated. Additionally, further work is required to scale up the juice-based odorant-binding protein system for use on a commercial scale.

A more recent approach involves the use of magnetic molecularly imprinted polymers to remediate wine with elevated IBMP levels. Molecular imprinting involves the creation of a polymer with selective pockets based on a molecular or biomolecular template. Literature shows that the polymerization of monomers in the presence of a molecular target leads to the formation of corresponding binding sites in the resulting polymer [80]. Liang et al. [81] prepared molecularly imprinted polymers using 2-methoxypyrazine as a template, and incorporated iron oxide nanoparticles as magnetic substrates in order to be able to remove the polymers after treatment with magnets. They also used non-imprinted magnetic polymers for comparison, and assessed both polymers in two different Sauvignon Blanc wines, each spiked with 30 ng/L IBMP. After 30 min of contact time they removed the polymers using a permanent magnet. The authors reported that both the non-imprinted and imprinted polymers adsorbed IBMP, with the imprinted polymers having a higher rate of adsorption (up to 45%, compared with up to 38% for the non-imprinted polymer). Furthermore, they showed that the polymers were washable and reusable for up to five cycles [81].

Further research from Liang et al. [82] investigated the addition of a putative imprinted magnetic polymer both pre- and post-fermentation as a corrective treatment for high concentrations of IBMP in Cabernet Sauvignon grape must, and compared its efficacy to both a non-imprinted magnetic polymer and a polylactic-acid-based film added post-fermentation. Sensory analyses of wines showed that treatment with the putative imprinted magnetic polymer was more effective than the polylactic acid film at decreasing "fresh green" aroma nuances, with no negative impact on the overall aroma profiles reported. Post-fermentation addition of a magnetic polymer removed up to 74% of the initial IBMP concentration, compared with 18% for polylactic acid. Adding the magnetic polymers pre-fermentation removed 20–30% less IBMP compared with post-fermentation addition, and also had less effect on other wine volatiles and color parameters [82]. Theoretically, the same general approach could be used for molecularly imprinting polymers that target IPMP and SBMP—the two main MPs involved in LBT—but the literature is currently lacking on that front.

4. Conclusions

Ladybug taint can appear in wine when Coccinellidae species—and especially *H. axyridis*—are incorporated into the grape processing stream after harvest. It is primarily caused by 2-*iso*propyl-3-methoxypyrazine, which originates in the hemolymph of the beetles, and for which the human detection threshold in wine is very low (0.3–2 ng/L). Ladybug taint is characterized by "asparagus", "bell pepper", "green beans", "potatoes", "herbaceous", "musty", and "earthy" descriptors, and varietal attributes are diminished. Several options are available in the vineyard for reducing the density of beetles to acceptable limits—approximately 1200–2000 beetles/ton—depending on beetle species and grape variety. Of these options, several insecticidal sprays and potassium metabisulfite are effective, although some sprays are limited by their pre-harvest intervals.

Several methods for the remediation of tainted grape juice and wine have been investigated, including the use of traditional fining agents, with mixed results. Juice clarification prior to fermentation and the use of oak chips in the wine reduce methoxypyrazine loads and mask ladybug taint, respectively, while thermovinification of juice is also effective. Silicone and polylactic acid also show significant potential for reducing methoxypyrazine levels. A challenge with all of these approaches is the relative lack of specificity for methoxypyrazines, meaning that non-target components of the juice/wine—including desirable aroma compounds—can be adversely affected. Two promising treatments with good specificity for methoxypyrazines are at varying stages of development/commercialization: an odorant-binding protein for use in juice, and molecularly imprinted magnetic polymers for use in either juice or wine. Ladybug taint affects wines from many international winemaking regions, and further research on both prevention and treatment is encouraged, given the invasive nature of *H. axyridis* and its widening global distribution.

Author Contributions: Conceptualization, G.J.P. and A.B.; methodology, G.J.P. and A.B.; resources, G.J.P. and A.B.; writing—original draft preparation, A.B.; writing—review and editing, G.J.P. and A.B.; project administration, G.J.P.; funding acquisition, G.J.P. All authors have read and agreed to the published version of the manuscript.

Funding: This research received no external funding.

Institutional Review Board Statement: Ethical review and approval were waived for this work, as it is a review paper, and did not directly require the use of human or animal research subjects.

Informed Consent Statement: Not applicable. See Institutional Review Board Statement.

Data Availability Statement: There are no data directly connected to this review paper.

Acknowledgments: We give thanks to Margaret Thibodeau, Brock University, for editorial assistance; and Wendy McFadden-Smith, Ontario Ministry of Agriculture, Food, and Rural Affairs, and Ian Brindle, Brock University, for advice.

Conflicts of Interest: The authors declare no conflict of interest. The funders had no role in the design of the study, in the collection, analyses, or interpretation of data, in the writing of the manuscript, or in the decision to publish the results.

References

1. Gordon, R.D. The Coccinellidae (Coleoptera) of America north of Mexico. *J. N. Y. Entomol. Soc.* **1985**, *93*, 33–35.
2. Koch, R.L.; Burkness, E.C.; Burkness, S.J.W.; Hutchison, W.D. Phytophagous preferences of the Multicolored Asian Lady Beetle (Coleoptera: Coccinellidae) for autumn-ripening fruit. *J. Econ. Entomol.* **2004**, *97*, 539–544. [CrossRef] [PubMed]
3. Vincent, C.; Pickering, G. Multicolored Asian ladybeetle, Harmonia axyridis (Coleoptera: Coccinellidae). In *Biological Control Programmes in Canada 2001–2012*; Mason, P.G., Gillespie, D.R., Eds.; CABI: Wallingford, UK, 2013; pp. 192–198. ISBN 9781780642574.
4. Coderre, D.; Lucas, É.; Gagné, I. The occurrence of Harmonia axyridis (Pallas)(Coleoptera: Coccinellidae) in Canada. *Can. Entomol.* **1995**, *127*, 609–611. [CrossRef]
5. Galvan, T.L.; Burkness, E.C.; Hutchison, W.D. Influence of berry Injury on Infestations of the Multicolored Asian Lady Beetle in wine grapes. *Plant Health Prog.* **2006**, *7*. [CrossRef]
6. Pickering, G.; Lin, J.; Riesen, R.; Reynolds, A.; Brindle, I.; Soleas, G. Influence of Harmonia axyridis on the sensory properties of white and red wine. *Am. J. Enol. Vitic.* **2004**, *55*, 153–159.
7. Pickering, G.J.; Spink, M.; Kotseridis, Y.; Brindle, I.D.; Sears, M.; Inglis, D. Morbidity of Harmonia axyridis mediates ladybug taint in red wine. *J. Food Agric. Environ.* **2008**, *6*, 133–138. [CrossRef]
8. Pickering, G.J.; Ker, K.; Soleas, G. Determination of the critical stages of processing and tolerance limits for Harmonia axyridis for "ladybug taint" in wine. *Vitis* **2007**, *46*, 85–90.
9. Pickering, G.J.; Lin, Y.; Reynolds, A.; Soleas, G.; Riesen, R.; Brindle, I. The influence of Harmonia axyridis on wine composition and aging. *J. Food Sci.* **2005**, *70*, S128–S135. [CrossRef]

10. Ross, C.F.; Weller, K. Sensory evaluation of suspected Harmonia axyridis -tainted red wine using untrained panelists. *J. Wine Res.* **2007**, *18*, 187–193. [CrossRef]
11. Cai, L.; Koziel, J.A.; O'Neal, M.E. Determination of characteristic odorants from Harmonia axyridis beetles using in vivo solid-phase microextraction and multidimensional gas chromatography–mass spectrometry–olfactometry. *J. Chromatogr. A* **2007**, *1147*, 66–78. [CrossRef]
12. Slabizki, P.; Legrum, C.; Meusinger, R.; Schmarr, H.-G. Characterization and analysis of structural isomers of dimethyl methoxypyrazines in cork stoppers and ladybugs (Harmonia axyridis and Coccinella septempunctata). *Anal. Bioanal. Chem.* **2014**, *406*, 6429–6439. [CrossRef] [PubMed]
13. Cudjoe, E.; Wiederkehr, T.B.; Brindle, I.D. Headspace gas chromatography-mass spectrometry: A fast approach to the identification and determination of 2-alkyl-3- methoxypyrazine pheromones in ladybugs. *Analyst* **2005**, *130*, 152. [CrossRef] [PubMed]
14. Kögel, S.; Gross, J.; Hoffmann, C.; Ulrich, D. Diversity and frequencies of methoxypyrazines in hemolymph of Harmonia axyridis and Coccinella septempunctata and their influence on the taste of wine. *Eur. Food Res. Technol.* **2012**, *234*, 399–404. [CrossRef]
15. Pickering, G.J.; Spink, M.; Kotseridis, Y.; Brindle, I.D.; Sears, M.; Inglis, D. The influence of Harmonia axyridis morbidity on 2-Isopropyl-3-methoxypyrazine in "Cabernet Sauvignon" wine. *Vitis* **2008**, *47*, 227–230. [CrossRef]
16. Kögel, S.; Botezatu, A.; Hoffmann, C.; Pickering, G. Methoxypyrazine composition of Coccinellidae-tainted Riesling and Pinot noir wine from Germany. *J. Sci. Food Agric.* **2015**, *95*, 509–514. [CrossRef]
17. Pickering, G.J.; Lin, Y.; Ker, K. Origin and remediation of Asian Lady Beetle (Harmonia axyridis) taint in wine. In *Crops: Growth, Quality and Biotechnology. III. Quality Management of Food Crops for Processing Technology*; Dris, R., Ed.; WFL Publisher: Helsinki, Findland, 2006; pp. 785–794. ISBN 952-91-8601-0.
18. Galvan, T.L.; Kells, S.; Hutchison, W.D. Determination of 3-Alkyl-2-methoxypyrazines in Lady Beetle-infested wine by solid-phase microextraction headspace sampling. *J. Agric. Food Chem.* **2008**, *56*, 1065–1071. [CrossRef]
19. Ross, C.F.; Rosales, M.U.; Fernandez-Plotka, V.C. Aroma profile of Niagara grape juice contaminated with multicoloured Asian lady beetle taint using gas chromatography/mass spectrometry/olfactometry. *Int. J. Food Sci. Technol.* **2010**, *45*, 789–793. [CrossRef]
20. Botezatu, A.I.; Kotseridis, Y.; Inglis, D.; Pickering, G.J. Occurrence and contribution of alkyl methoxypyrazines in wine tainted by Harmonia axyridis and Coccinella septempunctata. *J. Sci. Food Agric.* **2013**, *93*, 803–810. [CrossRef]
21. Allen, M.S.; Lacey, M.J.; Harris, R.L.N.; Brown, W.V. Contribution of methoxypyrazines to Sauvignon blanc wine aroma. *Am. J. Enol. Vitic.* **1991**, *42*, 109–112.
22. Lacey, M.J.; Allen, M.S.; Harris, R.L.N.; Brown, W.V. Methoxypyrazines in Sauvignon blanc grapes and wines. *Am. J. Enol. Vitic.* **1991**, *42*, 103–108.
23. Roujou de Boubée, D.; Van Leeuwen, C.; Dubourdieu, D. Organoleptic impact of 2-Methoxy-3-isobutylpyrazine on red Bordeaux and Loire wines. Effect of environmental conditions on concentrations in grapes during ripening. *J. Agric. Food Chem.* **2000**, *48*, 4830–4834. [CrossRef] [PubMed]
24. Roujou de Boubée, D.; Cumsille, A.M.; Pons, M.; Dubourdieu, D. Location of 2-methoxy-3-isobutylpyrazine in Cabernet Sauvignon grape bunches and its extractability during vinifica-tion. *Am. J. Enol. Vitic.* **2002**, *53*, 1–5.
25. Sala, C.; Busto, O.; Guasch, J.; Zamora, F. Contents of 3-alkyl-2-methoxypyrazines in musts and wines fromVitis vinifera variety Cabernet Sauvignon: Influence of irrigation and plantation density. *J. Sci. Food Agric.* **2005**, *85*, 1131–1136. [CrossRef]
26. Belancic, A.; Agosin, E. Methoxypyrazines in grapes and wines of Vitis vinifera cv. Carmenere. *Am. J. Enol. Vitic.* **2007**, *58*, 462–469.
27. Godelmann, R.; Limmert, S.; Kuballa, T. Implementation of headspace solid-phase-microextraction–GC–MS/MS methodology for determination of 3-alkyl-2-methoxypyrazines in wine. *Eur. Food Res. Technol.* **2007**, *227*, 449. [CrossRef]
28. Botezatu, A.; Kotseridis, Y.; Inglis, D.; Pickering, G.J. A survey of methoxypyrazines in wine. *J. Food Agric. Environ.* **2016**, *14*, 24–29. [CrossRef]
29. Ortega, C.; López, R.; Cacho, J.; Ferreira, V. Fast analysis of important wine volatile compounds: Development and validation of a new method based on gas chromatographic–flame ionisation detection analysis of dichloromethane microextracts. *J. Chromatogr. A* **2001**, *923*, 205–214. [CrossRef]
30. Ferreira, V.; López, R.; Cacho, J.F. Quantitative determination of the odorants of young red wines from different grape varieties. *J. Sci. Food Agric.* **2000**, *80*, 1659–1667. [CrossRef]
31. Romero, R.; Chacón, J.L.; García, E.; Martínez, J. Pyrazine contents in four red grape varietes cultivated in warm climate. *J. Int. Sci. Vigne Vin* **2006**, *40*, 203–207. [CrossRef]
32. Maga, J. Sensory and stability properties of added methoxypyrazines to model and authentic wines. In *Flavors and Off-Flavors '89, Proceedings of the 6th International Flavor Conference, Rethymnon, Crete, Greece, 5–7 July 1989*; Charalambous, G., Ed.; Elsevier Science Limited: Amsterdam, The Netherlands, 1990.
33. Kotseridis, Y.; Beloqui, A.A.; Bertrand, A.; Doazan, J.P. An analytical method for studying the volatile compounds of Merlot noir clone wines. *Am. J. Enol. Vitic.* **1998**, *49*, 44–48.
34. Pickering, G.J.; Karthik, A.; Inglis, D.; Sears, M.; Ker, K. Determination of ortho- and retronasal detection thresholds for 2-Isopropyl-3-Methoxypyrazine in wine. *J. Food Sci.* **2007**, *72*, S468–S472. [CrossRef] [PubMed]
35. Pickering, G.J.; Karthik, A.; Inglis, D.; Sears, M.; Ker, K. Detection thresholds for 2-Isopropyl-3-methoxypyrazine in Concord and Niagara grape juice. *J. Food Sci.* **2008**, *73*, S262–S266. [CrossRef]

36. Galvan, T.L.; Burkness, E.C.; Vickers, Z.; Stenberg, P.; Mansfield, A.K.; Hutchison, W.D. Sensory-based action threshold for Multicolored Asian Lady Beetle-related taint in winegrapes. *Am. J. Enol. Vitic.* **2007**, *58*, 518–522.
37. Ross, C.; Ferguson, H.; Keller, M.; Walsh, D.; Weller, K.; Spayd, S. Determination of ortho-nasal aroma threshold for Multicolored Asian Lady Beetle in Concord grape juice. *J. Food Qual.* **2007**, *30*, 855–863. [CrossRef]
38. Glemser, E.; McFadden-Smith, W.; Parent, J.-P. Evaluation of compounds for repellency of the multicoloured Asian lady beetle (Coleoptera: Coccinellidae) in vineyards. *Can. Entomol.* **2021**, 1–12. [CrossRef]
39. Koch, R.L. The multicolored Asian lady beetle, Harmonia axyridis: A review of its biology, uses in biological control, and non-target impacts. *J. Insect Sci.* **2003**, *3*, 32. [CrossRef] [PubMed]
40. Glemser, E.J.; Dowling, L.; Inglis, D.; Pickering, G.J.; Mcfadden-Smith, W.; Sears, M.K.; Hallett, R.H. A novel method for controlling Multicolored Asian Lady Beetle (Coleoptera: Coccinellidae) in vineyards. *Environ. Entomol.* **2012**, *41*, 1169–1176. [CrossRef] [PubMed]
41. Seko, T.; Yamashita, K.; Miura, K. Residence period of a flightless strain of the ladybird beetle Harmonia axyridis Pallas (Coleoptera: Coccinellidae) in open fields. *Biol. Control* **2008**, *47*, 194–198. [CrossRef]
42. Pickering, G.J.; Glemser, E.J.; Hallett, R.; Inglis, D.; McFadden-Smith, W.; Ker, K. Good bugs gone bad: Coccinellidae, sustainability and wine. *WIT Trans. Ecol. Environ.* **2011**, *167*, 239–251. [CrossRef]
43. Leroy, P.D.; Schillings, T.; Farmakidis, J.; Heuskin, S.; Lognay, G.; Verheggen, F.J.; Brostaux, Y.; Haubruge, E.; Francis, F. Testing semiochemicals from aphid, plant and conspecific: Attraction of Harmonia axyridis. *Insect Sci.* **2012**, *19*, 372–382. [CrossRef]
44. Pettersson, J.; Birkett, M.A.; Pickett, A. Pyrazines as Attractants for Insects of Order Coleoptera. WO1999037152 A1, 29 July 1999. Available online: https://www.google.com/patents/WO1999037152A1?cl=pt (accessed on 5 May 2021).
45. Riddick, E.W.; Brown, A.E.; Chauhan, K.R. Harmonia axyridis adults avoid catnip and grapefruit-derived terpenoids in laboratory bioassays. *Bull. Insectol.* **2008**, *61*, 81–90.
46. Riddick, E.W.; Aldrich, J.R.; De Milo, A.; Davis, J.C. Potential for modifying the behavior of the multicolored Asian lady beetle (Coleoptera: Coccinellidae) with plant-derived natural products. *Ann. Entomol. Soc. Am.* **2000**, *93*, 1314–1321. [CrossRef]
47. Riddick, E.W.; Aldrich, J.R.; Davis, J.C. DEET repels Harmonia axyridis (Pallas) (Coleoptera: Coccinellidae) adults in laboratory bioassays. *J. Entomol. Sci.* **2004**, *39*, 373–386. [CrossRef]
48. Botezatu, A.; Pickering, G. Ladybug (Coccinellidae) taint in wine. In *Managing Wine Quality. Vol. 2, Oenology and Wine Quality*; Reynolds, A.G., Ed.; Woodhead Publishing Limited: Cambridge, UK, 2010; pp. 418–429. ISBN 9781845694845.
49. Hashizume, K.; Umeda, N. Methoxypyrazine content of japanese red wines. *Biosci. Biotechnol. Biochem.* **1996**, *60*, 802–805. [CrossRef]
50. Sala, C.; Busto, O.; Guasch, J.; Zamora, F. Influence of vine training and sunlight exposure on the 3-Alkyl-2-methoxypyrazines content in musts and wines from the Vitis vinifera variety Cabernet Sauvignon. *J. Agric. Food Chem.* **2004**, *52*, 3492–3497. [CrossRef]
51. Sidhu, D.; Lund, J.; Kotseridis, Y.; Saucier, C. Methoxypyrazine analysis and influence of viticultural and enological procedures on their levels in grapes, musts, and wines. *Crit. Rev. Food Sci. Nutr.* **2015**, *55*, 485–502. [CrossRef]
52. Kotseridis, Y.S.; Spink, M.; Brindle, I.D.; Blake, A.J.; Sears, M.; Chen, X.; Soleas, G.; Inglis, D.; Pickering, G.J. Quantitative analysis of 3-alkyl-2-methoxypyrazines in juice and wine using stable isotope labelled internal standard assay. *J. Chromatogr. A* **2008**, *1190*, 294–301. [CrossRef]
53. Treloar, J.; Howell, C.S. Influence of yeast and malolactic bacteria strain choice on 3-isobutyl-2-methoxypyrazine concentration in Cabernet Sauvignon and Franc wines. In *Wine Making Workshop, Vegetable & Farm Market EXPO, Great Lakes Fruit, Grand Rapids, MI, USA*; 2006; Available online: www.glexpo.com/summaries/2006summaries/WineMaking2006.pdf (accessed on 5 May 2021).
54. Jiménez-Moreno, N.; Ancín-Azpilicueta, C. Sorption of volatile phenols by yeast cell walls. *Int. J. Wine Res.* **2009**, *1*, 11–18. [CrossRef]
55. Pickering, G.J.; Spink, M.; Kotseridis, Y.; Inglis, D.; Brindle, I.D.; Sears, M.; Beh, A.-L. Yeast strain affects 3-isopropyl-2-methoxypyrazine concentration and sensory profile in Cabernet Sauvignon wine. *Aust. J. Grape Wine Res.* **2008**, *14*, 230–237. [CrossRef]
56. Pickering, G.; Lin, J.; Reynolds, A.; Soleas, G.; Riesen, R. The evaluation of remedial treatments for wine affected by Harmonia axyridis. *Int. J. Food Sci. Technol.* **2006**, *41*, 77–86. [CrossRef]
57. Sajilata, M.G.; Savitha, K.; Singhal, R.S.; Kanetkar, V.R. Scalping of flavors in packaged foods. *Comp. Rev. Food Sci. Food Saf.* **2007**, *6*, 17–35. [CrossRef]
58. Blake, A.; Kotseridis, Y.; Brindle, I.D.; Inglis, D.; Sears, M.; Pickering, G.J. Effect of closure and packaging type on 3-Alkyl-2-methoxypyrazines and other impact odorants of Riesling and Cabernet Franc wines. *J. Agric. Food Chem.* **2009**, *57*, 4680–4690. [CrossRef]
59. Pickering, G.J.; Blake, A.J.; Soleas, G.J.; Inglis, D.L. Remediation of wine with elevated concentrations of 3-alkyl-2-methoxypyrazines using cork and synthetic closures. *J. Food Agric. Environ.* **2010**, *8*, 97–101.
60. Bakker, J.; Clarke, R.J. *Wine Flavour Chemistry*, 2nd ed.; Blackwell Publishing Ltd.: Chichester, UK, 2012; ISBN 9781444330427.
61. Roujou de Boubée, D. Research on the vegetal green pepper character in grapes and wines. *Rev. Oenol.* **2004**, *31*, 6–10.
62. Jackson, R.S. *Wine Science: Principles and Applications*, 3rd ed.; Academic Press: Burlington, MA, USA, 2008; ISBN 9780123736468.
63. Maza, M.; Álvarez, I.; Raso, J. Thermal and non-thermal physical methods for improving polyphenol extraction in red winemaking. *Beverages* **2019**, *5*, 47. [CrossRef]

64. Baggio, P. Flash extraction—what can it do for you? In Proceedings of the ASVO Proceedings: Managing the Best out of Difficult Vintages; 2017; pp. 29–31. Available online: https://www.dtpacific.com/dev/wp-content/uploads/2017/03/Art-ASVO-Flash-Bio-Thermo-Extraction-What-can-it-do-for-you.pdf (accessed on 1 April 2021).
65. Vinsonneau, E.; Escaffre, P.; Crachereau, J.; Praud, S. *Evaluation du Procédé de Vinification par "Flash Détente" Dans le Bordelais*; ITV France Bordeaux: Blanquefort, France, 2006.
66. Gómez-Plaza, E.; Cano-López, M. A review on micro-oxygenation of red wines: Claims, benefits and the underlying chemistry. *Food Chem.* **2011**, *125*, 1131–1140. [CrossRef]
67. Sáenz-Navajas, M.-P.; Henschen, C.; Cantu, A.; Watrelot, A.A.; Waterhouse, A.L. Understanding microoxygenation: Effect of viable yeasts and sulfur dioxide levels on the sensory properties of a Merlot red wine. *Food Res. Int.* **2018**, *108*, 505–515. [CrossRef]
68. Cejudo-Bastante, M.J.; Pérez-Coello, M.S.; Hermosín-Gutiérrez, I. Effect of wine micro-oxygenation treatment and storage period on colour-related phenolics, volatile composition and sensory characteristics. *LWT* **2011**, *44*, 866–874. [CrossRef]
69. Cejudo-Bastante, M.J.; Hermosín-Gutiérrez, I.; Pérez-Coello, M.S. Improvement of Cencibel red wines by oxygen addition after malolactic fermentation: Study on color-related phenolics, volatile composition, and sensory characteristics. *J. Agric. Food Chem.* **2012**, *60*, 5962–5973. [CrossRef]
70. Oberholster, A.; Elmendorf, B.L.; Lerno, L.A.; King, E.S.; Heymann, H.; Brenneman, C.E.; Boulton, R.B. Barrel maturation, oak alternatives and micro-oxygenation: Influence on red wine aging and quality. *Food Chem.* **2015**, *173*, 1250–1258. [CrossRef]
71. Ryona, I.; Reinhardt, J.; Sacks, G.L. Treatment of grape juice or must with silicone reduces 3-alkyl-2-methoxypyrazine concentrations in resulting wines without altering fermentation volatiles. *Food Res. Int.* **2012**, *47*, 70–79. [CrossRef]
72. Botezatu, A.; Pickering, G.J. Application of plastic polymers in remediating wine with elevated alkyl-methoxypyrazine levels. *Food Addit. Contam. A* **2015**, *32*, 1199–1206. [CrossRef] [PubMed]
73. Botezatu, A.; Kemp, B.; Pickering, G. Chemical and sensory evaluation of silicone and polylactic acid-based remedial treatments for elevated methoxypyrazine levels in wine. *Molecules* **2016**, *21*, 1238. [CrossRef] [PubMed]
74. Wilson, K. Applications of Radiation within the Wine Industry. Can. Bachelor's Thesis, McMaster University, Hamilton, ON, Canada, 2003.
75. Wilson, K.J.; Moran, G.; Boreham, D. Application of radiation within the wine industry. In Proceedings of the 12th Quadrennial Congress of the International Association for Radiation Research Incorporating the 50th Annual Meeting of Radiation Research Society, RANZCR Radiation Oncology Annual Scientific Meeting and AINSE Radiation Science Conference, Brisbane, Australia, 17–22 August 2003.
76. Cavaggioni, A.; Findlay, J.B.C.; Tirindelli, R. Ligand binding characteristics of homologous rat and mouse urinary proteins and pyrazine-binding protein of calf. *Comp. Biochem. Physiol. B* **1990**, *96*, 513–520. [CrossRef]
77. Pevsner, J.; Hou, V.; Snowman, A.M.; Snyder, S.H. Odorant-binding protein. Characterization of ligand binding. *J. Biol. Chem.* **1990**, *265*, 6118–6125. [CrossRef]
78. Inglis, D.; Beh, A.L.; Brindle, I.D.; Pickering, G.; Humes, E.F. Method for Reducing Methoxypyrazines in Grapes and Grape Products. U.S. Patent No. 8859026B2, 14 October 2014.
79. Pickering, G.; Inglis, D.; Botezatu, A.; Beh, A.; Humes, E.; Brindle, I. New approaches to removing alkyl-methoxypyrazines from grape juice and wine. In *Scientific Bulletin. Series F. Biotechnologies, Vol. XVIII*; The University of Agronomic Sciences and Veterinary Medicine: Bucharest, Romania, 2014; pp. 130–134.
80. Graham, S.P.; El-Sharif, H.F.; Hussain, S.; Fruengel, R.; McLean, R.K.; Hawes, P.C.; Sullivan, M.V.; Reddy, S.M. Evaluation of molecularly imprinted polymers as synthetic virus neutralizing antibody mimics. *Front. Bioeng. Biotechnol.* **2019**, *7*, 115. [CrossRef]
81. Liang, C.; Jeffery, D.W.; Taylor, D.K. Preparation of magnetic polymers for the elimination of 3-Isobutyl-2-methoxypyrazine from wine. *Molecules* **2018**, *23*, 1140. [CrossRef]
82. Liang, C.; Ristic, R.; Stevenson, R.; Jiranek, V.; Jeffrey, D. Green characters: Using magnetic polymers to remove overpowering green capsicum flavour from Cabernet Sauvignon wine. *Wine Vitic. J.* **2019**, *34*, 24–26.

Article

Influence of Triazole Pesticides on Wine Flavor and Quality Based on Multidimensional Analysis Technology

Ouli Xiao [1,2,†], Minmin Li [3,†], Jieyin Chen [2], Ruixing Li [3], Rui Quan [3], Zezhou Zhang [1,2], Zhiqiang Kong [2,3,*] and Xiaofeng Dai [1,2,*]

1. Feed Research Institute, Chinese Academy of Agricultural Sciences, Beijing 100081, China; xiaoouli123@163.com (O.X.); zhangzezhou7689@163.com (Z.Z.)
2. State Key Laboratory for Biology of Plant Diseases and Insect Pests, Institute of Plant Protection, Chinese Academy of Agricultural Sciences, Beijing 100193, China; chenjieyin@caas.cn
3. Key Laboratory of Agro-Products Quality and Safety Control in Storage and Transport Process, Ministry of Agriculture and Rural Affairs/Institute of Food Science and Technology, Chinese Academy of Agricultural Sciences, Beijing 100193, China; liminmin@caas.cn (M.L.); liruixing06@163.com (R.L.); qr802319@163.com (R.Q.)
* Correspondence: kongzhiqiang@caas.cn (Z.K.); daixiaofeng_caas@126.com (X.D.); Tel.: +86-10-62813566 (Z.K.); +86-10-62813566 (X.D.)
† These authors contributed equally to this work.

Academic Editors: Fernando M. Nunes, Fernanda Cosme and Luís Filipe Ribeiro
Received: 28 October 2020; Accepted: 26 November 2020; Published: 28 November 2020

Abstract: Triazole pesticides are widely used to control grapevine diseases. In this study, we investigated the impact of three triazole pesticides—triadimefon, tebuconazole, and paclobutrazol—on the concentrations of wine aroma compounds. All three triazole pesticides significantly affected the ester and acid aroma components. Among them, paclobutrazol exhibited the greatest negative influence on the wine aroma quality through its effect on the ester and acid aroma substances, followed by tebuconazole and triadimefon. Qualitative and quantitative analysis by solid-phase micro-extraction gas chromatography coupled with mass spectrometry revealed that the triazole pesticides also changed the flower and fruit flavor component contents of the wines. This was attributed to changes in the yeast fermentation activity caused by the pesticide residues. The study reveals that triazole pesticides negatively impact on the volatile composition of wines with a potential undesirable effect on wine quality, underlining the desirability of stricter control by the food industry over pesticide residues in winemaking.

Keywords: triazole pesticides; wine; fermentation; sensory analysis; flavor components

1. Introduction

Wine has become one of the three most globally popular alcoholic beverages because of its good flavor and taste and its unique health benefits. Indeed, a decreased risk of cardiovascular disease, improved immunity, and reduced mortality rate have been reported in moderate wine drinkers [1]. According to statistics, with grapes being the raw material in wine production, more than half of the global annual grape production is used for wine production [2]. However, because of their high fructose and glucose contents, grapes are susceptible to contamination by vine microbial pathogens during cultivation. Thus, chemical pesticides are applied to control diseases and pests during the whole grape cultivation cycle to obtain high-quality wine grapes [3]. Unfortunately, unsuitable agricultural practices associated with potential grape and wine contamination, are frequently observed during the application of these active materials [4]. Consequently, consumers are indirectly exposed to these pesticides, which pose potential risks to human health. Thus, there is an ever-increasing global concern about wine quality and food safety.

Triadimefon [1-(4-chlorophenoxy)-3,3-dimethyl-1-(1H-1,2,4triazol-1-yl)butan-2-one], tebuconazole [(RS)-1-(4-chlorophenyl)-4,4-dimethyl-3-(1H-1,2,4-triazol-1-ylmethyl)-pentan-3-ol], and paclobutrazol [(2RS,3RS)-1-(4-chlorophenyl)4,4-dimethyl-2-(1H-1,2,4-triazol-1-yl)pentan-3-ol] (Supplementary Figure S1) are triazole pesticides commonly used in grape cultivation [5]. These pesticides (e.g., triadimefon and tebuconazole) are mainly used as fungicides because of their broad-spectrum activity and protective, curative, and eradicating action against actinomycetes and basidiomycetes [6], such as *Streptomyces scabies*, a genus of actinomycetes, which causes scabs in tap root crops and potato tubers [7]. Moreover, the grape skin extract containing fungicide can obviously inhibit the germination of *Penicillium expansum*, *Penicillium chrysogenum*, and *Aspergillus niger* [8]. However, some triazole pesticides (e.g., paclobutrazol) are also used to regulate plant extension growth. The Joint FAO (Food and Agriculture Organization of the United Nations)/WHO (World Health Organization) Meeting on Pesticide Residues (JMPR), reported the data from 17 residue trials matching good agricultural practice (GAP), with the highest pesticide residue of 0.52 mg/kg in grapes. It was also reported that the use of triadimefon on grapes would contribute to high pesticide residues of 3.2 mg/kg [9]. Additionally, many triazole pesticides have been reported as potential endocrine disruptors with anti-androgen activity and inhibition of the enzymatic activity of cytochrome 3A4 (CYP3A4), which was the dominant form of CYP450 in the liver that mediates the 6b-hydroxylation of testosterone [10]. Thus, it is important to monitor the presence of pesticides and regulate their levels in grapes and wine to limit human health risks, and to use phytochemical biopesticides that are less toxic, least persistent, environmentally friendly, and safe to humans and non-target organisms. It was reported that several phytochemical biopesticides such as azadirachtin, nicotine, pyrethrins, rotenone, veratrum, annonins, rocaglamides, isobutylamides, etc. have been successfully commercialized in the past [11].

Maximum residue levels (MRLs) in plant-based products have been established by many countries and international organizations to regulate pesticide levels. The MRLs for pesticide residues in grapes are normally in the range of 0.01–5 mg/kg, depending on the pesticides [12]. For example, in China, the MRLs of triadimefon, tebuconazole, and paclobutrazol on grapes are 0.3, 2.0, and 0.5 mg/kg, respectively [13]. In addition, pesticides that penetrate plant tissue and contaminate the grape will lead to stimulated or sluggish wine fermentation [14]. Moreover, the residual pesticides on grapes affect the flavor characteristics of wine volatile compounds through fermentation [15]. In summary, these pesticide residues not only pose potential health risks to the consumers but also reduce the wine quality.

Red wine is fermented with the existing pericarp to extract more chemical components (e.g., anthocyanins, polyphenols, and volatile compounds). This is because the wine aroma originates from the substances produced by the fermentation of the grape itself and yeast. González-Álvarez et al. [16] reported that residual fungicides affected the contents of discriminant volatiles (4-vinylguaiacol, 3-methylbutanoic acid and acetates, and ethyl ester) produced by biosynthesis using fatty acids as precursors, thereby changing the flavor of white wine. Pesticides can also affect the activity of lactic acid bacteria and change the process of malolactic fermentation [17]. Notably, flavor substance is an important quality index to judge the quality of wine as well as an important factor that affects the purchase intention of consumers [18].

Instrumental analytical technologies have attracted much attention in the analysis and identification of volatile components in food because of their objectivity, high efficiency, sensitivity, and stability. Particularly, electronic nose and tongue technologies are intelligent smell and taste recognition systems that simulate the respective human physiological senses [19]. These techniques are commonly applied to the flavor analysis of various food products such as beverages (e.g., wine) and condiments [20]. The combination of electronic nose and tongue sensing makes the detection result more accurate. Solid-phase micro-extraction gas chromatography coupled with mass spectrometry

(SPME-GC-MS) and gas chromatography-ion mobility spectrometry (GC-IMS) have been widely used in food flavor analysis. Applications include the analysis of volatile components in different vintages [21], comparison of different fermentation properties of raspberry wine [22], etc. These instrumental analytical technologies can effectively separate and identify the volatile compounds responsible for the fragrance of the product. Moreover, the application of these techniques can effectively avoid the risk of exposing traditional human sensory evaluators to the chemical hazards present in contaminated target samples. To the best of our knowledge, literature on the influence of triazole pesticides on wine flavor and quality based on electronic sensory evaluation systems and GC-MS is scarce [15].

With these facts in mind, we aimed to investigate the effects of triadimefon, tebuconazole, and paclobutrazol on the flavor and quality of wine during fermentation. Sensor evaluation of wine treated with triazole pesticides was conducted by electronic nose and tongue analyses. Additionally, GC-IMS was used to analyze the differences in the volatile organic compounds (VOCs) of the triazole-pesticide-treated wine samples. SPME-GC-MS was further used to identify and quantify the flavor components present in the tested wine samples. The results from this study provide more accurate information on the wine flavor and quality changes induced by triazole pesticides.

2. Results and Discussion

2.1. Electronic Sensory Evaluation

As new bionic sensing technologies for smell and taste, electronic nose and tongue sensory systems are simple, fast, nondestructive, and repeatable. Thus, they can provide an alternative to smell and taste evaluation to effectively distinguish the differences between the test samples.

2.1.1. Electronic Nose Analysis

The taste quality of the blank and triazole-pesticide-treated samples was analyzed by electronic nose technology. Principal component analysis (PCA) was applied to the data analysis as a multivariate method of generating principal component (PC) variables by investigating the correlation among several variables [23], which was used to eliminate the correlation among original characteristic variables. According to the PCA analysis (Figure 1a), the first component (PC1) explained 99.76% of the total system variance. This indicated that the first principal components represent most of the valid information of the original data, which can be used to reflect the changes in the wine odor. The flavor difference between any two groups was >0.5 (Supplementary Table S1), indicating that there was a significant difference in flavor between the four groups. This may be due to the presence of pesticide residue during fermentation causing significant changes in the flavor profile of all the wine samples; similar results have been reported in the literature [24,25]. In addition to the influence of pesticide residues on yeast fermentation, wild yeast on grape surface has relatively high invertase activity, which may also affect the volatile composition and taste of grape. *Saccharomyces cerevisiae* × *S. kudriavzevii* hybrids are prized for their unique flavor profiles in beer and wine, because these hybrids have good enological properties, such as high glycerol content, decreased ethanol, improved taste, and a lower production of undesirable acetic acid [26]. On the contrary, because of the complexity of the yeast strain, hybrids and introgressed strains from *S. eubayanus* and *S. uvarum* could create an odor, which is considered a brewery contaminant [27]. Linear discriminant analysis (LDA) was used as a dimensionality approach that retains most of the information in the original data and finds the best linear fit that separates two or more groups of samples. From Figure 1b, we can see that the data collection points of the same group of samples were gathered in the same area, while the data collection points of the different groups of samples (three triazole-pesticide-treated experimental groups and control group) were scattered in different areas. Figure 1b also reveals that the volatile odors of the wine samples treated with different triazole pesticides were significantly different in discriminant function 1 (DF1) and discriminant function 2 (DF2), and the four wine samples could therefore be effectively discriminated.

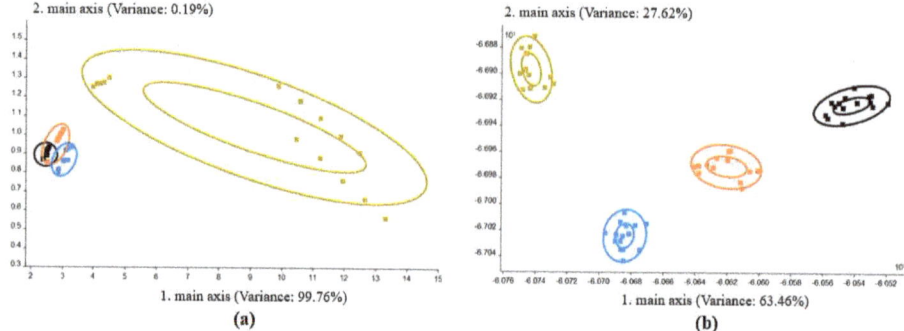

Figure 1. Results obtained by using the electronic nose. (a) Principal component analysis (PCA) and (b) linear discriminant analysis (LDA) of the electronic nose data for the control (CK; black), triadimefon-treated (SZT; orange), tebuconazole-treated (WZC; blue), and paclobutrazol-treated (DXZ; yellow) wine groups.

2.1.2. Electronic Tongue Analysis

The electronic tongue analysis system consists of seven flavor sensors, namely AHS (sour), CTS (salty), NMS (umami), and SCS, ANS, CPS, and PKS (general purpose) sensors. In this study, the seven flavor sensors responded to the taste of the four wine samples with different sensitivities. Thus, the taste and quality of the control and triazole-treated wine samples were compared and analyzed by electronic tongue technology. Discriminant factor analysis (DFA) is a useful pattern recognition technique for multivariate data analysis [28], which was used as a supervised linear pattern recognition algorithm for data classification. Electronic tongue DFA (Figure 2a) revealed that the four wine groups were significantly different, indicating marked differences in the tastes of the four groups (Supplementary Table S2). The data revealed significant differences between any two groups of each sample ($p < 0.01$), while the flavor of the DXZ wine was different from those of the other three groups. Figure 2b is the radar image of the taste characteristics of the wine treated with triazole fungicides and illustrates the similarity of the response values of each sensor. The data indicate that the most affected flavor was that of the DXZ wine, while the flavors of the SZT and WZC wines were less affected. Notably, paclobutrazol is a plant growth regulator that can regulate the growth and development of crops and induce stress resistance in plants [29]. Studies have shown that this triazole can regulate secondary metabolite contents such as Vitamin C (acerbity) and soluble sugars in fruits, thereby affecting their nutrition and quality [30].

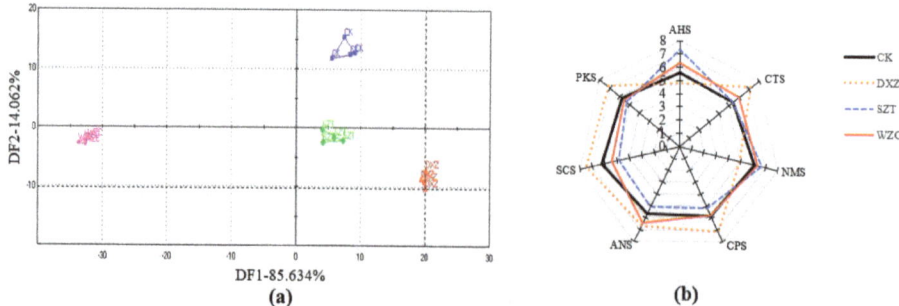

Figure 2. Results obtained by using the electronic tongue. (a) Discriminant factor analysis (DFA) diagram of the electronic tongue data for the control (CK; blue), triadimefon-treated (SZT; green), tebuconazole-treated (WZC; pink), and paclobutrazol-treated (DXZ; red) wine groups. (b) Radar image of the taste characteristics of the triazole-pesticide-treated wine samples.

2.2. GC-IMS Analysis

2.2.1. Effects of the Different Triazole Pesticide Treatments on the VOCs in Wine

The VOCs in the wine samples comprising the differently treated grapes are illustrated in Supplementary Figure S2. The color represents the concentration of the substance, whereby white indicates a low concentration and red indicates a high concentration. Additionally, a darker color implies a higher concentration. The contrast model was adopted to select the control (CK) spectra as reference, and those of the other samples were deducted from the reference. Thus, for two identical VOCs, the background after deduction was white, while red and blue backgrounds indicated that the substance concentrations were respectively higher and lower than that of the reference. The results reveal that the volatile component content of the SZT wine was only slightly different from that of the control group, while those of the WZC and DXZ wine samples were significantly different. Supplementary Figure S3 is a PCA analysis chart of all the samples, wherein the peak intensities of all the characteristic peaks are selected as characteristic variables for the PCA. This graph visually indicates the differences between the different samples. Thus, a short distance between the samples represents a small difference, while a long distance represents a significant difference. Supplementary Figures S2 and S3 therefore indicate that the VOCs in the WZC and DXZ wines were remarkably similar, while the CK wine was relatively more similar to the SZT wine.

The NIST and IMS databases built in the application software identified 40 signal peaks according to the retention time index of the standard substances and the standard drift time; 23 known components and 17 unknown components were determined by comparison with existing databases. Thus, to further investigate the changes in the main volatile substances, the signal peaks of all the VOCs were selected to form a fingerprint for comparison.

2.2.2. Comparison of the VOC Fingerprints in the Triazole-Pesticide-Treated and Control Wine Samples

Figure 3 presents the gallery plot (fingerprint) of the VOCs in the four groups of wines. Each row in the figure represents all the signal peaks selected in a wine sample, while each column represents the signal peaks of the same VOC in the different wine samples. The plot therefore provides the complete VOC information of each sample as well as the differences in the VOCs of the different samples. In Figure 3a, the component contents in region A were higher in the CK and SZT wines, and mainly comprised propionic aldehyde, isoamyl acetate, ethyl propionate, ethyl isobutyrate, and acetone. On the other hand, the component contents in region B were higher in the WZC and DXZ wines and included ethyl octanate, ethyl caproate, 1-butanol, and ethyl butyrate. In region C, the component contents of the control group were significantly different from those of the treatment groups, in which the contents of components no. 36 and no. 40 decreased after treatment, while the contents of components such as isobutyraldehyde increased after treatment.

Samples containing similar VOCs were also compared. Thus, the components in region D (Figure 3b) comprised a higher content in the control and included acetone, ethyl hexanoate, ethyl octoate, ethyl butyrate, 1-butanol, and isobutyl acetate. Higher levels were found in the SZT wine (region E), which included isobutyraldehyde, propionaldehyde, ethyl isobutyrate, isoamyl acetate, and ethyl propionate. Figure 3c reveals that there is little difference between the VOCs of the WZC and DXZ wines. Only the component contents in region F are slightly higher in the DXZ wines, which include isobutyl acetate, isopropyl ethyl butyrate, propyl acetate, isoamyl acetate, ethyl butyrate, and ethyl propionate.

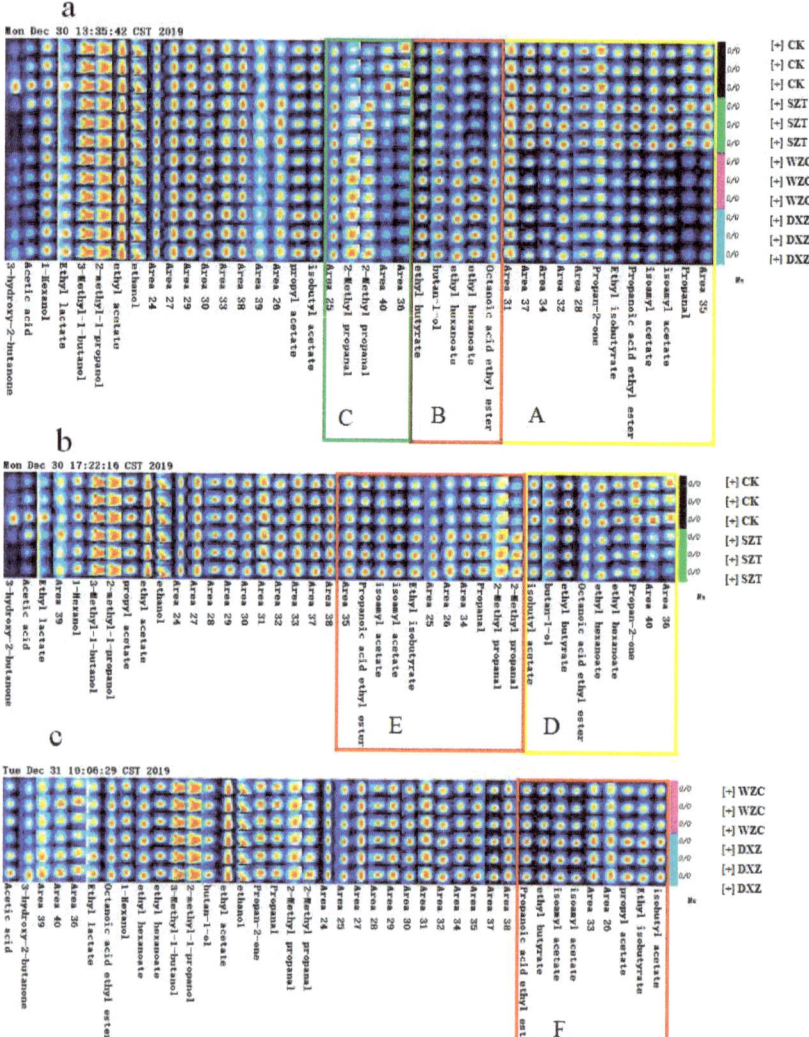

Figure 3. Analyzed by GC-IMS. (**a**) Comparison of gallery plot in all wine samples (CK, SZT, WZC, and DXZ wine samples). (**b**) Comparison of gallery plot between the control (CK) and triadimefon-treated (SZT) wine groups. (**c**) Comparison of gallery plot between the tebuconazole-treated (WZC) and paclobutrazol-treated (DXZ) wine groups. A, B, C, D, E, F represent regions where volatile compounds differ significantly.

These results indicated that the main volatile substances in the treated wine include ethyl hexanoate, isobutyl acetate, ethyl isobutyrate, propyl acetate, isoamyl acetate, ethyl propionate, ethyl butyrate, 1-esters (e.g., butanol and ethyl lactate), acetone, propionaldehyde, and isobutyraldehyde. The residues of pesticides may affect the uptake of microorganisms and delay the alcohol fermentation, but esters and aldehydes are still the main volatile components [24]. Similar results have also been reported by other research groups [31,32], which suggested that the ethyl esters produced during alcohol fermentation contribute to the typical fruit aroma of wine. On the other hand, alcohols do not affect the wine flavor quality due to their higher sensory thresholds [33]. Studies have also shown that changes in the esters' concentrations effect the quality of red wine by altering the fruity aroma [34]. In addition to these esters, other compounds that do not necessarily exhibit fruity aromas may have an important effect on the overall fruity aroma of the wine [35].

2.3. Head-Space SPME-GC-MS Analysis

The volatile components of the wine samples were further qualitatively and quantitatively analyzed by GC-MS. The results were compared with the spectra of unknown volatile compounds using the NIST11 database and semi-quantified by the internal standard method. A 20 µL/L cyclohexanone content in the wine samples was used as an internal standard to semi-quantify the content of the volatile substances in the different triazole-pesticide-treated wine samples.

2.3.1. GC-MS Qualitative Analysis

The total ion chromatogram of the volatile compounds of the four wine samples are displayed in Figure 4, whereby a total of 58 volatile substances were detected in the four differently treated wines (Table 1). These comprised 14 alcohols, accounting for 24.1% of the total volatile components; 33 esters (56.9%); six acids (10.3%); a ketone (1.7%); an aldehyde (1.7%); and other components [pentane, 2,4-di-tert-butylphenol, and 1,3-di-tert-butylbenzene; 5.3%]. Thus, the results revealed that alcohols, esters, acids, and to a lesser extent aldehydes and ketones were the main volatile components of the wine samples.

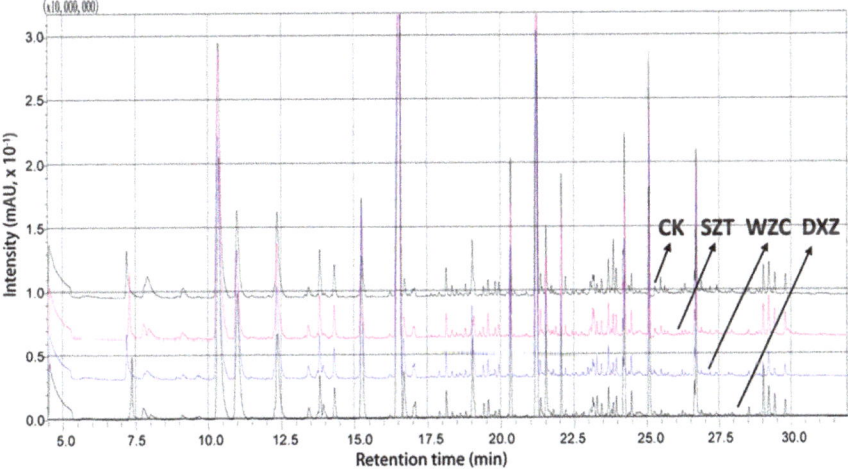

Figure 4. SPME-GC-MS total ion chromatogram of the volatile compounds for, from top to bottom, the blank (CK), triadimefon-treated (SZT), tebuconazole-treated (WZC), and paclobutrazol-treated (DXZ) wine samples.

Table 1. Composition of the wine volatile substances determined by SPME-GC-MS analysis.

Elution Order	Compound	Retention Index	Descriptor	CK	SZT	WZC	DXZ	Odor Threshold (mg/L)
1	Ethanol	463	Pungent, wine [a]	33.13	27.66	22.2	27.59	100 [d]
2	Isobutanol	597	Disagreeable, wine [a]	5.27	2.08	-	4.17	40 [d]
3	Pentane	518	Alkane [b]	2.48	2.41	7.09	2.33	4.1 [d]
4	Isoamyl acetate	820	Banana [a]	6.05	3.39	2.39	1.22	0.2 [d]
5	Isoamyl alcohol	697	Fusel oil, pungent [a]	72.82	80.02	73.73	60.65	30 [d]
6	Ethyl caproate	984	Fruit [a]	20.46	20.24	20.55	26.95	0.08 [d]
7	Cyclohexanone *	891	Peppermint [a]	20.0	20.0	20.0	20.0	20
8	Ethyl heptanoate	1083	Pineapple [b]	-	-	0.27	0.39	0.3 [d]
9	Ethyl lactate	848	Wine, cream [a]	-	-	0.11	1.92	8 [d]
10	Hexyl formate	981	Apple, unripe plum [b]	1.28	1.22	2.39	1.1	6.4 [d]
11	1-Hexanol	860	Sweet, green fruity [b]	2.77	2.89	1.23	2.34	8 [d]
12	Methyl octanoate	1083	Oranges, grapes [a]	13.33	14.8	16.33	18.78	0.2 [d]
13	1,3-Di-tert-butylbenzene	1334	Unknown	0.33	0.27	-	-	Unknown [e]
14	Ethyl caprylate	1183	Orange, oily [a]	85.87	92.28	88.98	78.35	0.51 [c]
15	Acetic acid	576	Pungent, vinegar [a]	2.7	1.99	2.53	0.56	22 [d]
16	Isopentyl hexanoate	1218	Banana, pineapple [b]	0.82	0.94	1.43	0.96	0.9 [d]
17	Methyl nonanoate	1183	Wine, coconut [a]	0.24	0.37	-	0.18	0.04 [d]
18	Propyl caprylate	1282	Spice [a]	0.47	-	0.85	0.35	Unknown [e]
19	Ethyl nonanoate	1282	Grape, rose, wine [b]	5.24	8.29	2.31	2.84	0.377 [d]
20	(R,R)-2,3-Butanediol	743	Unknown	1.3	1.36	-	1.25	Unknown [e]
21	Linalool	1082	Lily, citrus [a]	-	-	0.46	-	5 [d]
22	Isobutyl caprylate	1317	Unknown	0.78	0.76	0.86	0.78	Unknown [e]
23	1-Octanol	1059	Orange, rose [a]	1.36	1.91	1.8	1.33	0.13 [d]
24	2,3-Butanediol	743	Spices [b]	1.97	1.52	1.6	1.29	20 [d]
25	Methyl n-caprate	1282	Unknown	10.83	9.7	11.13	8.44	Unknown [e]
26	Ethyl caprate	1381	Coconut [a]	55.15	53.01	49.81	39.4	2.4 [c]
27	Methyl cis-4-decenoate	1290	Unknown	0.57	-	-	0.4	Unknown [e]
28	Isoamyl caprylate	1417	Fruity [b]	2.11	-	-	4.77	0.125 [d]
29	3-Methylbutyl octanoate	1417	Fruity, brandy [a]	6.73	7.6	7.7	6.3	0.125 [d]
30	2-Methyl butyric acid	811	Cheese, fruit [a]	-	-	-	0.31	5.9 [d]
31	Diethyl succinate	1151	Faint, pleasant [a]	0.37	0.28	3.09	0.44	200 [d]
32	3-Methylthiopropanol	912	Onion, meat [a]	0.33	0.28	0.37	0.26	4 [d]
33	Ethyl undecanoate	1481	Coconut [a]	0.41	0.27	-	0.08	Unknown [e]
34	2-Undecenal	1311	Fresh aldehyde [a]	0.15	-	-	-	0.001 [d]
35	1-Decanol	1258	Flower, fatty [a]	1.69	2	2.1	1.3	2.8 [d]
36	Citronellol	1179	Rose [b]	1.26	1.8	1.43	0.7	0.1 [d]
37	Methyl salicylate	1281	Ilex leaf [a]	0.96	1.59	1.12	1.45	0.071 [d]
38	Ethyl phenylacetate	1259	Rose [a]	-	1.03	1.16	0.96	0.65 [c]

Table 1. *Cont.*

Elution Order	Compound	Retention Index	Descriptor	Concentration (mg/L) CK	SZT	WZC	DXZ	Odor Threshold (mg/L)
39	Methyl laurate	1481	Fatty, floral [a]	2.93	2.36	2.73	2.11	Unknown [e]
40	Ethyl salicylate	1380	Wintergreen [b]	0.54	0.6	0.52	0.47	0.115 [d]
41	Phenethyl acetate	1259	Sweet [a]	4.17	3.25	1.41	0.87	1.8 [d]
42	Hexanoic acid	974	Stink, unpleasant [a]	-	-	-	2.28	0.42 [d]
43	Ethyl laurate	1580	Apricot [a]	14.43	12.47	12.93	9.13	0.5 [d]
44	Isoamyl decanoate	1615	Unknown	1.26	1.47	1.68	0.98	5 [d]
45	Phenylethyl alcohol	1136	Rose [b]	14.7	13.98	16.36	17.31	7.5 [c]
46	1-Dodecanol	1457	Fatty, waxy [a]	-	-	-	0.04	0.073 [d]
47	Methyl tetradecanoate	1580	Onion, honey, orris [a]	0.16	0.12	0.26	0.3	0.5 [d]
48	Strawberry furanone	1322	Fruit, caramel [a]	0.12	0.12	0.08	-	Unknown [e]
49	Ethyl myristate	1779	Essence [b]	1.27	1.22	1.29	1.86	0.5 [d]
50	Octanoic acid	1173	Mildly unpleasant [a]	11.39	11.82	12.25	12.62	10 [d]
51	Ethyl pentadecanoate	1878	Unknown	-	-	-	0.07	Unknown [e]
52	Nonanoic acid	1272	Light fat, coconut [a]	-	-	-	0.08	3.0 [d]
53	Myristyl alcohol	1656	Essence [a]	-	-	0.18	-	5 [d]
54	Methyl hexadecanoate	1878	Unknown	0.3	0.29	0.27	0.46	2 [d]
55	Methyl palmitoleate	1886	Unknown	0.11	-	0.19	0.06	Unknown [e]
56	Ethyl palmitate	1978	Mild, sweet [a]	1.5	1.94	2.04	2.16	2 [d]
57	n-Decanoic acid	1372	Unpleasant [a]	2.88	2.63	3.45	2.5	1 [d]
58	Ethyl 9-hexadecenoate	1986	Unknown	1.23	1.64	1.73	1.25	Unknown [e]
59	2,4-Di-tert-butylphenol	1555	Burning, sweet [a]	1.45	2.45	1.82	1.37	Unknown [e]

The volatile components of the wine samples were semi-quantified by the internal standard method; dashes "_" in the columns denote that the sample was not detected; CK, SZT, WZC, DXZ represent the control and triadimefon-, tebuconazole-, and paclobutrazol-treated wine; [a] Reference [36]; [b] Reference [37]; [c] Reference [38]; [d] Reference [39]; [e] Unknown odor value; * Internal standard.

In this study, 47, 45, 46, and 54 volatile compounds were detected in the CK, SZT, WZC, and DXZ wine samples, respectively. Among them, 36 volatile substances were common to the four wine samples, including 22 esters, three acids, nine alcohols, one phenol, and one alkane (Supplementary Figure S4). Of these, some were alcohols (isoamyl alcohol, 1-octanol, citronellol, and phenylethyl alcohol), esters (isoamyl acetate; ethyl caproate; methyl octanoate; ethyl caprylate; ethyl nonanoate; ethyl caprate; methyl salicylate; ethyl salicylate; ethyl laurate; ethyl myristate; and 3-methylbutyl octanoate), and fatty acids (n-decanoic acid and octanoic acid). Esters have been reported to be the main volatile substance contributors in wine [16]. This is consistent with the GC-IMS results attained in this study (Section 2.2.2).

2.3.2. GC-MS Quantitative Analysis

Alcohols

Figure 5 displays the alcohols present in the differently treated wines, with total alcohol contents in the CK, SZT, WZC, DXZ wine samples of 136.6, 135.5, 121.46, and 118.23 mg/L, respectively. Rapp and Mandery [40] proposed that a small amount of the major alcohols has a positive effect on the wine quality, while a total concentration of >300 mg/L endows the wine with an unpleasant taste. For the WZC and DXZ wines, the alcohol contents were significantly reduced, indicating that tebuconazole and paclobutrazol affect the flavor of wine by affecting the fermentation of *S. cerevisiae* [41]. It may be that the level of expression of genes involved in alcohol synthesis is affected, for example, phenylalanine metabolism, lysine degradation and biosynthesis in *S. cerevisiae* are inhibited. Linalool and myristyl alcohol, which have lily and citrus aromas, respectively, were also detected in the WZC wine; nevertheless, since their concentrations were below the odor thresholds, it was assumed that these two compounds did not affect the flavor of the wine. Isobutanol was not detected because the biosynthesis of valine, the precursor amino acid of alcohols, may be affected by the pesticide residues [42]. Additionally, pesticide treatment reduced the content of n-hexanol (C6 alcohol) in the WZC wines, while the higher alcohol and geraniol (terpene) contents were not affected. These results are consistent with those reported by Noguerol-Pato et al. [43]. The concentration of isoamyl alcohol (fusel oil, floral descriptor), 1-octanol (orange fragrance), citronellol (rose fragrance), and phenylethyl alcohol (rose fragrance) were higher than their respective odor thresholds, indicating a significant contribution to the wine flavor. Notably, in this study, the concentration of citronellol decreased under paclobutrazol treatment. This was not consistent with the citronellol concentration changes reported by Oliva et al. [44], which did not change following treatment with six fungicides. This difference was ascribed to the disparate mechanisms of paclobutrazol as a plant growth regulator.

Esters

The total ester contents in the CK, SZT, WZC, and DXZ wine samples were 232.84, 239.17, 233.55, and 210.69 mg/L, respectively. The contents in the SZT wine were slightly higher than those of the CK and WZC wine samples. On the other hand, after paclobutrazol treatment, the ester content was significantly reduced. The formation of acetate esters is highly dependent on enzyme activity. These enzymes are responsible for both the synthesis and the hydrolysis of medium-chain fatty acid ethyl esters [45]. The levels of ester were significantly reduced in paclobutrazol-treated wine, which may be related to reduced enzyme activity. The most abundant compound was ethyl caproate (78.35–85.87 mg/L), followed by ethyl caprylate (39.4–55.15 mg/L) and ethyl caprate (20.24–26.95 mg/L); the contents of the other esters were <20 mg/L. The concentrations of isoamyl acetate (banana fragrance); decanoic acid, ethylene ester (coconut fragrance); acetic acid, 2-phenylethal ester (sweet fragrance); and dodecanoic acid, ethylene ester (apricot fragrance) in the pesticide-treated groups were lower than those in the CK wines. On the other hand, the concentrations of isopentyl hexanoate, methyl salicylate, ethyl palmitate, ethyl phenylacetate, ethyl heptanoate, and ethyl lactate were higher or appeared in the treatment group, possibly due to the type of nitrogen composition that the pesticides may confer on the must [46]. Notably, although these esters had a fruity aroma, the threshold was not high or <1. In all the treated wine groups, the ethyl caproate, methyl octanoate, and ethyl caprylate concentrations were significantly higher than their odor thresholds (0.08, 0.2, and 0.51 mg/L, respectively) and significantly differed from the concentrations of the control group. However, a high ester concentration, with a strong fruit flavor, in the treated wine has a negative effect on the aromatic quality of the wine [42].

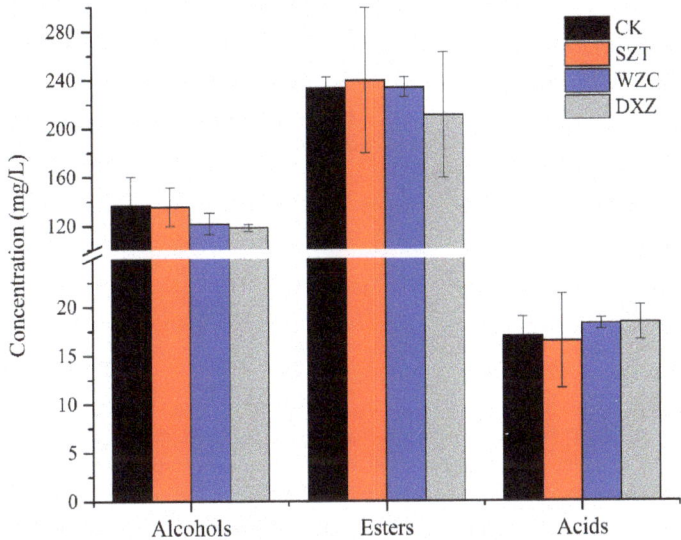

Figure 5. Comparison of relative volatile components of the four wine samples. Blank (CK), triadimefon-treated (SZT), tebuconazole-treated (WZC), and paclobutrazol-treated (DXZ) wine samples.

Diverse Volatile Compounds

The total acid substance contents in the CK, SZT, WZC, and DXZ wine samples were 16.97, 16.44, 18.23, and 18.35 mg/L, respectively. Triadimefon slightly influenced the concentration of n-decanoic acid, octanoic acid, and acetic acid. On the other hand, treatment with tebuconazole increased the n decanoic (unpleasant) and octanoic (smell of rancid butter) acid concentrations, while treatment with paclobutrazol increased the octanoic and hexanoic acid concentrations, which impart the wine with an unpleasant flavor. Only aldehyde, 2-undecenal was detected in the CK wines, giving the wine its fresh aldehyde flavor because of its polar threshold. Thus, treatment with the tested pesticides affected aldehyde synthesis, possibly because they promoted the synthesis of the corresponding alcohols during fermentation [47].

2.4. Combined Sensory, GC-IMS, and SPME-GC-MS Analysis

In this study, electronic nose and tongue technologies were used to investigate the flavor of the wine samples treated with different triazole pesticides. The results revealed significant changes in the overall flavor of the wines treated with triazole pesticides, which could be effectively discriminated by electronic nose and tongue technologies.

GC-IMS and GC-MS methods were further used to identify and quantify the volatile components of the wine samples after different treatments. Comparison of the GC-IMS fingerprints indicated that esters are important factors in determining the wine quality. Moreover, because their concentrations are usually higher than their threshold levels, they endow the wine with a fruity aroma [46]. Differences in the types and relative contents of the volatile substances were observed in the different samples. In all, 40 typical compounds were determined by GC-IMS, but there were still 17 compounds with no qualitative results due to the limited data of the library database. Based on the identified compounds, the volatile compounds in samples were mainly esters, alcohols, and aldehydes, which was consistent with the results of the SPME-GC–MS analysis. Of the 58 compounds identified by SPME-GC-MS technology, the main volatile components were esters, alcohols, acids, and some aldehydes, ketones, and alkanes. Paclobutrazol was the most influential pesticide on the volatile components of wine, significantly changing the concentrations of citronellol; isoamyl acetate, hexanoic acid; ethyl ester, octanoic acid; metallic ester, nonanoic acid; and ethical ester hexanoic acid. Tebuconazole and paclobutrazol reduced the ester and alcohol contents in the wine samples, while conversely, the triadimefon-treated samples retained most of their original wine flavor quality. This result was similar to that of the GC–IMS analysis. Oliva et al. [41] reported that fungicides of the triazole family can affect the amount of ethyl esters, acetates, acids, and ethyl acetate in wine. Moreover, they inhibit the synthesis of major sterols on fungal cell membranes, reducing the fermentation activity of *S. cerevisiae* and, consequently, affecting the synthesis of volatile compounds by other metabolic pathways [48,49].

3. Material and Methods

3.1. Materials and Reagents

Cyclohexanone analytical standards (purity ≥99.9%) were purchased from Dr. Ehrenstorfer (LGC Standards, Augsburg, Germany). Commercial triadimefon 20% emulsifiable concentrate was sourced from Jiangsu Sword Agrochemicals Co., Ltd. (Yancheng, China); tebuconazole 43% suspension was acquired from Qingdao Haina Biotechnology Co., Ltd. (Qingdao, China); and paclobutrazol 15% wet-table powder was obtained from Jiangsu Kesheng Group Co., Ltd. (Yancheng, China). Sodium chloride (NaCl) was provided by Sinopharm Chemical Reagent Co., Ltd. (Shanghai, China). *Saccharomyces cerevisiae* powder was purchased from Yantai Di Boshi brewing machine Co., Ltd. (Yantai, China). The fermentation tanks were bought from Hebei Chaoya glass products Co., Ltd. (Hebei, China). Ultra-pure water was produced using a Millipore purification system (Millipore, Bedford, MA, USA). The grapes were obtained from Changxinghongyuan grape professional cooperative of Liaoning province and did not contain the target pesticides.

3.2. Red Wine Processing

A total of 60 kg fresh and mature kyoho grapes was used, the brix was 16° and total acidity was 5 mg/g of berry, with the rotten grape particles and stems removed, and divided into 4 groups, and every 5 kg of grapes was selected and crushed into fermentation vessels; the pulp dregs, juice, and skin were all included in the fermentation tanks. The spiked levels of triadimefon, tebuconazole, and paclobutrazol were 0.3, 2.0, and 0.5 mg/kg, respectively, according to the corresponding maximum residue limits established by GB 2763-2019, and a different pesticide was added to each fermentation tank. Next, 30 mg/kg SO_2 and 1 g/kg yeast powder activated in 100 mL of 30 °C pure water were added into each fermentation tank, and the mixtures were stirred three times daily at the initial stage of the fermentation process. After 240 h at room temperature (25 °C), the skin residues were filtered out and secondary fermentation was performed during 192 h. Finally, the fermented grape musts were clarified and filtered by sieve to attain four groups of wine samples, bottled at 4 °C for further study; three repetitions were performed for each group. The four groups were labeled CK, representing the control group, and SZT, WZC, and DXZ representing the wines treated with triadimefon, tebuconazole, and paclobutrazol, respectively. The detailed processing procedure is illustrated in Figure 6.

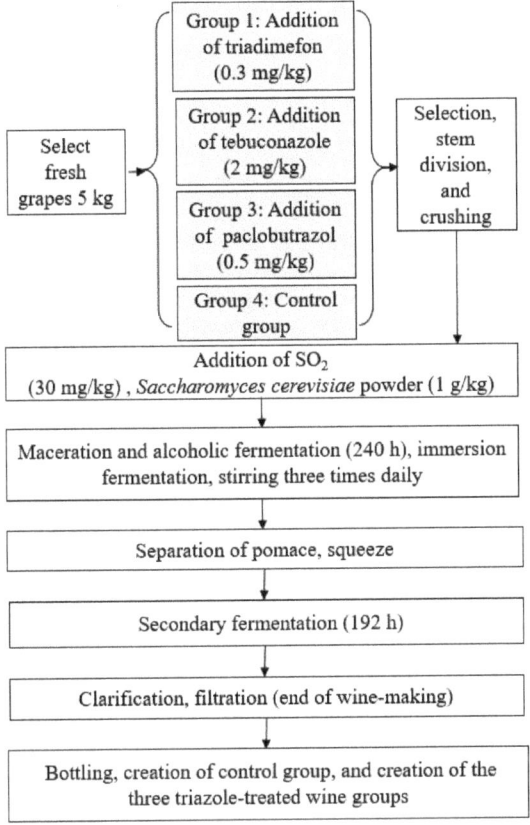

Figure 6. Wine making process adopted in the experiment.

3.3. Electronic Nose Detection

A PEN3.5 electronic nose (Airsense Analytics, GmBH, Schwerin, Germany) was employed for the volatile analysis. Thus, 1.0 mL each of the control and three triazole-pesticide-treated wine samples were separately placed into four 10 mL headspace sampling vials. Each sample was injected manually with a PEN3.5 electronic nose and allowed to stand for 10 min, under laboratory conditions of 25 ± 1 °C, until the volatile gases in the sample filled the headspace space of the vials. Electronic nose detection comprised respective sampling and cleaning times of 60 and 180 s and three parallel samples were set for each group.

3.4. Electronic Tongue Detection

An Alpha Astree II electronic tongue system connected with an LS16 auto sampler unit (Alpha MOS, Toulouse, France) was used for the interacting chemical substances in solution. Thus, 10 mL aliquots of the control and triazole-pesticide-treated wines were collected, filtered with 0.45 μm aqueous phase filter membranes, and placed into separate 100 mL beaker with a constant volume of 120 mL. Each beaker was then placed in the spot tongue detector sample position so that it was sitting just over the electrode surface. Respective collection and cleaning times of 120 and 10 s were employed. To eliminate the interference of the unstable factors of the initial detection response signal and obtain a stable detection signal, seven parallel measurements were conducted in the study. After obtaining the analysis data, the first three circles of data were discarded, and the last four circles of stable electronic tongue response data were selected as the analysis data.

3.5. GC-IMS Analysis of the Volatile Compounds

A commercial GC-IMS (FlavourSpec®; GAS, Dortmund, Germany) was used for flavor analysis. Thus, a 1 mL wine sample was first transferred into a 20 mL headspace bottle. Then, 100 µL aliquots were automatically injected under splitless mode at 85 °C, after incubating at 60 °C and 500 rpm incubation speed for 10 min. The separation was carried out on an MXT-WAX column (dimensions, 30 m × 0.53 mm × 1 µm) at 60 °C. Nitrogen (purity, >99.99%) was used as the carrier gas, and the linear pressure program was as follows: 2 mL/min for 2 min, ramped up to 100 mL/min over 18 min, and finally maintained at this rate for 10 min. The drift gas flow rate was 150 mL/min (nitrogen), and a 5 cm drift tube was operated at a constant voltage of 400 V/cm at 45 °C.

3.6. Head-Space SPME-GC-MS Analysis of the Volatile Compounds

A gas chromatography-mass spectrometer (Shimadzu GC-MSQP 2010plus) was employed for compound separation and analysis. Thus, 5 mL wine samples were accurately measured into 15 mL headspace bottles, and 1 g NaCl was added to each bottle to promote volatilization of the volatile components. Subsequently, 50 µL cyclohexanone internal standard solution (2 µL/mL) was added and the bottles were immediately sealed with a polypropylene cap comprising a PTFE-silicone rubber septum. The headspace volatile components of the wine samples were then extracted by 65 µm divinylbenzene/polydimethylsiloxane (DVB/PDMS). After equilibration of each sample at 50 °C for 20 min, the extraction head was inserted into a headspace bottle for 40 min. Next, the DVB/PDMS fiber was placed in a 250 °C gas chromatography-mass spectrometer inlet for 2 min to allow desorption. The instrument was equipped with a DB-WAX column (dimensions, 30 m × 0.25 mm × 0.25 µm). Analytical conditions were set as follows: an initial column chamber temperature of 40 °C, injection port temperature of 250 °C, spitless injection mode, and a carrier gas flow rate of 1.0 mL/min. The column chamber temperature was initially set at 40 °C for 3 min, then increased to 120 °C at 5 °C/min, and finally to 230 °C at 10 °C/min where it was held for 5 min. The ion source and transfer-line temperatures were 200 and 250 °C, respectively, and the signal was collected under scan mode in the scanning range 35–500 m/z. Peak identification was primarily achieved using Lab Solution software in retention index calibration mode with retention index calibrated in-house MS libraries, and compared to the NIST mass spectral search program (NIST 11).

3.7. Data Statistics

The wine fermentation experiments were carried out in triplicate, and PCA was used for sample discrimination and classification. LDA was based on the determination of the linear discriminant functions (DFs) to extract features by maintaining class separability. DFA was used to further expand the differences between the different groups and narrow the differences within each group based on principal component analysis. The stable intervals selected for the PCA and LDA analyses were in the range 48–52 s. These analyses were used to distinguish the smell characteristics of the four groups of wine samples. Discriminant factor analysis (DFA) was conducted to compare the differences in the taste characteristics of the four wine sample groups. The instrumental analysis software includes LAV (Laboratory Analytical Viewer, version 2.2.1—G.A.S. Dortmund, Germany) and three plugins (Reporter, Gallery Plot, and Dynamic PCA) as well as GC × IMS Library Search, which can be used for sample analysis from different perspectives.

4. Conclusions

The main objective of this study was to explore the influence of three different triazole pesticides on wine flavor. Analysis at the molecular sensing level was based on multidimensional analysis technology. The pesticide with the greatest influence on wine flavor composition was paclobutrazol, as the wines treated with this pesticide exhibited the most significant changes in the concentrations (relative to threshold) of the flavor components (citronellol, isoamyl acetate, ethyl caproate, methyl octanoate, ethyl caprylate, ethyl caprate, methyl cis-4-decenoate, ethyl laurate, ethyl myristate, and hexanoic acid). Treatment with the three pesticides mainly changed the concentrations of the esters, the main contributors to the significant differences in the wine flavors. The study demonstrated that pesticide residues on the grapes are transferred from the grape berries to the grape juice, thereby changing the flavor components and thus, the flavor quality of the final wine. Electronic nose and tongue analyses were also employed to assess the potential flavor differences caused by the triazole pesticides. Moreover, the combination of dynamic headspace sampling with GC-MS and GC-IMS allowed us to determine the compounds present in the volatile components of the wine. The results of the three tests were in agreement, which was conducive to the comprehensive analysis of the effects of the three triazole pesticides on wine flavor. Further, the application of these techniques can effectively avoid the risk of exposing traditional human sensory evaluators to chemical hazards present in target samples. Meanwhile, it is important to monitor the presence of pesticides in grapes and wines in order to improve wine flavor, and biopesticides, which are safe, less toxic, least persistent, and environmentally friendly to humans and non-target organisms, can be used as alternatives to chemical pesticides in the vineyard.

Supplementary Materials: The supplementary materials are available online, Figure S1: Structure of triadimefon (a), tebuconazole (b), paclobutrazol (c); Figure S2: GC-IMS spectra of volatile organic compounds in four groups of wine; Figure S3: PCA (Principal Component Analysis) diagram of all samples; Figure S4: Venn diagram of the common volatile components in the four wine samples. Table S1: Electronic nose principal component analysis table; Table S2: Taste differences of wines.

Author Contributions: Formal analysis, O.X., J.C., and Z.K.; data curation, O.X., R.L., and Z.Z.; writing—original draft preparation, O.X.; writing—review and editing, M.L., J.C., Z.K., and X.D.; methodology, M.L., R.Q., and Z.K.; validation, M.L. and R.Q.; software, R.L. and Z.K.; investigation, R.Q.; supervision, Z.K. and X.D.; project administration, X.D. and Z.K.; funding acquisition, Z.K. All authors have read and agreed to the published version of the manuscript.

Funding: This study was supported by the National Natural Science Foundation of China (31671942) and the Beijing Nova Program of Science and Technology (Z191100001119121).

Conflicts of Interest: The authors declare no conflict of interest.

References

1. Mortensen, E.L.; Jensen, H.H.; Sanders, S.A.; Reinisch, J.M. Better Psychological Functioning and Higher Social Status May Largely Explain the Apparent Health Benefits of Wine. *Arch. Intern. Med.* **2001**, *161*, 1844–1848. [CrossRef] [PubMed]
2. OIV. *OIV FOCUS 2017 Distribution of the World's Grapevine Varieties*; International Organisation of Vine and Wine: Paris, France, 2017; p. 54. ISBN 979-10-91799-89-8.
3. Flamini, R. Mass spectrometry in grape and wine chemistry. Part I: Polyphenols. *Mass Spectrom. Rev.* **2003**, *22*, 218–250. [CrossRef] [PubMed]
4. Cabras, P.; Angioni, A. Pesticide Residues in Grapes, Wine, and Their Processing Products. *J. Agric. Food Chem.* **2000**, *48*, 967–973. [PubMed]
5. Zhang, A.; Xie, X.; Liu, W. Enantioselective separation and phytotoxicity on Rice Seedlings of paclobutrazol. *J. Agric. Food Chem.* **2011**, *59*, 4300–4305. [CrossRef] [PubMed]
6. Singh, N. Mobility of four triazole fungicides in two Indian soils. *Pest Manag. Sci.* **2005**, *61*, 191–196. [CrossRef] [PubMed]
7. Lerat, S.; Simao-Beaunoir, A.-M.; Beaulieu, C. Genetic and physiological determinants of *Streptomyces scabies* pathogenicity. *Mol. Plant Pathol.* **2009**, *10*, 579–585. [CrossRef]

8. Corrales, M.; Fernández, A.; Pinto, M.G.V.; Butz, P.; Franz, C.M.; Schuele, E.; Tauscher, B. Characterization of phenolic content, in vitro biological activity, and pesticide loads of extracts from white grape skins from organic and conventional cultivars. *Food Chem. Toxicol.* **2010**, *48*, 3471–3476. [CrossRef]
9. FAO. *FAO Plant Production and Protection. Paper, 502(574), 1 from the Report of the Joint Meeting of the FAO Panel of Experts on Pesticide Residues in Food and the Environment and the WHO Core Assessment Group on Pesticide Residues*; Food and Agriculture Organization of the United Nations: Geneva, Switzerland, 2009.
10. Lv, X.; Pan, L.; Wang, J.; Lu, L.; Yan, W.; Zhu, Y.; Xu, Y.; Guo, M.; Zhuang, S. Effects of triazole fungicides on androgenic disruption and CYP3A4 enzyme activity. *Environ. Pollut.* **2017**, *222*, 504–512. [CrossRef]
11. Walia, S.; Saha, S.; Tripathi, V.; Sharma, K.K. Phytochemical biopesticides: Some recent developments. *Phytochem. Rev.* **2017**, *16*, 989–1007. [CrossRef]
12. Bakırcı, G.T.; Acay, D.B.Y.; Bakırcı, F.; Otles, S. Pesticide residues in fruits and vegetables from the Aegean region, Turkey. *Food Chem.* **2014**, *160*, 379–392. [CrossRef]
13. GB 2763-2019. National food safety standard-maximum residue limits for pesticides in food. In *Standardization administration of the People's Republic of China*; Agriculture Press: Beijing, China, 2019. (In Chinese)
14. Caboni, P.; Cabras, P. Chapter 2—Pesticides' Influence on Wine Fermentation. In *Advances in Food and Nutrition Research*; Elsevier: Amsterdam, The Netherlands, 2010; Volume 59, pp. 43–62.
15. Vitalini, S.; Ruggiero, A.; Rapparini, F.; Neri, L.; Tonni, M.; Iriti, M. The application of chitosan and benzothiadiazole in vineyard (Vitis vinifera L. cv Groppello Gentile) changes the aromatic profile and sensory attributes of wine. *Food Chem.* **2014**, *162*, 192–205. [CrossRef] [PubMed]
16. González-Álvarez, M.; González-Barreiro, C.; Cancho-Grande, B.; Simal-Gándara, J. Impact of phytosanitary treatments with fungicides (cyazofamid, famoxadone, mandipropamid and valifenalate) on aroma compounds of Godello white wines. *Food Chem.* **2012**, *131*, 826–836. [CrossRef]
17. Ruediger, G.A.; Pardon, K.H.; Sas, A.N.; Godden, P.W.; Pollnitz, A.P. Fate of Pesticides during the Winemaking Process in Relation to Malolactic Fermentation. *J. Agric. Food Chem.* **2005**, *53*, 3023–3026. [CrossRef] [PubMed]
18. Parker, M.; Capone, D.L.; Francis, I.L.; Herderich, M. Aroma Precursors in Grapes and Wine: Flavor Release during Wine Production and Consumption. *J. Agric. Food Chem.* **2017**, *66*, 2281–2286. [CrossRef] [PubMed]
19. Escuder-Gilabert, L.; Peris, M. Review: Highlights in recent applications of electronic tongues in food analysis. *Anal. Chim. Acta* **2010**, *665*, 15–25. [CrossRef] [PubMed]
20. Valente, N.I.P.; Rudnitskaya, A.; Oliveira, J.; Gaspar, E.M.; Gomes, M.T.S. Cheeses Made from Raw and Pasteurized Cow's Milk Analysed by an Electronic Nose and an Electronic Tongue. *Sensors* **2018**, *18*, 2415. [CrossRef] [PubMed]
21. Botelho, G.; Mendes-Faia, A.; Clímaco, M.C. Differences in Odor-Active Compounds of Trincadeira Wines Obtained from Five Different Clones. *J. Agric. Food Chem.* **2008**, *56*, 7393–7398. [CrossRef]
22. Li, H.; Jiang, D.; Liu, W.; Yang, Y.; Zhang, Y.; Jin, C.; Sun, S.Y. Comparison of fermentation behaviors and properties of raspberry wines by spontaneous and controlled alcoholic fermentations. *Food Res. Int.* **2020**, *128*, 108801. [CrossRef]
23. Yin, Y.; Zhao, Y. A feature selection strategy of E-nose data based on PCA coupled with Wilks Λ-statistic for discrimination of vinegar samples. *J. Food Meas. Charact.* **2019**, *13*, 2406–2416. [CrossRef]
24. Briz-Cid, N.; Castro-Sobrino, L.; Rial-Otero, R.; Cancho-Grande, B.; Simal-Gándara, J. Fungicide residues affect the sensory properties and flavonoid composition of red wine. *J. Food Compos. Anal.* **2018**, *66*, 185–192. [CrossRef]
25. Cynkar, W.; Dambergs, R.; Smith, P.; Cozzolino, D. Classification of Tempranillo wines according to geographic origin: Combination of mass spectrometry based electronic nose and chemometrics. *Anal. Chim. Acta* **2010**, *660*, 227–231. [CrossRef] [PubMed]
26. Peris, D.; Pérez-Torrado, R.; Hittinger, C.T.; Barrio, E.; Querol, A. On the origins and industrial applications of Saccharomyces cerevisiae × Saccharomyces kudriavzevii hybrids. *Yeast* **2018**, *35*, 51–69. [CrossRef] [PubMed]
27. Langdon, Q.K.; Peris, D.; Baker, E.P.; Opulente, D.A.; Nguyen, H.-V.; Bond, U.; Gonçalves, P.; Sampaio, J.P.; Libkind, D.; Hittinger, C.T. Fermentation innovation through complex hybridization of wild and domesticated yeasts. *Nat. Ecol. Evol.* **2019**, *3*, 1576–1586. [CrossRef]

28. Li, Y.; Jiang, J.H.; Chen, Z.P.; Xu, C.J.; Yu, R.Q. Robust linear discriminant analysis for chemical pattern recognition. *J. Chemom.* **1999**, *13*, 3–13. [CrossRef]
29. Bhattacherjee, A.K.; Singh, V.K. Uptake of soil applied paclobutrazol in mango cv. Dashehari and its persistence in soil, leaves and fruits. *Indian J. Plant Physiol.* **2015**, *20*, 39–43. [CrossRef]
30. Ahmed, W.; Tahir, F.M.; Rajwana, I.A.; Raza, S.A.; Asad, H.U. Comparative Evaluation of Plant Growth Regulators for Preventing Premature Fruit Drop and Improving Fruit Quality Parameters in 'Dusehri' Mango. *Int. J. Fruit Sci.* **2012**, *12*, 372–389. [CrossRef]
31. Ferreira, V.; Ortín, N.; Escudero, A.; López, R.; Cacho, J. Chemical Characterization of the Aroma of Grenache Rosé Wines: Aroma Extract Dilution Analysis, Quantitative Determination, and Sensory Reconstitution Studies. *J. Agric. Food Chem.* **2002**, *50*, 4048–4054. [CrossRef]
32. Torrens, J.; Urpí, P.; Riu-Aumatell, M.; Vichi, S.; López-Tamames, E.; Buxaderas, S. Different commercial yeast strains affecting the volatile and sensory profile of cava base wine. *Int. J. Food Microbiol.* **2008**, *124*, 48–57. [CrossRef]
33. Ferreira, V.; Lopez, R. The Actual and Potential Aroma of Winemaking Grapes. *Biomolecules* **2019**, *9*, 818. [CrossRef]
34. Lytra, G.; Tempère, S.; Marchand, S.; De Revel, G.; Barbe, J.-C. How do esters and dimethyl sulphide concentrations affect fruity aroma perception of red wine? Demonstration by dynamic sensory profile evaluation. *Food Chem.* **2016**, *194*, 196–200. [CrossRef]
35. Segurel, M.A.; Razungles, A.J.; Riou, C.; Salles, M.; Baumes, R.L. Contribution of Dimethyl Sulfide to the Aroma of Syrah and Grenache Noir Wines and Estimation of Its Potential in Grapes of These Varieties. *J. Agric. Food Chem.* **2004**, *52*, 7084–7093. [CrossRef] [PubMed]
36. Burdock, G.A. *Fenaroli's Handbook of Flavor Ingredients*, 6th ed.; CRC Press: Boca Raton, FL, USA, 2010.
37. Available online: http://www.flavornet.org/flavornet.html (accessed on 3 October 2020).
38. Lesschaeve, I.; Langlois, D.; Etiévant, P. Volatile Compounds in Strawberry Jam: Influence of Cooking on Volatiles. *J. Food Sci.* **1991**, *56*, 1393–1398. [CrossRef]
39. van Gemert, L.J. *Odour Thresholds: Compilations of Odour Threshold Values in Air, Water and Other Media*, 2nd ed.; Oliemans Punter & Co.: Zeist, The Netherlands, 2011.
40. Rapp, A.; Mandery, H. Wine aroma. *Cell. Mol. Life Sci.* **1986**, *42*, 873–884. [CrossRef]
41. Oliva, J.; Zalacain, A.; Payá, P.; Salinas, M.R.; Barba, A. Effect of the use of recent commercial fungicides [under good and critical agricultural practices] on the aroma composition of Monastrell red wines. *Anal. Chim. Acta* **2008**, *617*, 107–118. [CrossRef]
42. Oliva, J.; Navarro, S.; Barba, A.; Navarro, G.; Salinas, M.R. Effect of pesticide residues on the aromatic composition of red wines. *J. Agric. Food Chem.* **1999**, *47*, 2830–2836. [CrossRef] [PubMed]
43. Noguerol-Pato, R.; González-Rodríguez, R.; González-Barreiro, C.; Cancho-Grande, B.; Simal-Gándara, J. Influence of tebuconazole residues on the aroma composition of Mencía red wines. *Food Chem.* **2011**, *124*, 1525–1532. [CrossRef]
44. Oliva, J.; Martínez-Gil, A.M.; Lorenzo, C.; Cámara, M.; Salinas, M.R.; Barba, A.; Garde-Cerdán, T. Influence of the use of fungicides on the volatile composition of Monastrell red wines obtained from inoculated fermentation. *Food Chem.* **2015**, *170*, 401–406. [CrossRef] [PubMed]
45. Malcorps, P.; Cheval, J.M.; Jamil, S.; Dufour, J.P. A New Model for the Regulation of Ester Synthesis by Alcohol Acetyltransferase in *Saccharomyces cerevisiae* during Fermentation. *J. Am. Soc. Brew. Chem.* **1991**, *49*, 47–53. [CrossRef]
46. Garcia, M.A.; Oliva, J.; Barba, A.; Cámara, M.Á.; Pardo, F.; Díaz-Plaza, E.M. Effect of Fungicide Residues on the Aromatic Composition of White Wine Inoculated with Three *Saccharomyces cerevisiae* Strains. *J. Agric. Food Chem.* **2004**, *52*, 1241–1247. [CrossRef] [PubMed]
47. Perestrelo, R.M.D.S.; Fernandes, A.; Albuquerque, F.F.; Marques, J.C.; Câmara, J.S. Analytical characterization of the aroma of Tinta Negra Mole red wine: Identification of the main odorants compounds. *Anal. Chim. Acta* **2006**, *563*, 154–164. [CrossRef]

48. Kong, Z.; Li, M.; An, J.; Chen, J.; Bao, Y.; Francis, F.; Dai, X. The fungicide triadimefon affects beer flavor and composition by influencing *Saccharomyces cerevisiae* metabolism. *Sci. Rep.* **2016**, *6*, 33552. [CrossRef] [PubMed]
49. Munayyer, H.K.; Mann, P.A.; Chau, A.S.; Yarosh-Tomaine, T.; Greene, J.R.; Hare, R.S.; Heimark, L.; Palermo, R.E.; Loebenberg, D.; McNicholas, P.M. Posaconazole Is a Potent Inhibitor of Sterol 14α-Demethylation in Yeasts and Molds. *Antimicrob. Agents Chemother.* **2004**, *48*, 3690–3696. [CrossRef] [PubMed]

Sample Availability: Samples of the compound cyclohexanone are available from the authors.

Publisher's Note: MDPI stays neutral with regard to jurisdictional claims in published maps and institutional affiliations.

© 2020 by the authors. Licensee MDPI, Basel, Switzerland. This article is an open access article distributed under the terms and conditions of the Creative Commons Attribution (CC BY) license (http://creativecommons.org/licenses/by/4.0/).

Article

Commercial Mannoproteins Improve the Mouthfeel and Colour of Wines Obtained by Excessive Tannin Extraction

Alessandra Rinaldi [1,2,*], Alliette Gonzalez [1], Luigi Moio [1] and Angelita Gambuti [1]

1. Dipartimento di Agraria, Sezione di Scienze della Vigna e del Vino, Università degli Studi di Napoli Federico II, Viale Italia, Angolo Via Perrottelli, 83100 Avellino, Italy; a.gonzalessifuentes@studenti.unina.it (A.G.); moio@unina.it (L.M.); angelita.gambuti@unina.it (A.G.)
2. Biolaffort, 126 Quai de la Souys, 33100 Bordeaux, France
* Correspondence: alessandra.rinaldi@unina.it

Abstract: In the production of red wines, the pressing of marcs and extended maceration techniques can increase the extraction of phenolic compounds, often imparting high bitterness and astringency to finished wines. Among various oenological products, mannoproteins have been shown to improve the mouthfeel of red wines. In this work, extended maceration (E), marc-pressed (P), and free-run (F) Sangiovese wines were aged for six months in contact with three different commercial mannoprotein-rich yeast extracts (MP, MS, and MF) at a concentration of 20 g/hL. Phenolic compounds were measured in treated and control wines, and sensory characteristics related to the astringency, aroma, and colour of the wines were studied. A multivariate analysis revealed that mannoproteins had a different effect depending on the anthocyanin/tannin (A/T) ratio of the wine. When tannins are strongly present (extended maceration wines with A/T = 0.2), the MP conferred mouthcoating and soft and velvety sensations, as well as colour stability to the wine. At A/T = 0.3, as in marc-pressed wines, both MF and MP improved the mouthfeel and colour of Sangiovese. However, in free-run wine, where the A/T ratio is 0.5, the formation of polymeric pigments was allowed by all treatments and correlated with silk, velvet, and mouthcoat subqualities. A decrease in bitterness was also obtained. Commercial mannoproteins may represent a way to improve the mouthfeel and colour of very tannic wines.

Keywords: mannoproteins; mouthfeel; astringency; subquality; colour; pressing; extended maceration; Sangiovese

1. Introduction

The key step in the production of red wine is the maceration of the solid parts of the berries during fermentation. In this step, important phenomena occur in which the phenolic compounds of the grapes are involved: the release of part of them from the skins and seeds into the must, reactions among themselves and with other metabolites of fermentation, and the absorption on grape pomaces and yeast lees [1]. The phenolic compounds constitute a wide group of compounds, and, among them, the most important in winemaking are the anthocyanins extractable from the skins and the proanthocyanidins (namely condensed tannins) extractable from the skins and seeds [2]. Grape maceration is a critical point in red wine production, as an excessive extraction of tannins and/or low extraction of anthocyanins or loss of part of them during the process can determine an unbalanced ratio between these classes of phenolic compounds. This may cause defects such as astringency and bitterness, which reduce the commercial value of wines [3]. Apart from the specific varietal composition in anthocyanins and tannins of the berry, numerous factors can modulate the extraction of these two important groups of compounds during the initial stages of winemaking [4,5]. One of the technological practices that often determine an excessive extraction of tannins is the prolonged maceration after the end of alcoholic fermentation [6]. This practice is usually applied to obtain wines richer in phenolic compounds and with a longer shelf-life, but sometimes, these wines are too rich in phenolic compounds responsible for bitterness and astringency [7]. Fining treatments with high doses of animal and vegetable proteins are necessary to diminish the content of flavanols and proanthocyanidins and decrease the undesired mouthfeel sensations elicited by these compounds [8,9]. Fining practices, on the other hand, can impoverish the aroma of wines, so the commercial value of these products may be low anyway. Unbalanced red wines are also produced by the marc-pressing of wines shortly after the end of maceration. Usually, the free-run juice is used to produce higher quality wines that are richer in compounds easily extracted from the grape skin and seeds such as anthocyanins and lower molecular weight tannins, which are characterised by more pleasant mouthfeel sensations [10]. The corresponding marc-pressed wines are lower quality wines, because they are richer in bitter and astringent compounds, such as flavanols and proanthocyanidins, which are extracted from the skin and seeds during the pressing of the marcs [11]. However, sometimes, these marc-pressed wines can be rich in aromatic compounds, and with appropriate treatment, they could have a higher commercial value [12,13].

In addition to fining agents capable of precipitating phenolic compounds such as albumin, gelatin, and some vegetable proteins, which can be used to improve the mouthfeel properties of wines too rich in tannins, mannoproteins can also be used to improve the in-mouth characteristics of red wines [14].

Mannoproteins represent major polysaccharides found in wine, because they are released from the yeast cell wall during alcoholic fermentation and wine ageing [15,16]. In recent decades, several commercial mannoproteins have been added to wine, because they confer favourable oenological properties, such as the decrease of astringency, the improvement of mouthfeel sensation [14,17], the increase of colour [17], and protein and tartrate stability [18].

Mannoproteins or yeast products rich in mannoproteins are used for various types of wines, such as still white wine [19] and red wine [17], as well as white and rosé sparkling wines [20]. However, they are rarely used to treat wines that are very rich in astringent and bitter tannins, as those obtained by excessive extractive procedures like prolonged maceration and marc-pressing.

The effectiveness of commercial mannoproteins on protein stabilisation, phenolic compounds, and the chromatic and sensory properties of wine depend on the structural characteristics of mannoproteins [19]. Recently, Manjon et al. [21] also showed that the formation of salivary protein–mannoprotein systems, mainly involved in altering astringency sensations, depends on the structural characteristics and hydrophobicity of mannoproteins. Therefore, the possible use of commercial mannoproteins to modulate the astringency and bitterness attributes of tannin-rich red wines should consider different preparations. On the other hand, the contact time between mannoproteins and red wine is important to reach the colloidal state able to induce a significant variation in the perceived sensations [17]. In this study, three commercial mannoproteins were tested to remediate the excessive astringency and bitterness of red wines produced by prolonged maceration and marc-pressing, considering the contemporary effect on chromatic wine characteristics and aroma compounds.

2. Results

2.1. The Content in BSA-Reactive Tannins and Vanillin-Reactive Flavans

The BSA-reactive tannins and vanillin-reactive flavans in extended maceration (E), marc-pressed (P), and free-run (F) Sangiovese wines were measured before (t0) and after the ageing on mannoproteins for six months.

In Figure 1a, the prolonged contact between grape solids and wine during the extended maceration process extracted around 1500 mg/L of BSA-reactive tannins (E-t0), compared to 600 mg/L of marc-pressed wine (P-t0) and 500 mg/L of free-run (F-t0) wine. After six months, the control wines E-C, P-C, and F-C showed a significantly lower concentrations of proanthocyanidins with respect to t0. In E wines, the treatments with MP and MS induced a greater reduction in BSA-reactive tannins; in P wines, only MF was efficient in reducing these compounds; in F wines, there were no differences between the control and treated wines, except for MS, which showed a higher tannin concentration than the control but a lower one than before ageing.

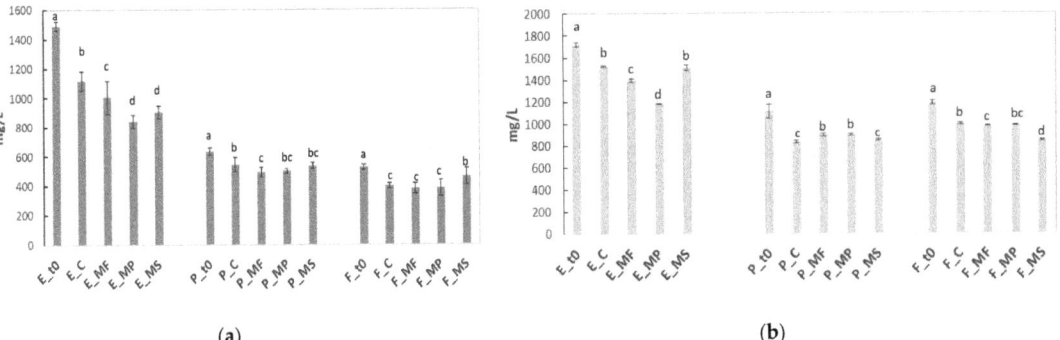

Figure 1. BSA-reactive tannins (a) and vanillin-reactive flavans (b) after six months of ageing on mannoproteins at 20 g/hL in extended maceration (E), marc-pressed (P), and free-run (F) Sangiovese wines. C represents the control wine; MF, MP, and MS are the mannoprotein treatments. t0 represents the wine before ageing. According to Fisher's LSD analysis, histograms with different letters indicate significant differences for each wine typology ($p < 0.05$).

Extended maceration also resulted in a high extraction of vanillin-reactive flavans (Figure 1b). The content in E-t0 was around 1700 mg/L compared to 1200 mg/L in P-t0 and F-t0 wines. The vanillin-reactive flavans in the control wines decreased concerning the t0 in all the wines after six months. The decrease was also observed after the treatments with MF and MP in E wines and MF and MS in F wines, with MS being the most effective. In contrast, in the P-MF and P-MP wines, the contents of flavans were higher than in the control, probably due to reduced precipitation over time.

2.2. The Effect of Mannoproteins on the Colour of Wines

We evaluated the impact of mannoproteins after ageing on colour by measuring the colour intensity; hue; total anthocyanins; polymeric pigments content; and CIELab coordinates (L^*, a^*, b^*, and ΔE), as shown in Table 1.

Table 1. The colour parameters of Sangiovese wines (E = extended maceration, P = marc-pressed, and F = free-run) aged on mannoproteins MF, MP, and MS for six months.

Wine Typology	Samples	Total Anthocyanins mg/L	Colour Intensity (420 + 520 + 620) a.u. [†]	Hue	Polymeric Pigments (520) a.u. [†]	L^*	a^*	b^*	ΔE
Extended Maceration (E)	E-C	256.99 ± 0.84 a	13.70 ± 0.02 a	0.75 ± 0.00 b	5.87 ± 0.07 c	70.2 ± 0.2 c	37.6 ± 0.3 a	19.7 ± 0.3 a	-
	E-MF	252.13 ± 1.82 b	13.05 ± 0.02 c	0.75 ± 0.00 b	6.06 ± 0.09 b	71.5 ± 0.1 b	36.6 ± 0.0 b	19.3 ± 0.5 a	1.79 b
	E-MP	199.63 ± 0.44 d	11.98 ± 0.06 d	0.76 ± 0.01 a	6.08 ± 0.02 b	73.6 ± 0.1 a	34.1 ± 0.3 c	17.5 ± 0.6 b	5.34 a
	E-MS	228.97 ± 0.51 c	13.21 ± 0.05 b	0.75 ± 0.01 b	6.28 ± 0.04 a	71.4 ± 0.4 b	37.1 ± 0.4 ab	19.2 ± 0.3 a	1.48 b
Marc-Pressed (P)	P-C	174.51 ± 2.86 c	11.06 ± 0.47 ab	0.72 ± 0.01 b	3.77 ± 0.01 b	68.2 ± 0.4 b	28.8 ± 0.1 a	11.2 ± 0.2 a	-
	P-MF	175.74 ± 0.28 bc	10.76 ± 0.23 b	0.72 ± 0.00 b	4.02 ± 0.01 a	66.7 ± 0.5 c	29.2 ± 0.2 a	11.2 ± 0.4 a	1.58 b
	P-MP	178.92 ± 2.30 b	10.57 ± 0.16 b	0.73 ± 0.00 b	3.97 ± 0.09 ab	70.2 ± 0.4 a	26.8 ± 0.5 b	10.4 ± 0.3 b	2.98 a
	P-MS	192.03 ± 1.57 a	11.46 ± 0.13 a	0.75 ± 0.01 a	3.89 ± 0.20 ab	66.7 ± 0.2 c	29.0 ± 0.2 a	10.9 ± 0.5 a	1.58 b
Free-Run (F)	F-C	223.97 ± 4.98	12.63 ± 0.01 b	0.68 ± 0.00 b	4.42 ± 0.06 d	62.6 ± 0.3 b	49.3 ± 0.1 b	25.8 ± 0.1 c	-
	F-MF	219.71 ± 19.70	13.10 ± 0.26 a	0.70 ± 0.01 a	4.70 ± 0.05 b	62.8 ± 0.0 ab	49.7 ± 0.0 a	25.7 ± 0.1 c	0.52 b
	F-MP	207.78 ± 3.26	11.86 ± 0.01 c	0.69 ± 0.00 ab	4.54 ± 0.01 c	63.1 ± 0.0 a	49.4 ± 0.0 b	26.6 ± 0.1 b	0.91 b
	F-MS	218.82 ± 6.96	12.88 ± 0.07 ab	0.70 ± 0.00 a	4.80 ± 0.06 a	61.9 ± 0.2 c	48.1 ± 0.2 c	27.4 ± 0.2 a	2.11 a

[†] a.u. = absorbance unit. According to Fisher's LSD analysis, values ± standard deviation (SD) with different letters indicate significant differences for each wine typology ($p < 0.05$).

In E wines, the total anthocyanin content decreased after ageing on mannoproteins, mainly due to the MP treatment. Colour intensity and redness (a*) were also reduced in E-MP wine. However, a higher amount of polymeric pigments was formed, indicating that wine still had a red colour with violet hues, as shown by the b* coordinate (yellow–blue). The lightness (L*) increased in all treatments, indicating a more vivid colour. The ΔE represents the difference in colour between the control and treated wines, and the value 5.34 of E-MP showed that this wine had a different colour, easily detectable by the human eye. In Table 1, the decrease of total anthocyanins in P-MS and P-MP was lower than in P-C. Moreover, the P-MS showed a higher colour intensity and hue. Polymeric pigments were mainly formed in P-MF, although no differences were observed with the control in a* and b*. The P-MP was the wine with a colour difference (ΔE = 2.98) detectable to untrained eyes, probably due to the lower lightness and higher blue nuances than the other wines. A different effect of mannoproteins can be observed in free-run wines on the colour parameters. The total anthocyanins did not differ significantly in free-run wines (F), while the colour intensity was higher in F-MF and lower in F-MP than in F-C. A slight increase in hue was observed in F-MS and F-MF. Yet, the polymeric pigments were significantly increased after the treatment with all mannoproteins in free-run wines. In F-MF, a high redness was also observed. However, no evident colour differences were denoted in the treated wines.

2.3. The Effect of Mannoproteins on the Mouthfeel of Wines

After six months of ageing on mannoproteins, the mouthfeel profile of Sangiovese wines was evaluated using 16 attributes of astringency (Supplementary Table S1), which were analysed by the CATA analysis. The significant terms ($p < 0.01$) were plotted for each wine typology in Figure 2. Figure 2a shows the subqualities of the wines obtained by extended maceration (E-C) and treated with mannoproteins (E-MS, E-MP, and E-MF).

The explained inertia was 99.52% by the first two coordinates and permitted the separation of the samples according to their astringency attributes. E-C was characterised principally by dry, hard, green terms, i.e., high astringency felt with bitterness and acidity. The primary sensation of E-MS was corduroy, a feeling of a slight wrinkling of the soft palate that can be felt by tongue movements. E-MF was instead very similar to the other wines and did not differ from either the control or treated wines. E-MP wine resulted in soft and mouthcoating sensations, indicating that the MP represented the most suitable treatment to improve the mouthfeel of extended maceration wine. E-MP was also the wine with the lowest content of BSA-reactive tannins and vanillin-reactive flavans.

Figure 2b showed the CATA plot of the marc-pressed wines (P) after six months of ageing on the MS, MF, and MP mannoproteins. The first and second dimensions explained 98.62% of the total inertia and allowed a clear separation between the P-C and P-MS from the P-MP and P-MF. Green, dry, adhesive, and aggressive sensations characterised the control P-C, and the P-MS differed from the latter for the corduroy and pucker terms. In contrast, the treatment with MF and MP mannoproteins conferred positive subqualities to the wines: velvet, soft, full-body, and persistent. For the marc-pressed wines, the most evident effect on mouthfeel was similarly obtained with the MP and MF mannoproteins, although the BSA-reactive tannins and vanillin-reactive flavans did not show noticeable variations after these treatments.

Figure 2c showed the mouthfeel profile of Sangiovese free-run wines after six months of ageing on mannoproteins using the CATA analysis. The corduroy term highly characterised F-C. Even if each wine was different from the control, the treated wines were similar in their mouthfeel profiles. In particular, F-MF was persistent, indicating that the overall sensation associated with the aftertaste lasted long in the mouth. F-MP was principally velvety and mouthcoating, while F-MS was perceived as full-bodied.

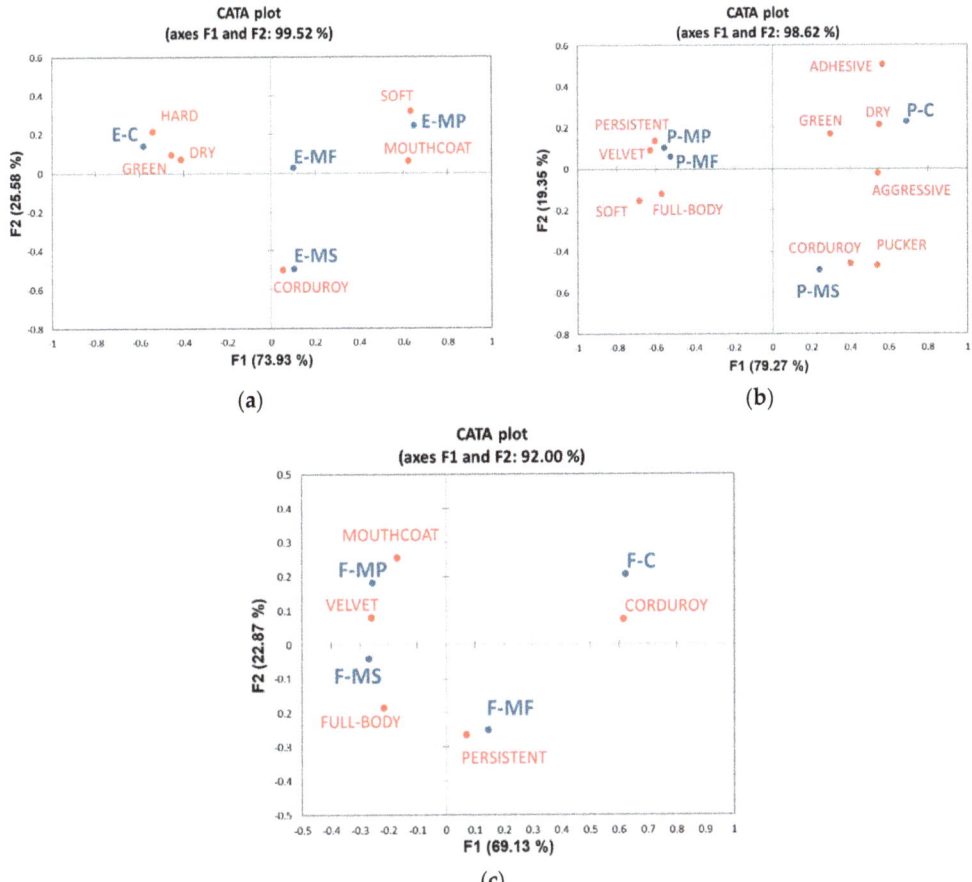

Figure 2. The mouthfeel profile of Sangiovese wines after six months of ageing with mannoproteins by the CATA analysis. The significant terms ($p < 0.01$) were plotted for each wine typology: (**a**) extended maceration (E), (**b**) marc-pressed (P), and (**c**) free-run (F) wines. C represents the control wine; MF, MP, and MS are the mannoprotein treatments at 20 g/hL.

2.4. Relationships between Subqualities, Colour Parameters, and Phenolic Content of Wines

It is essential to consider how the colour evolves together with the variation in mouthfeel during ageing with mannoproteins. For this reason, relationships between different variables such as the astringency subqualities (silk, velvet, dry, corduroy, adhesive, aggressive, hard, soft, mouthcoat, rich, full-body, green, grainy, satin, pucker, and persistent); colour parameters (colour intensity, hue, a*, b*, L*, and polymeric pigments); and phenolic content of the wine (total anthocyanins, BSA-reactive tannins, and vanillin-reactive flavans) were studied. A Multiple Factor Analysis (MFA) was carried out on each wine typology to characterise and find relationships between the variables and factors, as shown in Table 2.

For extended maceration wines (E), the first two dimensions (F1 and F2) of the MFA accounted for 93.4% of the variance of the experimental data, representing 75.9% and 17.5% of the variance, respectively. The F1 and F2 of the MFA for marc-pressed wines (P) accumulated 61.2% and 35% and, for free-run wines (F), 50.8% and 34.2%, totalling 96.3% and 85% of the initial variability, respectively. The eigenvalue of the first dimension indicated that it could be considered a significant direction in explaining the dispersion of analytical parameters (colour and phenolic content) and the frequency table of the CATA terms (astringency subqualities), being E > P > F. The first dimension of the MFA of E was positively correlated with lightness (L*) and hue and with the subqualities velvet, soft, mouthcoat, satin, and persistent. The factor score related to the treatment MP was positively loaded on F1, indicating that the wine showed positive subqualities, a more vivid colour, and a higher hue than other wines. On the same factor F1, the control wine was negatively projected, characterised by dry, adhesive, hard, aggressive, green, grainy, and pucker terms, which are correlated with the content BSA-reactive tannins, total anthocyanins, flavans, colour intensity, redness (a*), and yellowness–blueness direction (b*). The colour and the phenolic content similarly contributed to F1 (33.7%). The factor F2 was formed by the contribution of the variables subquality and colour by 55.2% and 30.1%. The highest factor score of MS was loaded on F2, indicating an association with polymeric pigments (of the variable colour) and silk, full-body, and corduroy (of the variable subqualities). This suggests that the MS treatment can impart colour stability and interesting mouthfeel characteristics to extended maceration wine. MF, on the other hand, was associated with F3 and aromatic richness (rich); however, it did not differ from the other wines.

Table 2. Relationships from the Multiple Factor Analysis between the variables: astringency subqualities (silk, velvet, dry, corduroy, adhesive, aggressive, hard, soft, mouthcoat, rich, full-body, green, grainy, satin, pucker, and persistent); colour parameters (colour intensity, hue, a*, b*, L*, and polymeric pigments); and phenolic content of the wine (total anthocyanins, BSA-reactive tannins, and vanillin-reactive flavans) and factors: C = control and mannoprotein treatments = MF, MP, and MS for extended maceration (E), marc-pressed (P), and free-run (F) wines.

	Extended Maceration Wines (E)			Marc-Pressed Wines (P)			Free-Run Wines (F)		
	F1	F2	F3	F1	F2	F3	F1	F2	F3
Eigenvalue	2.9	0.7	0.3	2.5	1.4	0.2	2.2	1.5	0.7
Variability (%)	75.9	17.5	6.6	61.2	35.0	3.7	50.8	34.2	15.0
Cumulative %	75.9	93.4	100.0	61.2	96.3	100.0	50.8	85.0	100.0
Correlations									
Subquality									
Silk	0.270	0.928	0.258	−0.838	0.354	0.416	0.132	−0.828	0.545
Velvet	0.984	−0.178	−0.010	−0.962	0.220	0.161	0.493	−0.868	0.054
Dry	−0.860	−0.471	−0.197	0.710	−0.703	0.051	0.101	0.623	−0.776
Corduroy	−0.064	0.834	−0.548	0.942	0.330	−0.058	−0.713	0.695	0.092
Adhesive	−0.942	−0.136	−0.308	0.464	−0.883	0.077	0.525	0.087	−0.846
Hard	−0.753	−0.600	−0.269	0.772	−0.495	0.398	−0.367	0.819	−0.441
Aggressive	−0.996	−0.088	0.013	0.867	−0.444	−0.226	−0.140	0.990	−0.033
Soft	0.917	−0.398	0.009	−0.823	0.551	−0.138	0.795	−0.587	−0.153
Mouthcoat	0.991	−0.097	−0.097	−0.869	0.491	0.060	0.143	−0.919	−0.368
Rich	0.439	0.389	0.810	−0.280	0.785	0.552	0.959	−0.192	−0.206
Green	−0.847	−0.488	−0.210	0.626	−0.776	0.076	−0.156	0.456	−0.876
Grainy	−0.948	0.102	0.302	0.765	−0.428	−0.482	−0.686	0.727	−0.038
Satin	0.970	−0.245	0.001	−0.393	0.795	−0.462	−0.994	−0.055	0.096
Pucker	−0.977	−0.207	−0.055	0.976	0.211	−0.052	−0.450	0.879	−0.157
Full-body	0.365	0.812	0.455	−0.826	0.506	−0.250	0.679	−0.342	0.650
Persistent	0.842	0.162	0.514	−0.993	0.098	−0.058	−0.341	−0.314	0.886
Colour parameter									
L*	0.997	−0.072	0.028	−0.776	−0.522	0.353	−0.853	−0.507	−0.122
a*	−0.957	0.287	−0.051	−0.752	−0.647	−0.127	−0.959	−0.191	0.210
b*	−0.961	0.232	0.147	0.378	0.846	0.376	0.941	−0.270	−0.202
Hue	0.935	−0.330	0.128	0.304	0.952	−0.046	0.971	−0.134	0.199
Colour Intensity	−0.988	0.146	−0.060	0.899	0.348	−0.267	0.177	0.670	0.721
Polymeric pigments	0.486	0.862	−0.143	−0.854	0.483	−0.193	0.714	−0.118	0.691
Phenolic content									
Total Anthocyanins	−0.930	0.002	0.368	0.454	0.889	0.066	−0.031	0.979	0.204
BSA-reactive tannins	−0.923	−0.342	0.176	0.976	−0.179	0.121	0.952	0.296	−0.076
Vanillin-reactive flavans	−0.908	0.373	−0.192	−0.962	0.269	0.038	−0.913	−0.370	0.169
Factor scores									
C	−2.006	−0.729	−0.215	1.486	−1.660	0.038	−1.106	1.611	−0.836
MF	−0.106	0.284	0.897	−1.641	−0.080	−0.523	−1.087	0.048	1.260
MP	2.768	−0.702	−0.169	−1.467	0.242	0.573	−0.469	−1.882	−0.615
MS	0.016	1.248	−0.429	1.769	1.763	−0.073	2.465	0.317	0.097
Contributions (%)									
Subquality	32.6	55.2	63.6	33.0	26.7	33.4	20.3	49.6	53.3
Colour parameter	33.7	30.1	5.2	27.6	43.7	60.3	43.2	13.5	41.3
Phenolic content	33.7	14.7	31.2	39.4	29.5	6.3	36.5	36.8	5.4

For the marc-pressed wines (P), the first dimension of the MFA contrasted the terms such as aggressive, pucker, and corduroy (positively projected on F1) with the subquality descriptors silk, velvet, soft, mouthcoat, full-body, and persistent (negatively projected on F1). Similarly, the BSA-reactive tannins and colour intensity were opposed to flavans and polymeric pigments. The MP and MS were equally correlated to F1, as their factor scores were −1.47 and −1.64, and highly differed from MS (1.2). The phenolic content and subquality variables contributed by 39.4% and 33% to the observations on F1. F2 correlated positively with the total anthocyanins, hue, and b* and negatively with the control, whose factor score was −1.66. The control wine was then associated with dry, adhesive, and green subqualities, which were mainly correlated with the factor F2.

In free-run wines, 50.8% of the experimental data variability was explained by F1, followed by 34.2% by F2 and 15% by F3 (Table 2). On F1, the BSA-reactive tannins, hue, polymeric pigments, and b* were correlated with the soft, rich, and full-body subqualities and the MS-treated wine. The colour parameters contributed mainly to this factor (43.2%). Conversely, F2 correlated with the total anthocyanin, colour intensity, aggressive, grainy, and pucker terms. The control was characterised by these variables, as shown by its factor score (−1.66). Opposite to the second factor, the MP correlated with the silk, velvet, and mouthcoat subqualities. Finally, MF was loaded on F3 and was mainly represented by subquality variables (53.3%), resulting in silk, persistent instead of dry, adhesive, green, and aggressive. Polymeric pigments also characterised this wine, positively correlated with F3 (0.691), although to a lesser extent than F1 (0.714).

2.5. The Effect of Mannoproteins on Aroma and Odour Descriptors

In addition to mouthfeel, mannoproteins influenced the aroma and odour of the wines differently depending on the wine typology. In Figure 3a, the main effect of mannoproteins on taste was exerted in a wine with high flavan and tannin content (E).

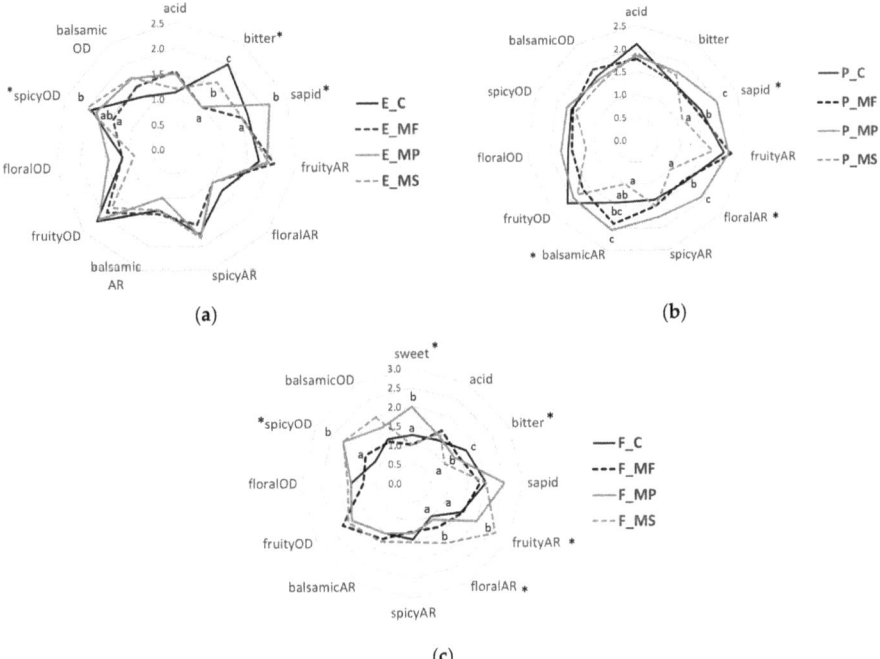

Figure 3. The spider charts representing the tastes, aromas (AR), and odours (OD) of (**a**) extended maceration (E), (**b**) marc-pressed (P), and (**c**) free-run (F) Sangiovese wines after six months of ageing. C represents the control wine. Mannoprotein treatments: MF, MP, and MS at 20 g/hL. The asterisk (*) indicates a statistical difference between wines with different letters according to Duncan's test ($p < 0.05$).

The MP and MF were able to reduce the bitterness of extended macerated wine. Additionally, MP contributed to an increase in sapidity. The spicy odour was instead reduced in E-MF. For the marc-pressed wines (Figure 3b), a significant improvement was obtained by the MP mannoprotein. The P-MP revealed an increased floral and balsamic aroma and a high sapidity compared to other wines. In free-run wines (Figure 3c), mannoproteins modulated the flavour, increasing the fruity and floral aromas (F-MS), as well as the spicy odour (F-MS and F-MP). The three mannoproteins determined a reduction in bitterness, MS being the most effective. However, an increase in the sweetness perception was felt only in F-MP wine.

3. Discussion

Mannoproteins represent a natural oenological product that aims to improve the sensory characteristics of red wine, such as mouthfeel, astringency, bitterness, and colour. Several works have shown that treatments with mannoproteins increase the perception of sweetness and roundness sensation, body, persistence, aroma intensity, and odour complexity and reduces the astringency, bitterness, and aggressive green tannins [19,22–24]. In this work, we used Sangiovese wines with high tannin contents obtained by different winemaking processes, such as extended maceration and marc-pressing. After six months of ageing, significant differences were observed, with mannoproteins showing distinct behaviours according to wine typology. After this contact period, the MP precipitated vanillin-reactive flavans and BSA-reactive tannins, probably due to its high content in peptides. Peptides, having a high affinity towards high and low molecular weight proanthocyanidins, induce them to precipitate, reducing the final concentration in the treated wine. This result is more evident when the wine is richer in these compounds (E > P > F). A decrease in bitterness and an increase in sapidity in E-MP was also detected and could be due to the masking effect of the sapid peptide, which is a part of the formulation of MP. This peptide (Hsp12p), belonging to the heat shock proteins family, exhibited a sweet taste [25], thus conferring sapidity and reducing wine bitterness [16]. Furthermore, the treatment with MP seemed the most suitable for extended maceration wines, as it improved the wine's mouthfeel by granting mouthcoating and soft and velvety sensations. This result is in accordance with previous work on Sangiovese, a wine rich in tannins and flavans, in which the MP was able to enhance the body, structure, and roundness of the wine [16].

From the multivariate analysis, the content of phenolic compounds in the treated wines was found to be differently correlated with the astringency subqualities and colour parameters according to the wine typology and mannoprotein treatment. In particular, it has been shown that the ratio of anthocyanins/tannins (A/T) in wines affects the formation of polymeric pigments during ageing [26]. When the wine has a high phenolic content (total anthocyanins, BSA-reactive tannins, and vanillin-reactive flavans) and the tannins are in excess with respect to anthocyanins (A/T = 0.2), as in the case of extended maceration wine, the decrease of phenolic compounds observed after six months of ageing with MP leads to a decreased astringency, felt as a dryness, hardness and unripeness. It equally resulted in the development of positive subqualities (velvet, soft, mouthcoat, satin, and persistent). Previous studies also showed that the addition of mannoproteins significantly modifies the mouthfeel and structural properties of red wines, leading to a reduction in astringency [22,27]. The decrease in phenolic content showed a more significant influence on the subqualities than on the formation of polymeric pigments. However, the colour stability of extended maceration wines was promoted by mannoproteins, which may allow multiple interactions between proanthocyanidins and anthocyanins, as observed by others [24,27].

In marc-pressed wines, where the content of proanthocyanidins was also in excess compared to anthocyanins (A/T ratio = 0.3), the high tannin content again correlated with negative astringency subqualities (dry, aggressive, hard, and pucker), as also reported for pressed wine fractions [11]. The higher the decrease in tannins detected after ageing in contact with MP, the more positive subqualities such as velvet, soft, mouthcoat, and full-body were obtained by the applied treatments (MP and MF). Moreover, the improvement in mouthfeel was correlated with polymeric pigment formation. Regarding the latter compounds, the sensory perception of the polymeric pigments such as velvety and mouth-coating was also observed during ageing [28]. Although the MF and MP mannoproteins similarly affected the wines, they differed more in their effects on colour and aroma than on mouthfeel. Specially, MF favoured more the formation or a smaller loss of polymeric pigments during ageing and then colour stability. In contrast, MP influenced the aroma revelation (more floral and balsamic) and sapidity.

In the free-run wines, the total anthocyanins accounted for half of the BSA-reactive tannin content and A/T = 0.5. Unlike the other wine typologies (P and E wines showed A/T < 0.5), the tannins were correlated with positive subqualities. The mannoproteins probably had some protective effects on the precipitation and depolymerisation of tannins when more anthocyanins were present, promoting the formation of stable macrostructures, which are less reactive towards salivary proteins and less astringent [29]. It is likely that BSA-reactive tannins remain in solution because: (i) the complexes between polymeric tannins and anthocyanins are stable [30], and (ii) mannoproteins contribute to the further stabilization of the complexes [31]. These hypotheses can be supported by the fact that the formation of polymeric pigments resistant to the action of SO_2 was observed after all mannoprotein treatments. Concurrently, an improvement in mouthfeel was observed with MS > MF > MP. A reduction in bitterness was also observed in treated wines. An effect of polysaccharides on the bitterness was also previously reported [17,23]. From the MFA, a significant correlation between the positive subqualities, polymeric pigments formation, and decrease of flavans was found. This means that condensed flavans, when A/T = 0.5, are principally involved with mannoproteins in the formation of polymeric pigments, which are also characterised by an improved mouthfeel. Alcalde-Eon et al. [32] already proposed an additional mechanism that can explain these data, in which the steric hindrance caused by mannoproteins can protect the flavanol from precipitation and stabilise the interaction with anthocyanins. Ultimately, mannoproteins can improve the aroma of wine [17,33,34], because free-run MS-treated wine was perceived as more floral, fruity, and spicy than the control. Mechanisms involving orthonasal perceptions could explain this result.

4. Materials and Methods

4.1. Wine Samples

Sangiovese wines were industrially produced in a winery located in the Chianti DOCG area (Toscana, Italy) during the 2016 vintage. Vinification was based on the following protocol: grapes (18 tons) were destemmed and crushed, the resulting must treated with potassium metabisulfite (40 mg/kg) and inoculated with 20 g/hL of yeast (F83 Laffort, Bordeaux, France); the fermentation/maceration lasted 12 days at 25 °C, during which yeast-assimilable nitrogen (YAN), in the form of diammonium phosphate (containing ≈0.12% of thiamine hydrochloride), was added with the inoculum and then again on the third and sixth days of fermentation, to a total concentration of 30 g/hL. The wine was then separated into three fractions: (i) extended maceration wine (E), which prolonged the skin contact for an additional 15 days, (ii) a devatted fraction of free-run wine (F), and (iii) a pressed fraction at 1.5 bar of marc-pressed wine (P). After completing the skin contact, E wine was pressed and, similarly to F and P, was transferred to 53-L carboys. After the addition of pectolytic enzymes (3 g/hL), the wines were inoculated with lactic bacteria (LF16 Direct, Laffort, Bordeaux, France) at 1 g/hL. Potassium metabisulfite (6 g/hL) was then added to the wines conserved under N_2 in stainless-steel tanks (15 L) before commencing the experiments in October 2017.

4.2. Yeast Mannoprotein Products

Mannoproteins MF, MS, and MP were supplied by Laffort (Bordeaux, France). According to the manufacturer, MF is a specific yeast cell wall mannoprotein from *Saccharomyces cerevisiae* used for the colloidal stabilisation of wine and to improve the mouthfeel. MS is a specific mannoprotein (MP40—Patent 2726284) naturally present in wines and used to inhibit potassium bitartrate crystallisation. MP is a yeast cell wall extract composed of mannoproteins rich in a sweet peptide fraction (Patent EP 1850682) and vegetal polysaccharides (gum arabic). These products are qualified for use in oenology and comply with Regulation (EC) n° 606/2009.

Before ageing (t0), the E, P, and F wines were treated with MF, MP, and MS at 20 g/hL in duplicate. We used the concentration of 20 g/hL as the average dose recommended by the manufacturer (10–30 g/hL) to compare the products at the same concentration. The base parameters of the wines at t0 are shown in Supplementary Table S2. The control wine (C) was not treated. Two independent bottles (750 mL) were considered for each treatment and were stored in a cellar for six months. After this period, the wines were filtered under vacuum with Whatman® glass microfiber filters (64 g/m^2) (GE Healthcare, Chicago, IL, USA) before analysis.

4.3. Wine Sensory Evaluation

Sangiovese wines were evaluated in duplicate by 13 trained assessors (comprising five women between the ages of 35–50 and eight men between the ages of 25–44 years), as previously described [9,28]. Two tasting evaluations of four anonymous samples were conducted on each session. They were presented in balanced random order at room temperature (18 ± 2 °C) in black tulip-shaped glasses coded with 3-digit random numbers. The assessors were instructed to pour the whole sample in their mouth, hold it for 8 s, expectorate, and answer a check-all-that-apply (CATA) question with 16 sensory attributes of astringency. The attributes were the following: silk, velvet, dry, corduroy, adhesive, aggressive, hard, soft, mouthcoat, rich, full-body, green, grainy, satin, pucker, and persistent, defined in Supplementary Table S1. Judges waited for 4 min before rinsing twice for 10 s with mineral water (Sorgesana, pH ≈ 7) and then waited at least 30 s before drinking the following sample. The serving order design was a juxtaposition of Latin squares balanced for carryover effects [35]. The panel also evaluated the taste (sweet, acid, bitter, sapid); odour; and aroma (floral, fruity, spicy, balsamic) of the wines using a 5-point scale.

4.4. Chemical Analyses

All spectrophotometric determinations were performed using a Spectrophotometer Shimadzu UV-1800 model. Wine colour intensity (CI), given by the sum of the absorbances at 420, 520, and 620 nm and hue (420/520 Abs) were analysed using the Glories method [36]. CIELab allows the specification of colour perception in terms of a three-dimensional space. The L*-axis is known as the lightness and ranges from 0 (black) to 100 (white). The other two coordinates a* and b* represent redness–greenness and yellowness–blueness, respectively. CIELab coordinates were determined by Panorama software (Shimadzu, Milan, Italy). The total colour difference (ΔE) between two samples (treated wine and control) was obtained using the following expression: $\Delta E = [(\Delta L^*)^2 + (\Delta a^*)^2 + (\Delta b^*)^2]^{1/2}$, in CIELab units [37]. Vanillin-reactive flavans were determined according to Di Stefano and Guidoni [38]. Total anthocyanins, polymeric pigments (LPP + SPP) as a measure of the colour stability of the wine, and BSA-reactive tannins were determined by the Harbertson et al. method [39]. Briefly, in this method, pH changes allow the evaluation of polymeric pigments by combining the analysis of the supernatant obtained after protein precipitation using bovine serum albumin (Sigma, Merck Life Science, Milano, Italy) for the tannin analysis (BSA-reactive tannins) and the bisulfite bleaching of the pigments in wine. For the determination of the total anthocyanins, 500 mL of wine diluted in a buffer solution (5-g/L potassium bitartrate,12% EtOH, and pH adjusted to 3.3 with HCl) were added to 1 mL of a buffer solution (200-mM maleic acid, 170-mM NaCl, and pH adjusted to 1.8 with NaOH) and incubated for 5 min. Total anthocyanins were determined by reading the absorbance of this solution at 520 nm. All analyses were carried out in duplicate on each bottle, for a total of four replicates.

4.5. Data Analysis

As a one-way ANOVA analysis, Fisher's Least Significant Differences (LSD) procedure was used to distinguish the means of the phenolic and colour variables over four replicates. Sensory attributes (taste, odour, and aroma) were evaluated using Duncan's test. Differences of $p < 0.05$ were considered significant. CATA responses were elaborated by the CATA analysis for each wine typology, and the most significant astringency subqualities ($p < 0.01$) were projected as explanatory variables in the CATA plot. A Multiple Factor Analysis (MFA) was performed on the frequency table containing responses to the CATA question, the phenolic content, and the colour parameters of the wines to investigate the relationships between the data from the chemical analyses and responses to the CATA question as separate groups of variables. Elaborations were carried out by means of XLSTAT software (Addinsoft, XLSTAT 2021).

5. Conclusions

Depending on the A/T ratio, each Sangiovese wine could necessitate a specific mannoprotein to improve the mouthfeel and/or colour. If there is a strong excess in tannins (extended maceration wines with A/T = 0.2), ageing with MP at 20 g/hL can be preferred, because it confers positive subqualities and colour stability to the wine. When A/T = 0.3, as in marc-pressed wines, both MF and MP can improve the mouthfeel and colour of Sangiovese. However, in free-run wine, where the A/T ratio is 0.5, the polymeric pigment formation was enabled by all treatments and correlated with an improved mouthfeel sensation. Further experiments on the role of the A/T ratio will be carried out to better understand the mechanisms involved in these phenomena. In addition, the bitterness was reduced by mannoproteins. For free-run wines, the ideal treatment can be represented by the MS, as it also showed a significant effect on the aroma revelation. In all cases, the formation or preservation of polymeric pigments by mannoproteins during ageing can be associated with positive subqualities, like velvet, soft, and mouthcoat.

Supplementary Materials: The following is available online. Table S1: Definitions of the attributes used to characterise the mouthfeel of wines. Table S2: Analyses of the base parameters of extended maceration (E), marc-pressed (P), and free-run (F) Sangiovese wines before ageing (t0).

Author Contributions: Conceptualisation, A.R.; methodology, A.R.; formal analysis, A.G. (Alliette Gonzalez); investigation, A.R., and A.G. (Angelita Gambuti); data curation, A.R.; writing—original draft preparation, A.R.; writing—review and editing, A.G. (Angelita Gambuti); and supervision, L.M. All authors have read and agreed to the published version of the manuscript.

Funding: This research received no external funding.

Institutional Review Board Statement: This study was conducted according to the guidelines of the Declaration of Helsinki. Panellists gave their informed consent, and their privacy rights were always observed.

Informed Consent Statement: Informed consent was obtained from all subjects involved in the study.

Data Availability Statement: Data are available upon request.

Acknowledgments: The authors thank Laffort Italia for supplying the products, the Tuscan winery, and the students involved in the internship agreement with the University of Naples.

Conflicts of Interest: The authors declare no conflict of interest.

Sample Availability: Samples of the mannoproteins are available from the authors.

References

1. Casassa, L.F.; Harbertson, J.F. Extraction, Evolution, and Sensory Impact of Phenolic Compounds during Red Wine Maceration. *Annu. Rev. Food Sci. Technol.* **2014**, *5*, 83–109. [CrossRef] [PubMed]
2. Waterhouse, A.L. Wine Phenolics. *Ann. N. Y. Acad. Sci.* **2002**, *957*, 21–36. [CrossRef]

3. Sáenz-Navajas, M.P.; Tao, Y.S.; Dizy, M.; Ferreira, V.; Fernández-Zurbano, P. Relationship between Nonvolatile Composition and Sensory Properties of Premium Spanish Red Wines and Their Correlation to Quality Perception. *J. Agric. Food Chem.* **2010**, *58*, 12407–12416. [CrossRef] [PubMed]
4. Setford, P.C.; Jeffery, D.W.; Grbin, P.R.; Muhlack, R.A. Factors Affecting Extraction and Evolution of Phenolic Compounds during Red Wine Maceration and the Role of Process Modelling. *Trends Food Sci. Technol.* **2017**, *69*, 106–117. [CrossRef]
5. Nel, A.P.; van Rensburg, P.; Lambrechts, M.G. The Influence of Different Winemaking Techniques on the Extraction of Grape Tannins and Anthocyanins. *S. Afr. J. Enol. Vitic.* **2016**, *35*, 304–320. [CrossRef]
6. Sacchi, K.L.; Bisson, L.F.; Adams, D.O. A Review of the Effect of Winemaking Techniques on Phenolic Extraction in Red Wines. *Am. J. Enol. Vitic.* **2005**, *56*, 197–206.
7. Frost, S.C.; Blackman, J.W.; Ebeler, S.E.; Heymann, H. Analysis of Temporal Dominance of Sensation Data Using Correspondence Analysis on Merlot Wine with Differing Maceration and Cap Management Regimes. *Food Qual. Prefer.* **2018**, *64*, 245–252. [CrossRef]
8. Gambuti, A.; Rinaldi, A.; Moio, L. Use of Patatin, a Protein Extracted from Potato, as Alternative to Animal Proteins in Fining of Red Wine. *Eur. Food Res. Technol.* **2012**, *235*, 753–765. [CrossRef]
9. Rinaldi, A.; Errichiello, F.; Moio, L. Alternative Fining of Sangiovese Wine: Effect on Phenolic Substances and Sensory Characteristics. *Aust. J. Grape Wine Res.* **2021**, *27*, 128–137. [CrossRef]
10. McRae, J.M.; Schulkin, A.; Kassara, S.; Holt, H.E.; Smith, P.A. Sensory Properties of Wine Tannin Fractions: Implications for in-Mouth Sensory Properties. *J. Agric. Food Chem.* **2013**, *61*, 719–727. [CrossRef]
11. Rinaldi, A.; Louazil, P.; Iturmendi, N.; Moine, V.; Moio, L. Effect of Marc Pressing and Geographical Area on Sangiovese Wine Quality. *Lebenson. Wiss. Technol.* **2020**, *118*. [CrossRef]
12. Lukić, I.; Milicević, B.; Banović, M.; Tomas, S.; Radeka, S.; Persurić, D. Characterization and Differentiation of Monovarietal Grape Marc Distillates on the Basis of Varietal Aroma Compound Composition. *J. Agric. Food Chem.* **2010**, *58*, 7351–7360. [CrossRef] [PubMed]
13. Genovese, A.; Lamorte, S.A.; Gambuti, A.; Moio, L. Aroma of Aglianico and Uva Di Troia Grapes by Aromatic Series. *Food Res. Int.* **2013**, *53*, 15–23. [CrossRef]
14. Vidal, S.; Francis, L.; Williams, P.; Kwiatkowski, M.; Gawel, R.; Cheynier, V.; Waters, E. The Mouthfeel Properties of Polysaccharides and Anthocyanins in a Wine like Medium. *Food Chem.* **2004**, *85*, 519–525. [CrossRef]
15. Charpentier, C.; Feuillat, M. Yeast Autolysis G. In *Wine Microbiology and Biotechnology*; Fleet, H., Ed.; Harwood Academic Publishers: Chur, Switzerland, 1993.
16. Domizio, P.; Liu, Y.; Bisson, L.F.; Barile, D. Use of Non-Saccharomyces Wine Yeasts as Novel Sources of Mannoproteins in Wine. *Food Microbiol.* **2014**, *43*, 5–15. [CrossRef] [PubMed]
17. Rinaldi, A.; Coppola, M.; Moio, L. Aging of Aglianico and Sangiovese Wine on Mannoproteins: Effect on Astringency and Colour. *Lebenson. Wiss. Technol.* **2019**, *105*, 233–241. [CrossRef]
18. Gonzalez-Ramos, D.; Cebollero, E.; Gonzalez, R. A Recombinant Saccharomyces Cerevisiae Strain Overproducing Mannoproteins Stabilizes Wine against Protein Haze. *Appl. Environ. Microbiol.* **2008**, *74*, 5533–5540. [CrossRef]
19. Ribeiro, T.; Fernandes, C.; Nunes, F.M.; Filipe-Ribeiro, L.; Cosme, F. Influence of the Structural Features of Commercial Mannoproteins in White Wine Protein Stabilization and Chemical and Sensory Properties. *Food Chem.* **2014**, *159*, 47–54. [CrossRef]
20. Pérez-Magariño, S.; Martínez-Lapuente, L.; Bueno-Herrera, M.; Ortega-Heras, M.; Guadalupe, Z.; Ayestarán, B. Use of Commercial Dry Yeast Products Rich in Mannoproteins for White and Rosé Sparkling Wine Elaboration. *J. Agric. Food Chem.* **2015**, *63*, 5670–5681. [CrossRef]
21. Manjón, E.; Brás, N.F.; García-Estévez, I.; Escribano-Bailón, M.T. Cell Wall Mannoproteins from Yeast Affect Salivary Protein-Flavanol Interactions through Different Molecular Mechanisms. *J. Agric. Food Chem.* **2020**, *68*, 13459–13468. [CrossRef] [PubMed]
22. Guadalupe, Z.; Palacios, A.; Ayestarán, B. Maceration Enzymes and Mannoproteins: A Possible Strategy to Increase Colloidal Stability and Colour Extraction in Red Wines. *J. Agric. Food Chem.* **2007**, *55*, 4854–4862. [CrossRef] [PubMed]
23. Loira, I.; Vejarano, R.; Morata, A.; Ricardo-da-Silva, J.M.; Laureano, O.; González, M.C.; Suárez-Lepe, J.A. Effect of Saccharomyces Strains on the Quality of Red Wines Aged on Lees. *Food Chem.* **2013**, *139*, 1044–1051. [CrossRef] [PubMed]
24. Del Barrio-Galán, R.; Pérez-Magariño, S.; Ortega-Heras, M.; Guadalupe, Z.; Ayestarán, B. Polysaccharide Characterization of Commercial Dry Yeast Preparations and Their Effect on White and Red Wine Composition. *Lebenson. Wiss. Technol.* **2012**, *48*, 215–223. [CrossRef]
25. Marchal, A.; Marullo, P.; Moine, V.; Dubourdieu, D. Influence of Yeast Macromolecules on Sweetness in Dry Wines: Role of the Saccharomyces Cerevisiae Protein Hsp12. *J. Agric. Food Chem.* **2011**, *59*, 2004–2010. [CrossRef]
26. Gambuti, A.; Picariello, L.; Rinaldi, A.; Moio, L. Evolution of Sangiovese Wines with Varied Tannin and Anthocyanin Ratios during Oxidative Aging. *Front. Chem.* **2018**, *6*, 63. [CrossRef] [PubMed]
27. Escot, S.; Feuillat, M.; Dulau, L.; Charpentier, C. Release of Polysaccharides by Yeasts and the Influence of Released Polysaccharides on Colour Stability and Wine Astringency. *Aust. J. Grape Wine Res.* **2001**, *7*, 153–159. [CrossRef]
28. Rinaldi, A.; Moio, L. Effect of Enological Tannin Addition on Astringency Subqualities and Phenolic Content of Red Wines. *J. Sens. Stud.* **2018**, *33*, e12325. [CrossRef]
29. Saucier, C.; Glories, Y.; Roux, D. Interactions Tanins-Colloides: Nouvelles Avancées Concernant La Notion de « bons » et de « mauvais » tanins. *Revue des Œnologues* **2000**, *94*, 7–8.

30. Maury, C.; Sarni-Manchado, P.; Lefebvre, S.; Cheynier, V.; Moutounet, M. Influence of Fining with Different Molecular Weight Gelatins on Proanthocyanidin Composition and Perception of Wines. *Am. J. Enol. Vitic.* **2001**, *52*, 140–145.
31. Rodrigues, A.; Ricardo-Da-Silva, J.M.; Lucas, C.; Laureano, O. Effect of Commercial Mannoproteins on Wine Colour and Tannins Stability. *Food Chem.* **2012**, *131*, 907–914. [CrossRef]
32. Alcalde-Eon, C.; Pérez-Mestre, C.; Ferreras-Charro, R.; Rivero, F.J.; Heredia, F.J.; Escribano-Bailón, M.T. Addition of Mannoproteins and/or Seeds during Winemaking and Their Effects on Pigment Composition and Color Stability. *J. Agric. Food Chem.* **2019**, *67*, 4031–4042. [CrossRef]
33. Chalier, P.; Angot, B.; Delteil, D.; Doco, T.; Gunata, Z. Interactions between Aroma Compounds and Whole Mannoprotein Isolated from Saccharomyces Cerevisiae Strains. *Food Chem.* **2007**, *100*, 22–30. [CrossRef]
34. Jones, P.R.; Gawel, R.; Francis, I.L.; Waters, E.J. The Influence of Interactions between Major White Wine Components on the Aroma, Flavour and Texture of Model White Wine. *Food Qual. Prefer.* **2008**, *19*, 596–607. [CrossRef]
35. Macfie, H.J.; Bratchell, N.; Greenhoff, K.; Vallis, L.V. Designs to Balance the Effect of Order of Presentation and First-Order Carry-over Effects in Hall Tests. *J. Sens. Stud.* **1989**, *4*, 129–148. [CrossRef]
36. Glories, Y. La Couleur Des Vins Rouges. 1° e 2° Partie. *Connaissance de la Vigne et du Vin* **1984**, *18*, 253–271.
37. Ayala, F.; Echávarri, J.F.; Negueruela, A.I. A New Simplified Method for Measuring the Colour of Wines. I. Red and rosé wines. *Am. J. Enol. Vitic.* **1997**, *48*, 357–363.
38. Di Stefano, R.; Guidoni, S. La Determinazione Dei Polifenoli Totali Nei Mosti e Nei Vini. *Vignevini* **1989**, *16*, 47–52.
39. Harbertson, J.F.; Picciotto, E.A.; Adams, D.O. Measurement of Polymeric Pigments in Grape Berry Extract Sand Wines Using a Protein Precipitation Assay Combined with Bisulfite Bleaching. *Am. J. Enol. Vitic.* **2003**, *54*, 301–306.

Article

Aromatic Higher Alcohols in Wine: Implication on Aroma and Palate Attributes during Chardonnay Aging

Antonio G. Cordente *,[†], Damian Espinase Nandorfy [†], Mark Solomon, Alex Schulkin, Radka Kolouchova, Ian Leigh Francis and Simon A. Schmidt

The Australian Wine Research Institute, P.O. Box 197, Glen Osmond, SA 5064, Australia; damian.espinasenandorfy@awri.com.au (D.E.N.); mark.solomon@awri.com.au (M.S.); alex.schulkin@awri.com.au (A.S.); radka.kolouchova@awri.com.au (R.K.); Leigh.francis@awri.com.au (I.L.F.); simon.schmidt@awri.com.au (S.A.S.)
* Correspondence: toni.garciacordente@awri.com.au; Tel.: +61-8-8313-6600; Fax: +61-8-8313-6601
[†] These authors contributed equally to this work.

Abstract: The higher alcohols 2-phenylethanol, tryptophol, and tyrosol are a group of yeast-derived compounds that have been shown to affect the aroma and flavour of fermented beverages. Five variants of the industrial wine strain AWRI796, previously isolated due to their elevated production of the 'rose-like aroma' compound 2-phenylethanol, were characterised during pilot-scale fermentation of a Chardonnay juice. We show that these variants not only increase the concentration of 2-phenylethanol but also modulate the formation of the higher alcohols tryptophol, tyrosol, and methionol, as well as other volatile sulfur compounds derived from methionine, highlighting the connections between yeast nitrogen and sulfur metabolism during fermentation. We also investigate the development of these compounds during wine storage, focusing on the sulfonation of tryptophol. Finally, the sensory properties of wines produced using these strains were quantified at two time points, unravelling differences produced by biologically modulating higher alcohols and the dynamic changes in wine flavour over aging.

Keywords: amino acid; yeast; wine; sulfur; aroma; aging; QDA

1. Introduction

Saccharomyces cerevisiae performs a wide range of industrial fermentations that functionally depend upon its ability to convert sugars to ethanol and carbon dioxide efficiently. While performing this primary function, *S. cerevisiae* also produces a range of secondary metabolites, such as esters, volatile fatty acids, higher alcohols, and volatile sulfur compounds (VSCs), which contribute substantially to the flavour and aroma of wine [1], beer [2], and sake [3]. Of these fermentation compounds, higher alcohols and esters are the most abundant groups [1].

Higher alcohols, also known as fusel alcohols, are compounds with more than two carbon atoms. These alcohols are derived from yeast amino acid metabolism via the Ehrlich pathway [4]. Amino acids assimilated by the Ehrlich pathway include the aliphatic or branched-chain (leucine, valine and isoleucine) and aromatic (phenylalanine, tyrosine and tryptophan) amino acids, as well as the sulfur-containing amino acid methionine. The Ehrlich pathway consists of three steps: the initial transamination of the amino acid to the corresponding α-keto acid analogue, decarboxylation to an aldehyde, and reduction to the corresponding higher alcohol by an alcohol dehydrogenase [4]. While the metabolism of aliphatic and aromatic amino acids via the Ehrlich pathway has been extensively studied in yeast, little is known about the branch of the pathway involved in methionine catabolism. However, some similarities exist between the catabolism of methionine and that of the aromatic amino acids. The aminotransferases Aro8p and Aro9p, which catalyse the first step of the Ehrlich pathway for aromatic amino acids, also play an essential role in methionine transamination [5,6]. Similarly, the broad-substrate phenylpyruvate decarboxylase Aro10p is also involved in the second step of the Ehrlich pathway for methionine [7,8].

Except for 2-phenylethanol (2-PE), which is derived from phenylalanine and associated with a 'rose-like' odour quality [9], the individual contribution of higher alcohols to wine aroma is not considered to be pleasant, particularly at higher concentrations [10]. For example, the higher alcohols derived from branched-chain amino acids (2-methylpropanol, 3-methylbutanol and 2-methylbutanol) are associated with 'solvent' and 'fusel' aroma descriptors, while 3-methylthio-1-propanol (or methionol), derived from methionine, imparts a 'boiled or cooked potato' aroma in wine [10]. Aside from these negative associations, higher alcohols may also contribute to wines' overall 'vinous' aroma as part of the 'aroma buffer' [11]. Of the many higher alcohols, tyrosol (TyrOH) and tryptophol (TOL) (from tyrosine and tryptophan, respectively) have not been associated with any aroma descriptors. However, there is growing evidence that they influence in-mouth sensory properties, especially the taste of some fermented beverages, as both compounds have been associated with bitterness in wine, sake, and beer [12–14]. Recently, it has been reported that TOL can react with sulfur dioxide (SO_2), which is widely used in winemaking as an additive due to its antimicrobial and antioxidant effects, to yield the tryptophol-2-sulfonate (TOL-SO_3H) adduct [15]. This reaction is favoured by the presence of small amounts of oxygen in wine [15], and the equilibrium towards the formation of TOL-SO_3H from TOL seems to be increased by bottle aging, particularly for white wines [16]. Although the effect of TOL-SO_3H on the taste and mouthfeel properties of fermented beverages remains unclear, it has recently been linked with bitterness [17].

Higher alcohols are substrates for acetate ester production, a reaction catalysed by yeast alcohol acetyltransferases. Many acetate esters are associated with 'fruity' and 'floral' aromas in wine [18]. For example, both 2-PE and 3-methylbutanol ('solvent') can be esterified and converted into 2-phenylethyl acetate (2-PEA) ('rose', 'fruity', 'honey') and 3-methylbutyl acetate ('banana'), respectively. Acetate esters are important contributors to the aroma of young wines, as their concentration tends to decrease post-fermentation with wine storage due to non-enzymatic, acid-catalysed reactions in the wine matrix [19,20].

Due to the ability of 2-PE and 2-PEA to impart floral notes, these compounds present an opportunity to shape wine style. The ability to influence the production of 2-PE and 2-PEA in white wines would be beneficial because, although naturally occurring, they are usually present at relatively low concentrations and are unlikely to impart a definitive character [1,21]. The concentration of amino acids in grape must is a crucial factor influencing the production of higher alcohols by yeast: increased concentration of a specific precursor amino acid will usually increase the concentration of the corresponding higher alcohol [22]. Therefore, one widely used strategy to increase the formation of 2-PE is to select yeast strains that overproduce the precursor phenylalanine. Obtaining phenylalanine-overproducing yeast can be achieved by selecting strains resistant to toxic fluorinated analogues of phenylalanine. Such a strategy has been successfully used to generate industrial *S. cerevisiae* strains that improve the organoleptic properties of sake, wine, and bread [23–25].

Conversely, and with the exception of a small group of polyfunctional thiols associated with 'fruity' and 'tropical' characters [26], most VSCs generated by yeast are considered to be off-flavours in wine, particularly the 'rotten-egg'-imparting compound hydrogen sulfide (H_2S). Similarly, a range of compounds derived from methionine by either enzymatic or non-enzymatic reactions are also associated with negative attributes in wine, such as methanethiol (MeSH) ('sewage and rubber' aromas), methional ('cooked potato' aroma) and methyl thioacetate (MeSAc) ('sulfurous and cheesy' aromas) [6,8,27]. While the formation of H_2S by yeast has been extensively studied, and commercial low H_2S producing strains have been generated [28], little is known about the formation of odoriferous VSCs derived from methionine, and few strategies exist to reduce their formation during fermentation.

This paper explores the influence of compounds derived from the metabolism of both aromatic and sulfur-containing amino acids in the sensory profile of Chardonnay wine over time. For this purpose, we characterised five variants of a commercially available wine yeast strain (AWRI796), which were previously isolated because of their high 2-PE production phenotype. Here we show that these variants influence not only the production of 2-PE and 2-PEA but also the concentration of the higher alcohols TOL, TyrOH, and methionol, as well as other odoriferous VSCs derived from the amino acid methionine. The progression of these compounds and the equilibrium shift between TOL-SO_3H and TOL during wine ageing are also reported. Formal sensory analysis was conducted on the Chardonnay wines produced by each strain at two time points, and the links between the resulting changes to chemical composition and sensory properties are presented and explored.

2. Results

2.1. Pilot-Scale White Winemaking of Five 2-Phenylethanol Overproducing Strains

Previously, we isolated a range of variants from the commercial wine yeast AWRI796 that were resistant to toxic fluorinated analogues of phenylalanine. These variants were shown to overproduce 2-PE and 2-PEA to different extents in laboratory-scale fermentations [23]. A single variant, AWRI2940, was further characterised in a Chardonnay pilot-scale winemaking study, where its high 2-PE production phenotype was validated. Sensory evaluation of the wines showed that while an increased 'floral' aroma was the attribute most affected by AWRI2940, this variant also produced wines that were noted as having a more 'bitter' taste and 'astringent' mouthfeel than the parent strain [23]. Compositional analysis showed that AWRI2940 produced a higher concentration of TOL and a lower concentration of TyrOH than the parent strain, compounds that have been associated with bitterness in different alcoholic beverages [12,13,29,30].

The effects of variations in 2-PE, TyrOH and TOL concentration on wine sensory properties were determined using five mutants of AWRI796 together with parent strain (control) to produce Chardonnay wine. The variant strains harbour distinct mutations in two of the enzymes involved in aromatic amino acid metabolism: Tyr1p and Aro4p (summarised in Table S1). The 19 L Chardonnay ferments were conducted in triplicate and were complete after 31 days. Post-fermentation analysis of the volatiles confirmed the 2-PE overproduction phenotype for all five variants (Figure 1 and Table S2). Fermentation with strain AWRI2940 resulted in a 15-fold increase in the concentration of 2-PE relative to the parent AWRI796. The strains AWRI2965, AWRI2969 and AWRI4124 showed a more moderate 2-PE overproduction phenotype (between 7- and 8-fold increase). AWRI2936 was the lowest 2-PE producer of all the variants (3-fold increase). In all five mutants of AWRI796, the relative increases of 2-PEA were even higher than those observed for 2-PE, ranging from 6- to nearly 40-fold (Figure 1). As expected, the concentration of 2-PE and 2-PEA was highly correlated (r = 0.983, $p < 0.0001$, Figure S1).

All five yeast variants, particularly AWRI2940, produced more TOL than AWRI796 (Figure 1). TOL was also positively correlated with the 2-PE concentration. The relationship was linear for the parent and the four variants that produced low to moderate concentrations of TOL ($R^2 = 0.87$, $p < 0.0001$), while an exponential model was a better fit when all strains were included ($R^2 = 0.97$) (Figure 2). In contrast to TOL production, the four Tyr1p variants produced lower concentrations of TyrOH than the parent AWRI796, while the Aro4p variant (AWRI2965) produced substantially more (9-fold increase) (Figure 1). While there was no evidence for strain-based differences in the concentration of branched-chain amino acid derived higher alcohols, there was evidence for strain-dependent differences in the concentration of the respective acetate esters (Table S2). Notably, the concentration of ethyl acetate, associated with a 'nail polish remover' aroma, was not affected by any of the 2-PE overproducing variants.

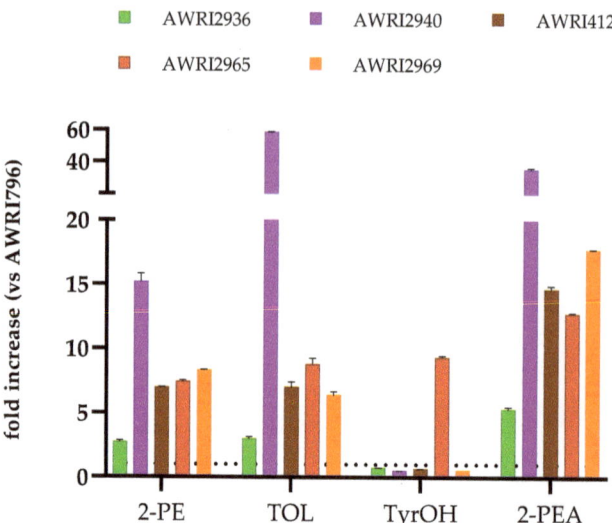

Figure 1. Production of the higher alcohols 2-PE, TOL and TyrOH, and 2-PEA, in Chardonnay wines by five variants of the wine strain AWRI796 carrying mutations in Aro4p (AWRI2965) or Tyr1p (AWRI2936, 2940, 4124, and 2969). Results are expressed as the average fold change in the concentration of these metabolites relative to the control strain AWRI796 (indicated with a dashed line) after alcoholic fermentation. Error bars show the standard deviation of three independent fermentations.

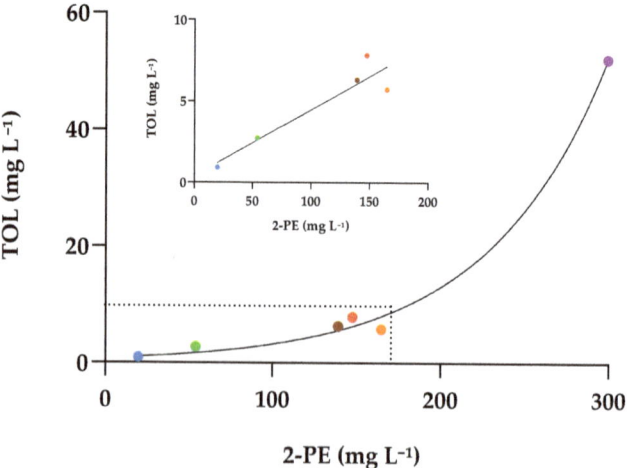

Figure 2. Relationship between 2-PE and TOL concentrations (mg L^{-1}) at the end of alcoholic fermentation in Chardonnay wines. The wines were fermented by the parent strain AWRI796 (•), and five variants carrying mutations in Aro4p (AWRI2965 (•)) or Tyr1p (AWRI2936 (•), AWRI2940 (•), AWRI4124 (•) and AWRI2969 (•)). The area under the dotted lines (inset) highlights the linear relationship between these two higher alcohols at lower concentrations.

2.2. Effect of Wine Aging on TOL/TOL-SO$_3$H Equilibrium

We assessed the effect of bottle storage duration on the concentration of aromatic higher alcohols, focusing on the sulfonation of TOL to yield the TOL-SO$_3$H adduct. At the end of alcoholic fermentation, only wines made with the highest TOL producer (AWRI2940) contained TOL-SO$_3$H (0.26 mg L^{-1}), representing a 0.3% molar conversion of TOL into TOL-SO$_3$H. The post-fermentation concentration of total SO$_2$ in the wines averaged 21 mg L^{-1}, with no detectable free SO$_2$ (Table S2). On completion of alcoholic fermentation, 80 mg L^{-1} of SO$_2$ was added to the wines, followed by a lengthy period of cold-stabilisation (two months). Before bottling, free SO$_2$ concentration was adjusted to between 35 and 40 mg L^{-1} (Table S3). Conversion of a substantial percentage of TOL into its sulfonated adduct was evident after three months in-bottle: for the low and moderate TOL producers (<10 mg L^{-1} TOL) the yield of TOL-SO$_3$H was at least 80%, while a 26% yield was observed in wines made using the high TOL producer (AWRI2940) (Figure 3).

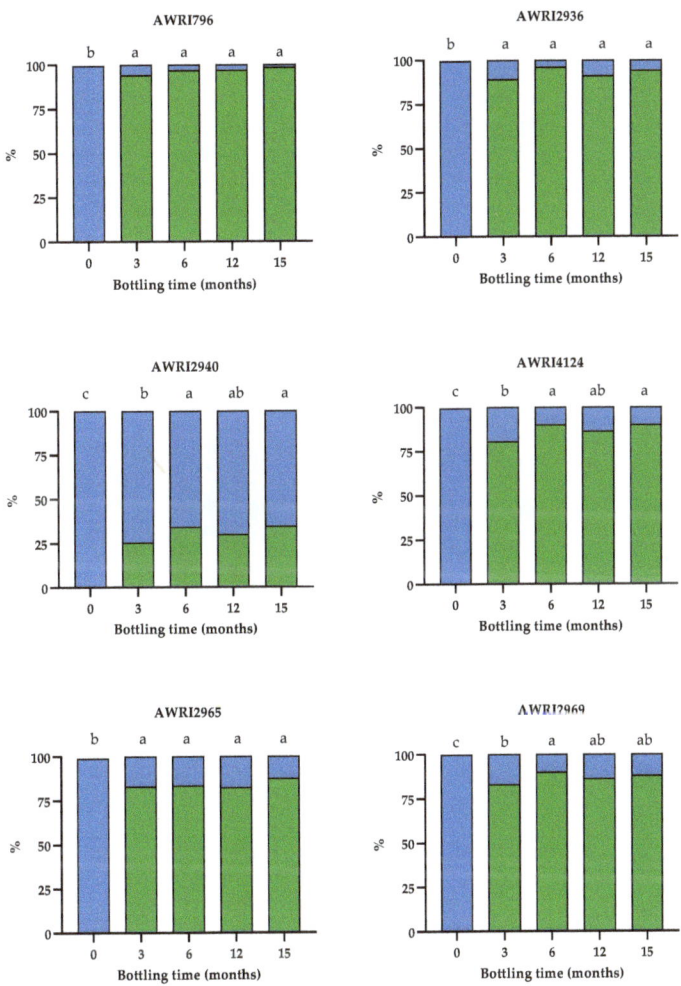

Figure 3. Relative molar percentages of TOL (blue) and TOL-SO$_3$H (green) in relation to bottling age in Chardonnay wines made with the parent AWRI796 and five variants carrying mutations in Aro4p (AWRI2965) or Tyr1p (AWRI2936, 2940, 4124, and 2969). Means with the same letters are not significantly different from each other (Tukey's test, alpha = 0.05).

After six months in-bottle, an equilibrium between TOL and TOL-SO$_3$H species was reached, with longer storage times (12 and 15 months) having little or no effect on the conversion of the higher alcohol into the sulfonated adduct (Figure 3). The maximum molar % yield of TOL-SO$_3$H inversely correlated with the concentration of TOL at the end of alcoholic fermentation (r = −0.998, $p < 0.0001$) (Figure S2). For the control strain (AWRI796 strain), which only produced 0.89 mg L^{-1} of TOL, almost 100% of the available TOL had been converted into TOL-SO_3H. For the four moderate TOL producers, the % yield of TOL-SO_3H ranged from 88 to 94%, while for the high TOL producer (AWRI2940) the yield was only 36% (Figure 3).

We sought to determine whether TOL-SO_3H formation was a significant contributor to SO_2 loss during bottle storage. On the day of bottling, the concentration of free SO_2 averaged 39 mg L^{-1} across all wines (Table S3). In the case of AWRI2940, we would expect a decrease in SO_2 concentration of 7.6 mg L^{-1} related to adduct formation assuming an equimolar reaction between SO_2 and TOL. For the low and moderate producers, we would expect less than 3 mg L^{-1} of SO_2 to be consumed. Measured decreases in SO_2 concentration were similar across all the wines after three months in-bottle: decreases of between 14 and 19 mg L^{-1} in free SO_2 concentration and between 24 and 30 mg L^{-1} in total SO_2 concentration were observed. Longer storage times resulted in further losses of SO_2 with respect to pre-bottling concentrations, with the data offering no evidence for an effect of strain on the loss of free or total SO_2 during wine aging (Figure 4). No significant correlation between loss of SO_2 and TOL-SO_3H concentration was observed at any time point assessed (Figure S3).

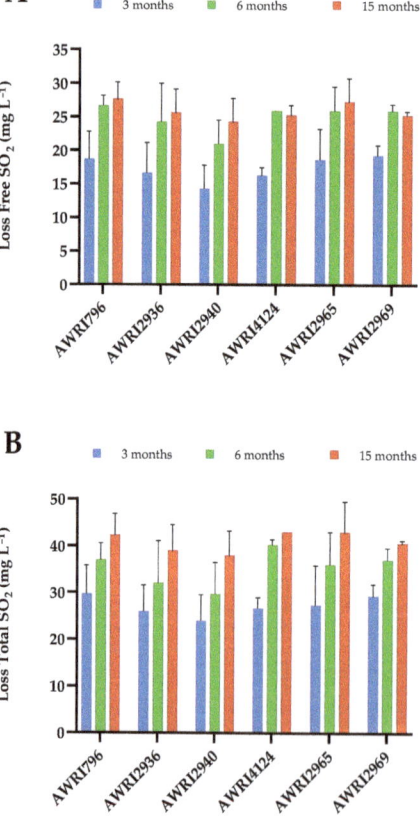

Figure 4. Loss in free (**A**) and total (**B**) SO_2 concentrations (mg L^{-1}) during storage in Chardonnay wines made with the parent AWRI796 and five variants carrying mutations in Aro4p (AWRI2965) or Tyr1p (AWRI2936, 2940, 4124, and 2969). Wines analysed after 3 (blue), 6 (green), and 15 (red) months

of storage at 15 °C were compared to the respective SO_2 concentrations on the day of bottling (which averaged 39 and 110 mg L^{-1} of free and total SO_2, respectively). Results are expressed as the mean and standard deviation of three independent fermentations. No differences between strains were seen at any of the three storage times assessed (Tukey's test, alpha = 0.05).

2.3. Effect of Wine Ageing on Volatile Sulfur Compounds

The effects of alterations to aromatic amino acid metabolism in yeast on the formation of several VSCs derived from methionine were also investigated in the Chardonnay wines after 3 and 15 months in-bottle. In yeast, two competing pathways are involved in methionine degradation: the Ehrlich and demethiolation pathways [31]. In the Ehrlich pathway, methionine is transaminated and decarboxylated to methional, and subsequently reduced to the higher alcohol methionol [4,6]. Methionine is also the precursor of MeSH, the production of which can occur enzymatically via demethiolation [8], or non-enzymatically [32]. Therefore, the question arises: how do alterations to aromatic amino acid metabolism change the contribution of these two methionine degradation pathways to wine composition?

After three months, the concentration of methionol was greater in wines made with the high 2-PE producing strains (AWRI4124, AWRI2969 and AWRI2940) than those made with the parent strain (AWRI796), whereas the low 2-PE producer and the high TyrOH-producing strains (AWRI2936 and AWRI2965, respectively) accumulated similar concentrations of methionol to the parent (Figure 5). The concentrations of methional were lowest in wines produced by strains AWRI2940 and AWRI2965. Similar trends were observed in the concentration of compounds derived from the demethiolation pathway; the concentrations of MeSH and its thioacetate, MeSAc, were both lower in wines made with AWRI2940 and AWRI2965.

After 15 months in-bottle, the concentration of most VSCs had increased in all wines relative to their concentration at three months. Nevertheless, a similar strain profile was evident in the VSC concentrations of wines sampled at both time points. Wines made with strain AWRI2965 still had a significantly lower concentration of MeSH and methional than the parent strain (AWRI796) while showing a slightly elevated concentration of dimethylsulfide (DMS). No statistical evidence supporting differences between the strains in the concentration of H_2S was observed at any time point (Figure 5).

To further confirm the effects that mutations in either Tyr1p or Aro4p might have on methionine catabolism, two of the variants were used to ferment a synthetic grape medium (SGM) with grape-like concentrations of methionine [33]. The two strains used were characterised by moderate (AWRI2965:Aro4) and high (AWRI2940:Tyr1) 2-PE production, thus reflecting different levels of activation of the Ehrlich pathway (Figure S4). Results in SGM reflected those found in the Chardonnay wines (Figure 6). Fermentation with both variants resulted in a lower concentration of MeSH and methional relative to the parent AWRI796. Conversely, AWRI2940 produced substantially more of the Ehrlich pathway end product methionol than the other strains.

2.4. Quantitative Descriptive Sensory Analysis

Sensory descriptive analysis of the wines was performed twice, after 3 and 15 months in-bottle. The analyses were compared to assess the effect of bottle storage on wine sensory attributes. The mean scores for a subset of sensory attributes are summarised in Tables 1 and 2. The sensory differences between the samples were relatively subtle after three months, indicated by the small ANOVA F-ratios (Table S4). Statistical evidence ($p < 0.05$) supports differences between strains in three attributes rated by the panel: 'yellow colour intensity', 'floral aroma' and 'grassy flavour' (Table 1 and Table S5). Wines from strain AWRI2940 (the highest 2-PE producer) were rated highest in both 'floral aroma' and 'yellow colour', but none of the variants were rated higher than the parent (AWRI796) in 'floral aroma'. Weak evidence ($p < 0.15$) alluded to possible trends among the strains for the attributes 'grassy aroma', 'cooked vegetable/potato aroma', 'sweetness', 'bitterness', 'stone fruit flavour' and 'flint flavour'. In particular, wines made with AWRI796 and AWRI4124 were rated highly in the attributes 'cooked vegetable/potato aroma' and 'sweetness', while there was a trend for 'bitterness' to be rated lowest for the low TOL-/TOL-SO$_3$H-producing strains AWRI2936 and AWRI796 compared to the other strains.

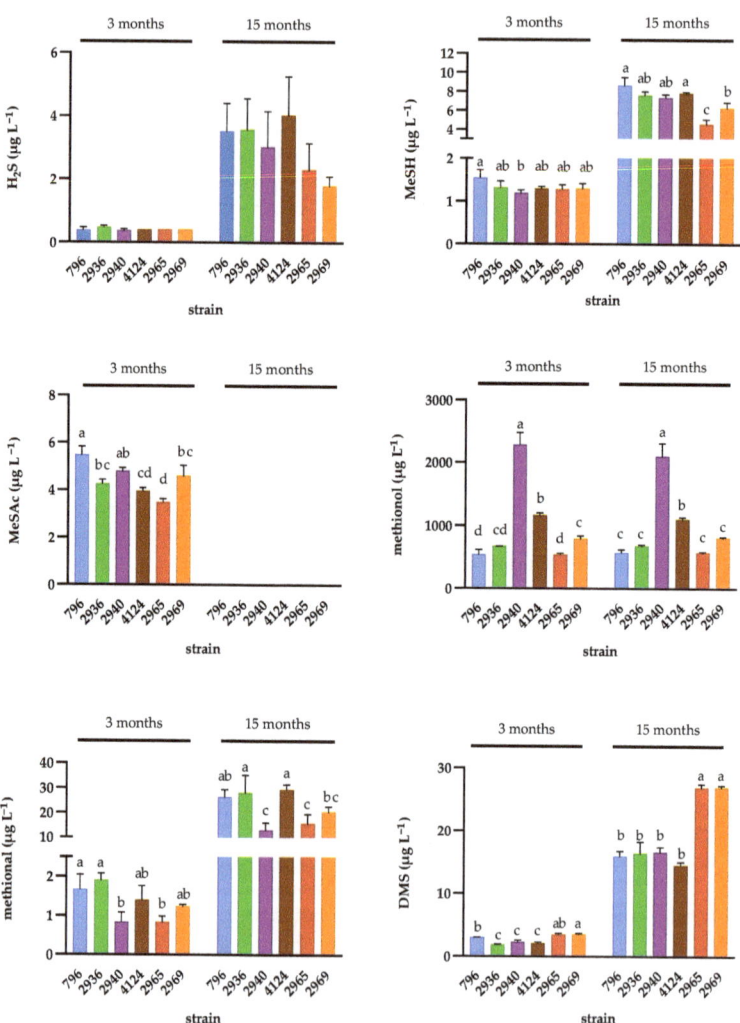

Figure 5. Concentrations (in μg L^{-1}) of VSCs in Chardonnay wines made with the parent AWRI796 and five variants carrying mutations in Aro4p (AWRI2965) or Tyr1p (AWRI2936, 2940, 4124, and 2969) at two different time points during wine storage. The results are expressed as the mean and standard deviation of three independent replicates. ANOVAs were conducted separately for each time point. Means with the same letters are not significantly different from each other (Tukey's test, alpha = 0.05). Concentrations of MeSAc at the 15 month time point were below the limit of quantification of the technique (<5 μg L^{-1}).

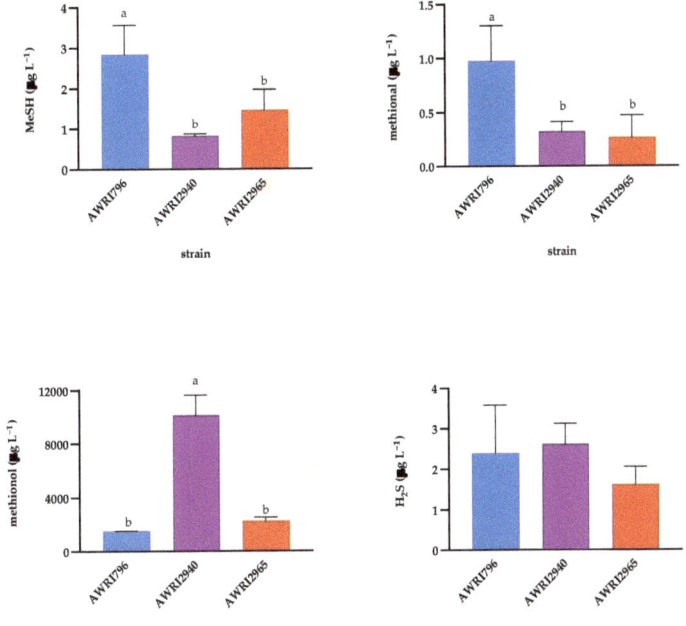

Figure 6. Concentrations (in µg L^{-1}) of VSCs after fermentation of a synthetic grape medium (SGM). Fermentations were carried out with the parent AWRI796, and two variants carrying mutations in either Aro4p (AWRI2965) or Tyr1p (AWRI2940). The results are expressed as the mean and standard deviation of three independent replicates. Means with the same letters are not significantly different from each other (Tukey's test, alpha = 0.05). No DMS and MeSAc were detected.

After 15 months in-bottle, differences in wine sensory profiles were more apparent, highlighted by more robust statistical evidence and larger F-ratios for more attributes in the ANOVA (Table S6). Seven attributes were influenced by the strains (Table 2 and Table S7). There was very strong evidence ($p < 0.001$) that strains produced wines with different intensities of 'yellow colour', 'cooked vegetable/potato aroma', 'sourness' and 'sweetness'. There was evidence ($p < 0.05$) for differences in 'pungency', 'rose aroma' and 'stone fruit flavour'. Notably, at this time point, judges chose to rate a specific 'rose aroma' quality rather than the more general 'floral' aroma attribute used to describe the wines at three months. Wine from strain AWRI2940 was notably higher in 'rose aroma', as well as in 'yellow colour' and 'sourness', while lower in 'cooked vegetable/potato' and 'sweetness'. Strains AWRI2936, AWRI2965, and AWRI4124 were rated with intermediate values for 'rose aroma' (Table 2), compared to the lowest (AWRI796 parent) and highest (AWRI2940) strains. Wines made with yeast strains AWRI796 and AWRI4124 were rated highly in the 'cooked vegetable/potato aroma'.

Table 1. Mean scores for a subset of appearance, aroma, and palate attributes of the yeast strains after 3 months in-bottle.

Yeast	Appearance	Aroma						Palate					
	Yellow Colour	Stone Fruit	Pungent	Floral	Grassy	Flint	Cooked Veg/Potato	Sweetness	Bitterness	Astringency	Stone Fruit F	Grassy F	Flint F
AWRI796	3.85	3.56	4.87	4.82	2.96	1.61	1.62	1.95	4.90	3.43	3.82	3.31	1.69
AWRI2936	4.06	3.53	4.99	4.61	2.38	1.61	1.35	1.87	4.92	3.59	3.82	3.12	1.66
AWRI2940	4.47	3.70	5.05	5.28	2.67	1.31	1.08	1.46	5.19	3.57	3.47	3.53	1.48
AWRI2965	4.03	3.66	5.02	4.68	2.92	1.90	1.28	1.42	5.26	3.61	3.32	3.33	2.02
AWRI2969	4.03	3.81	4.78	4.68	2.78	1.57	1.32	1.48	5.21	3.59	3.31	3.54	1.55
AWRI4124	4.09	3.78	4.85	4.42	2.72	1.40	1.86	1.69	5.21	3.64	3.76	3.36	1.75
HSD	0.37	ns	ns	0.79	ns	ns	ns	ns	ns	ns	ns	0.41	ns

F: Flavour. HSD ($p = 0.05$) values included for the significant attributes ($p < 0.05$), ns: not significantly different.

Table 2. Mean scores for a subset of appearance, aroma, and palate attributes of the yeast strains after 15 months in-bottle.

Yeast	Appearance	Aroma						Palate					
	Yellow Colour	Stone Fruit	Pungent	Rose	Grassy	Flint	Cooked Veg/Potato	Sweetness	Bitterness	Astringency	Stone Fruit F	Sourness	Rose F
AWRI796	4.05	3.24	4.41	3.72	2.31	1.43	2.38	1.44	3.44	2.80	3.31	5.04	2.07
AWRI2936	4.13	3.45	4.56	3.94	2.22	1.45	1.66	1.64	3.42	2.64	3.66	5.09	2.19
AWRI2940	4.79	3.28	4.37	4.55	2.21	1.16	1.06	0.78	3.78	2.86	2.96	5.46	2.77
AWRI2965	4.19	3.51	4.20	4.03	2.11	1.42	1.14	1.46	3.47	2.83	3.42	5.10	2.24
AWRI4124	4.40	3.19	4.47	4.06	2.11	1.61	2.36	1.37	3.47	2.84	3.29	5.29	2.31
HSD	0.27	ns	0.33	0.75	ns	ns	0.72	0.49	ns	ns	0.61	0.29	ns

F: Flavour. HSD ($p = 0.05$) values included for the significant attributes ($p < 0.05$), ns: not significantly different.

2.5. Relationships between Chemical Composition and Sensory Data

Partial least squares regression (PLS-R) was used to investigate the relationships between wine composition and sensory attributes for both time points at which the wines were evaluated (Figure 7). Chemical compounds and sensory attributes situated together in Figure 7C,D are covariant, and attributes toward the outside of the plots were well modelled. Compounds that were significant contributors to the overall model are indicated (sig analytes), while the magnitude of their regression coefficient can identify compounds most implicated in a specific attribute.

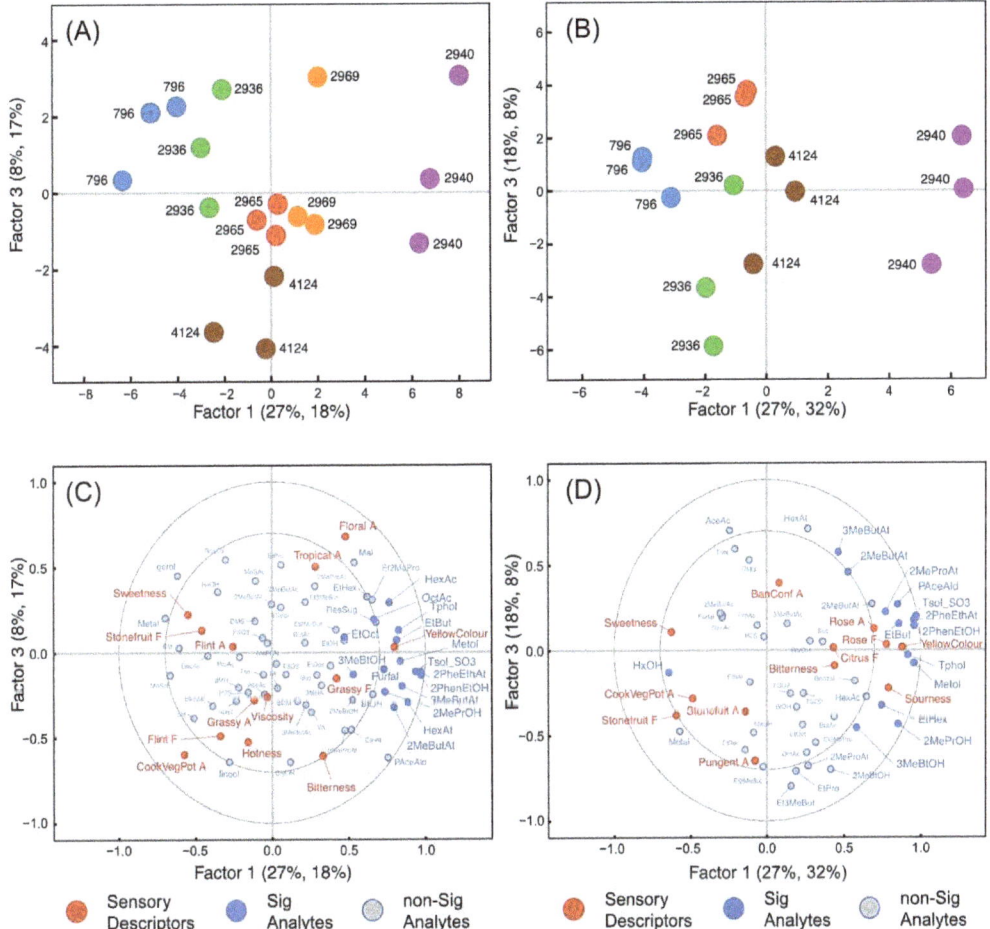

Figure 7. Factor 1 and 3 for the scores (**A,B**) and loadings (**C,D**) plots from PLS regression models for Chardonnay wines after 3 months (**A,C**) and 15 months (**B,D**) of aging generated using important sensory attributes (Y variables, red) and com-positional compounds (X variables, blue) for each fermentation replicate of the winemaking treatments. Compounds significant to each PLS model are indicated by filled blue circles and compounds with a lesser contribution to the model are shown as open circles. The proportion of the X-variance explained by the two factors is denoted by the first value in parentheses, the proportion of Y-variance by the second value. Chemical composition and labels are summarised in Tables S10 and S11.

Both the 3-month and 15-month PLS-R models indicated three optimum factors, explaining 50% and 54% of the variance of the sensory data, respectively. Figure 7 shows the scores (A,B) and loadings (C,D) plots for Factors 1 and 3. In both models, higher alcohols (2-PE, TOL, TOL-SO$_3$H, methionol), and the compounds 2-PEA, ethyl butanoate, 2- and 3-methylbutyl acetate, 2-methylpropanol, 3-methylbutanol were significant contributors to the sensory differences and were heavily loaded on Factor 1.

Of the sensory attributes at the 3-month time-point, 'yellow colour' (R^2 calibration 0.69 and R^2 validation 0.51), 'floral aroma' (R^2 calibration 0.67 and R^2 validation 0.20), 'cooked vegetable/potato aroma' (R^2 calibration 0.75 and R^2 validation 0.27), 'sweetness' (R^2 calibration 0.56 and R^2 validation 0.28), 'viscosity' (R^2 calibration 0.40 and R^2 validation 0.20), and 'stone fruit flavour' (R^2 calibration 0.59 and R^2 validation 0.30) were relatively well modelled but not so well predicted, indicated by the low R^2 validation. TOL and TOL-SO$_3$H were significantly associated with 'yellow colour'. No compounds were significant for 'floral aroma' but two monoterpenes, *cis*-rose oxide and α-terpineol, and several esters had moderately sized positive regression coefficients (values > 0.04) for this attribute, while 2-PE and 2-PEA were only weakly positively associated. 'Cooked vegetable/potato aroma' was associated with several sulfur compounds, notably H$_2$S and MeSH. 'Bitterness' was most strongly related to volatile acidity and 2-phenylacetaldehyde, with TyrOH and TOL-SO$_3$H weakly associated. TOL-SO$_3$H and volatile acidity, in addition to 2-PE and 2-PEA, were significantly negatively associated with 'sweetness'.

In the 15-month PLS model, similar links between compounds and sensory attributes emerged, but this model was generally stronger and predicted most attributes well. From the three-factor model, 'yellow colour' (R^2 calibration 0.83 and R^2 validation 0.67), 'stone fruit aroma' (R^2 calibration 0.55 and R^2 validation 0.22), 'rose aroma' (R^2 calibration 0.51 and R^2 validation 0.07), 'cooked vegetable/potato aroma' (R^2 calibration 0.76 and R^2 validation 0.59), 'sweetness' (R^2 calibration 0.39 and R^2 validation 0.23), 'stone fruit flavour' (R^2 calibration 0.60 and R^2 validation 0.34), 'rose flavour' (R^2 calibration 0.62 and R^2 validation 0.21), and 'sourness' (R^2 calibration 0.69 and R^2 validation 0.48) were relatively well modelled, while 'banana confection aroma', 'citrus flavour', 'pungent aroma', and 'bitterness' were not modelled well. Similar to the model at three months, TOL and TOL-SO$_3$H were again significantly associated with 'yellow colour', together with several other compounds. Of the many compounds associated with 'rose aroma and flavour', 2-PE and 2-PEA had the largest positive regression coefficients, while the volatiles 2-phenylacetaldehyde, methionol, hexyl acetate, 2- and 3-methylbutyl acetate, ethyl butanoate, 2-methylpropyl acetate, ethyl 2-methyl propanoate, H$_2$S, and non-volatiles TOL and TOL-SO$_3$H were also associated with these attributes. There was strong evidence for an association between 'cooked vegetable/potato aroma' and the compounds methional (regression coefficient 0.09) and MeSH (regression coefficient 0.07); although the association with H$_2$S was weak, it had a relatively high regression coefficient of 0.04. There was strong evidence for a negative association between the compounds TOL and TOL-SO$_3$H and 'sweetness', and a positive association with 'sourness'. Although 'bitterness' was not well modelled, TOL and TOL-SO$_3$H were positively associated with this attribute, with relatively high regression coefficients.

3. Discussion

In this study, we characterised the chemical and sensory profiles of a group of five variants derived from the commercial wine strain AWRI796 in a pilot-scale winemaking trial in Chardonnay. These five yeast strains' aromatic higher alcohol profiles were previously related to specific mutations found in two aromatic amino acid biosynthesis pathway genes, *ARO4* and *TYR1* [22]. The product of *ARO4* catalyses the first step in aromatic amino acid biosynthesis. It was shown that the Aro4p^{Q166R} mutation was responsible for the overproduction of 2-PE by AWRI2965 during laboratory-scale fermentation and the intracellular accumulation of the aromatic amino acids tryptophan, phenylalanine, and tyrosine [23]. We confirmed that AWRI2965 accumulates 2-PE in addition to high quantities of TyrOH while also producing slightly elevated concentrations of TOL compared to the strain from which it was derived (AWRI796). Together, these results indicate that the overall effect of Aro4p^{Q166R} could be to redirect pathway flux towards the accumulation of aromatic amino acids, and their respective higher alcohols, through the Ehrlich pathway.

The other four strains characterised in this work harboured point mutations in *TYR1*, a gene encoding a prephenate dehydrogenase enzyme catalysing the penultimate step in tyrosine biosynthesis. The higher alcohol profile of the *TYR1* mutant strains differed relative to the profiles of both AWRI2965 and AWRI796. While the *TYR1* mutant strains overproduced both TOL and 2-PE, they also produced significantly lower concentrations of TyrOH than the parent. It has been shown that Tyr1p mutations result in reduced formation of tyrosine and overproduction of 2-PE [23,25]. These results are compatible with a decrease in prephenate dehydrogenase activity, limiting tyrosine production, and consequently TyrOH biosynthesis. The constraint on the tyrosine branch of the aromatic amino acid biosynthesis pathway results in metabolic overflow into the tryptophan and phenylalanine branches, causing the increased production of these amino acids and their respective higher alcohols [23,25].

We observed dynamic changes between TOL and its sulfonated adduct, TOL-SO$_3$H, during wine storage, an event associated with wine aging and promoted by small amounts of oxygen [15,16]. Recently, small amounts of TOL-SO$_3$H were detected during laboratory-scale fermentation of a Chardonnay must with a yeast strain that produced high concentrations of TOL and SO$_2$, indicating that sulfonation of TOL can also occur in anaerobic conditions [17]. Similarly, in our Chardonnay study, trace amounts of TOL-SO$_3$H were found just after alcoholic fermentation, but only in the wines made with the highest TOL producer, AWRI2940. At this stage, the concentration of total SO$_2$ in the wines averaged 21 mg L^{-1}, with no detectable free SO$_2$. Subsequently, after three months of storage at 15 °C, during which the wines had been in contact with high concentrations of externally added free SO$_2$, a substantial amount of the initial TOL produced by yeast was converted into its sulfonated adduct.

Our results demonstrate that the time required to reach equilibrium between TOL and its sulfonated adduct, typically between 3 and 6 months, is independent of the initial amount of TOL. Longer storage times have little or no effect on the yield of TOL-SO$_3$H. In wines made with low and moderate TOL producers, the maximum observed conversion percentage (>85%) is consistent with that reported in a range of commercial white wines of different ages [16]. The molar ratio between free SO$_2$ and TOL was between 13 and 132 in wines with low to moderate TOL concentrations favouring the formation of TOL-SO$_3$H. The high conversion yields are therefore not unexpected. In the wines made with the highest TOL producer, AWRI2940, this molar ratio was as low as 1.8, explaining the lower conversion yield. The TOL concentration found in Chardonnay wines fermented with the low and moderate TOL-producing strains was in the range of those reported in the literature for commercial white wines, typically below 5 mg L^{-1} [16,34,35]. At the TOL concentration found in commercial white wines, and even considering a total conversion into its sulfonated adduct, the expected loss of SO$_2$ due to this reaction would be less than 2 mg L^{-1}. A survey of a large number of commercial bottled Australian white wines [36] found an average of 31 mg L^{-1} of free SO$_2$ in the most recent vintages assessed, a concentration similar to that found in our Chardonnay wines at bottling (39 mg L^{-1}). These data suggest that the conversion of TOL into TOL-SO$_3$H makes only a limited contribution to the losses of SO$_2$ observed during wine development, which are primarily driven by reaction with dissolved O$_2$ in wines after bottling and by sulfonation of other species [16,36]. This conclusion is supported by the lack of a correlation between TOL-SO$_3$H concentration and SO$_2$ loss in wine observed here.

Wines made with AWRI2940 accumulated exceptionally high concentrations of both TOL and TOL-SO$_3$H. This high accumulation allowed us to assess the possible effects of TOL and TOL-SO$_3$H on white wine sensory properties. After 15 months in-bottle, wines made with AWRI2940 were rated higher in 'sourness' and lower in 'sweetness', and tended to be more bitter for some assessors even though these wines had slightly elevated fructose and glycerol concentrations (both sweet compounds individually) relative to the control. The PLS-R results suggest that both TOL and TOL-SO$_3$H may impart a degree of 'bitterness' to these wines while decreasing 'sweetness' and increasing 'sourness' ratings, probably through well-established taste–taste interactions [37], especially for wines produced with AWRI2940.

To our knowledge, no sensory recognition or detection thresholds for TOL-SO$_3$H in a wine matrix have been published, but Van Gemert [38] lists a wide range for the detection threshold of TOL in beer (between 10 and >414 mg/L). Less clear is the possible effect of TyrOH on the sensory properties of white wines. Despite the potential of this higher alcohol to induce a 'bitter finish' in alcoholic beverages such as sake and beer [12,14,30], concentrations of TyrOH between its detection and recognition thresholds have been suggested to have a positive 'taste-sharpening' effect in sake [14,39]. Consequently, some effort has been devoted to breeding sake strains that overproduce this higher alcohol [14,40]. Different sensory thresholds for TyrOH appear in the literature: while a high threshold value (346 mg L^{-1}) eliciting bitter taste was reported in water [29], lower and different detection thresholds have been reported for beer (between 20 and 100 mg L^{-1}) [12]. Even though typical concentrations of TyrOH (20–30 mg L^{-1}) have been reported to impart bitterness in wine [13], to the best of our knowledge, these claims are not supported by documented research. Sapis and Ribereau-Gayon [34] added 50 mg L^{-1} of TyrOH to white wine and observed no influence on sensory properties, and only when unrealistically high concentrations were added (500 mg L^{-1}) did TyrOH seem to depress the overall 'flavour quality' of wine. The fact that our wines made with AWRI2965, with a concentration of TyrOH exceeding 120 mg L^{-1}, were not rated differently to the control wine in palate attributes such as 'astringency', 'bitterness' or 'sourness', suggest a limited effect of this higher alcohol on white wine in-mouth sensory properties.

Over time in-bottle, the sensory differences among the wines produced by each yeast became larger. Although the magnitude of the strain effect on 'floral/rose aroma' was similar across the two time points (Tables S4 and S6), an aroma quality shift was indicated, with the more general 'floral aroma' attribute of the first study replaced with a specific odour quality of 'rose aroma' in the second study. The appearance of 'rose aroma' may result from a decrease in the concentration of other compounds such as monoterpenes and acetate esters through acid hydrolysis, which likely also contributed similar floral and fruit nuances to the young wines, and that could have overshadowed the specific 'rose aroma'. Over time in-bottle we observed a pronounced decrease, between 45% to 70% depending on the strain, in the concentration of some acetate esters (Tables S8 and S9). The concentration of 2-PEA also decreased with time (between 40–55%) in wines made with the yeast variants; however, the concentration of this compound remained at levels that would likely transmit the 'rose aroma' character to the wines even after 15 months of aging. In wines made using the parent strain, 2-PEA concentration decreased to a level unlikely to strongly direct 'rose aroma'. Practically, this finding may provide a method to extend the shelf-life of floral wine styles.

Another aspect related to the effect of aging is the formation of aldehydes from oxidation of the analogous higher alcohols. These aldehydes can have a much lower sensory threshold than their corresponding higher alcohols; for example, the sensory threshold for methional is about 1000-times lower than that of methionol [41,42]. Therefore, even limited oxidation can affect wine aroma substantially. In particular, methional has been linked with the 'cooked vegetable' off-flavour observed in oxidised wine [42]. Even though a general increase in the concentration of methional with storage time was observed, wines made with the variants AWRI2965 and AWRI2940 still showed lower concentrations of this compound than those made with the parent strain, and had lower scores in 'cooked vegetable/potato' aroma. Interestingly, although AWRI2940 produced an elevated concentration of methionol (also with a 'cooked vegetable/potato' aroma), this did not seem to contribute to the 'vegetal aroma' rated in wines made with this strain. The lack of a correlation between methionol and 'vegetal aroma' could be due to the masking effect of high concentrations of 2-PE present in the wines made with AWRI2940.

The production of methionol and other negative VSCs has been shown to be affected by the branch of the Ehrlich pathway responsible for the production of aromatic higher alcohols [6]. Recently, it has been shown that the deletion of *ARO8* in a wine strain, encoding for the aromatic transaminase Aro8p, decreased methionol formation after fermentation of a synthetic grape must [6]. The decrease in methionol formation in *ARO8* mutants indicates that by blocking the aromatic Ehrlich pathway, the catabolism of methionine might also be impaired.

Conversely, a highly active Ehrlich pathway may result in higher catabolism of methionine to methionol. The aromatic transaminase Aro9p, as well as the decarboxylase Aro10p, have been shown to be involved in the catabolism of both aromatic amino acids and methionine [5,7,8], and the expression of both *ARO9* and *ARO10* genes are highly induced by aromatic amino acids [43] and the end product TOL [44]. In this work, upregulation of the first step of the Ehrlich pathway was confirmed in two variants (AWRI2940 and 2965) by using an *ARO9*-promoter-BFP reporter gene (Figure S4). Therefore, we can hypothesise that the overproduction of aromatic amino acids and/or TOL in the different variants used in our study might lead to an activation of these two critical steps in the Ehrlich pathway and an increase in the catabolism of methionine to form methionol. The Ehrlich pathway-mediated catabolism of methionine, in turn, may limit the amount of methionine available for the formation of MeSH, methional, and MeSAc by the competing demethiolation pathway and/or by non-enzymatic reactions.

The idea that there is competition for methionine between the Erhlich and demethiolation pathways is supported by the VSC profile observed in the pilot-scale wines made with the three homozygous Tyr1p mutants, particularly with the most active strain AWRI2940. Again, AWRI2965, which harbours a mutation (Aro4p) earlier in the amino acid biosynthetic pathway, behaved somewhat differently than the Tyr1p variants. Higher concentrations of methionol were not observed in the wines produced using AWRI2965 but decreased concentrations of other undesirable VSCs such as MeSH, MeSAc or methional were observed. Interestingly, AWRI2965 also produced slightly higher concentrations of DMS in these conditions. Even though DMS is also associated with reductive off-odours and 'vegetal' aromas, at low concentrations (about 25 µg L^{-1}) it can be described as contributing 'blackcurrant' and 'red fruit' aromas, and it is considered to enhance the bouquet of some wine styles [45,46].

4. Materials and Methods

4.1. Microorganisms and Culture Conditions

The commercial diploid wine strain AWRI796, and its five variants AWRI2936, AWRI2940, AWRI2965, AWRI4124, and AWRI2969, were obtained from The Australian Wine Research Institute (AWRI) culture collection (Table S1). These variants had been isolated previously using toxic analogues of the amino acid phenylalanine, as described in [23]. Yeast cultures were maintained on solid YPD agar plates (2% glucose, 2% peptone, 1% yeast extract, and 2% agar).

4.2. Laboratory-Scale Fermentation in a Synthetic Grape Medium

Laboratory-scale fermentations were performed in triplicate in a synthetic grape medium (SGM) [47], with a concentration of 6 mg L^{-1} of methionine. SGM was filtered through 0.22 µm Stericup filters (Millipore). Yeast starter cultures were prepared by growing cells aerobically in YPD medium for 24 h to stationary phase at 22 °C. Then, 1×10^6 cells mL^{-1} were inoculated into 50% diluted SGM medium and grown for another 48 h at 22 °C. The acclimatised cells were inoculated into 100 mL of SGM at a density of 1×10^6 cells mL^{-1}. Fermentations were conducted at 17 °C in 100 mL glass bottles (Schott Duran), fitted with stir bars and stirred at 200 rpm using a magnetic stirrer. The lids of the bottles were fitted with selective H_2S detector tubes (Komyo, Kitagawa, Japan) to measure the release of H_2S during fermentation. Fermentation progress was followed by CO_2 weight loss, measured every 24 h. After fermentation, the wines were cold settled at 4 °C for 5 days and sampled for different volatile and non-volatile compound analyses.

4.3. Pilot-Scale Winemaking

The pilot-scale winemaking trial with Chardonnay juice was performed by the Wine Innovation Cluster (WIC) winemaking services, according to a standardised white winemaking protocol. Hand-harvested Chardonnay grapes from the McLaren Vale region (South Australia, Australia) were used. The basic chemical parameters of the Chardonnay juice were: 12.7 °Baumé, yeast assimilable nitrogen 217 mg L^{-1}, and pH 3.29. A concentration of 25 mg L^{-1} of SO_2 was added to the grape must at the crusher. Yeast strains were grown for 48 h in filter-sterilised neutral grape concentrate (Tarac Technologies, Nuriootpa, Australia), which had been previously diluted to ~6 °Baume and pH adjusted to 3.5. Cells were inoculated at a density of approximately 2×10^6 cells mL^{-1} in 19 L of the Chardonnay juice, and fermentation was conducted at 15 °C in 20 L stainless steel kegs in triplicate. When °Baumé was below 3, wines were moved to 20 °C. Irrespective of the starter culture used, wines got stuck around 1 °Baumé (day 16–18 of fermentation). Ferments were then rescued at day 26 by the addition of the commercial yeast Lalvin EC 1118 (Lallemand, Adelaide, SA, Australia). Once alcoholic fermentation had finished (day 31), wines were sulfured with 80 mg L^{-1} of SO_2, and cold-stabilised for approximately 2 months at 0 °C. Before bottling, SO_2 concentration was adjusted to between 35–40 mg L^{-1} of free SO_2. Screw-cap sealed bottled wines (375 mL) were stored in the dark at a constant temperature of 15 °C.

4.4. Targeted Analyses of Volatile Compounds

Targeted analyses of fermentation-derived compounds (higher alcohols, acids, and esters) were performed by Metabolomics Australia (Adelaide) by GC-MS using a stable isotope dilution assay [48] at the end of fermentation, as well as 3 and 15 months post-bottling.

Analysis of monoterpenoids (linalool, cis-rose oxide, α-terpineol, nerol, geraniol) and C_{13}-norisoprenoids (β-damascenone and β-ionone) was performed at 3 months post-bottling by GC-MS on an Agilent 6890 gas chromatograph equipped with a Gerstel MPS2 autosampler and coupled to an Agilent 5973N mass selective detector. Sample preparation was as follows: 10 mL of wine was transferred into a 20 mL crimp-cap, headspace-SPME vial (Grace Davison) with 3 g of NaCl followed by 50 µL of a combined d_4-β-damascenone, d_3-α-ionone and d_3-β-ionone internal standard solution. Instrument control was performed with Agilent G1701EA Revision E.02.02 ChemStation software. The gas chromatograph was fitted with an Agilent DB-5ms 30 m × 0.25 mm × 0.5 um. Helium (Ultra High Purity) was used as the carrier gas with linear velocity 46 cm/s, flow rate 1.6 mL/min in constant flow mode. The oven temperature was started at 40 °C, held at this temperature for 2 min, then increased to 190 °C at 8 °C/min and held at this temperature for 5.25 min. The vial and its contents were heated to 60 °C for 10 min in the heater/agitator with the agitator on for 5 s and off for 2 s at 500 r.p.m. A Supelco grey 2 cm SPME fibre was exposed to the sample during this heating time through the septum. The fibre was then injected into a split/splitless inlet in splitless mode. The analytes were desorbed into a Supelco 0.75 mm ID sleeveless SPME liner for 10 min, which was held at 200 °C. The purge flow to the split vent was 50 mL/min at 2.1 min with the septum purge flow turned off. The mass spectrometer quadrupole temperature was set at 150 °C, the source was set at 230 °C and the transfer line was held at 250 °C. EMV Mode was set to Gain Factor = 1.00 and spectra were recorded in SIM mode.

4.5. Analysis of Volatile Sulfur Compounds (VSCs) and Aldehydes

The VSCs H_2S, MeSH, DMS, diethyl sulfide, dimethyl disulfide, diethyl disulfide, ethanethiol, carbon disulfide, MeSAc, and ethyl thioacetate were quantified using an Agilent 355 sulfur chemiluminescence detector coupled to an Agilent 6890A gas chromatograph (Forest Hill, Melbourne, VIC, Australia), as described previously [49]. Reference standards of the different compounds were of the highest purity as supplied by Sigma-Aldrich (Castle Hill, Sydney, NSW, Australia) and Lancaster Synthesis (Jomar Bioscience, Adelaide, SA, Australia). Sodium hydrosulfide hydrate and sodium thiomethoxide were used as standards for H_2S and MeSH, respectively. Ethylmethyl sulfide and propyl thioacetate were used as internal standards. Analytes were identified by comparison of their retention times with those of the corresponding pure reference compounds.

Analysis of the sulfur-containing compounds methionol and methional, as well as 2-methylpropanal, 3-methylbutanal, furfural, 5-methylfurfural, benzaldehyde, and 2-phenylacetadehyde was performed by GC-MS/MS, as described in [50]. Aldehydes were determined after derivatisation directly in the wine with O-(2,3,4,5,6-pentafluorobenzyl)hydroxylamine hydrochloride (Sigma-Aldrich). Reference standards for these compounds of the highest purity were purchased from Sigma-Aldrich. Isotopically labelled analogues for furfural, methionol, methional, benzaldehyde, and 2-phenylacetaldehyde were used as internal standards for accurate quantification of these compounds. For the quantitation of 2-methylpropanal and 3-methylbutanal, d_5-benzaldehyde was used as an internal standard. Similarly, d_4-furfural was used for the determination of 5-methylfurfural. With the exception of d_4-furfural (CDN Isotopes, Sydney, NSW, Australia), the synthesis of the other isotopically labelled standards was carried out in-house as described in [50].

VSCs and aldehydes were analysed 3 and 15 months post-bottling.

4.6. Analysis of Principal Non-Volatile Compounds

The concentrations of sugars, ethanol, glycerol, and organic acids (acetic, malic, and succinic) were measured by HPLC using a Bio-Rad HPX-87H column, as described previously [51]. Reference standards of the highest purity were obtained from Sigma-Aldrich.

TyrOH, TOL, and TOL-SO_3H were analysed on an Agilent 1200SL HPLC using a Phenomenex Kinetex PFP column (2.6 µm particle size, 2.1 mm × 150 mm) at different time points (end of alcoholic fermentation, and then 3, 6, 12, and 15 months post-bottling). The injection volume was 5 µL. The column was eluted at 45 °C with a gradient of 0.1% formic acid in Milli-Q water (A) and 0.1% formic acid in acetonitrile (B) at a flow rate of 0.4 mL min^{-1}. The gradient was as follows: an initial isocratic hold (0% B) for 8 min, then gradient to 5% B over 32 min, gradient to 25% B over 9 min, then gradient to 80% B over 3 min, held isocratically at 80% B for 3 min, and dropped to 0% B and held for another 15 min. Absorbance at 280 nm was monitored with an Agilent 1260 Series G7117C DAD, while fluorescence was monitored at excitation and emission wavelengths of 280 and 350 nm, respectively, with an Agilent 1260 Series G7121B FLD. Quantification of TOL and TyrOH was performed using the absorbance detector, while TOL-SO_3H was quantified using the fluorescence detector. Reference standards for TyrOH and TOL were obtained from Sigma-Aldrich. TOL-SO_3H was synthesised as previously described by Arapitsas, Guella and Mattivi [16] with modifications. Briefly, TOL solution (2.5 g in 200 mL EtOH) was slowly poured into a potassium metabisulfite solution (5 g in 500 mL H_2O) with stirring and reacted at room temperature for 48 h. The reaction product was dried under a vacuum (30 °C) and dissolved in H_2O. The product was purified using preparative HPLC with a Dionex UltiMate 3000 system, a C_{18} Synergi Hydro RP column (250 × 21.2 mm, 4 µm pore size, Phenomenex, Lane Cove, Australia), and solvent system of 100% H_2O (A) and 100% acetonitrile (B). Gradient: 0–10 min 0% B, 10–25 min 50% B, 25–35 min 100% B, 8 mL/min flow rate. The structure of TOL-SO_3H was confirmed using HRMS and NMR (400 MHz, Bruker, Germany) with samples in D_2O at 300 K. Results were processed using Topspin software. M-H mass (m/z): 240.0325; Chemical shifts for ^{1}H-NMR (400 MHz, D_2O) and ^{13}C-NMR (D_2O) concur with those previously reported [16].

4.7. Sensory Evaluation

Quantitative descriptive analysis (QDA) sensory studies [52] were conducted on the Chardonnay wines at two time points (3 months and 15 months post-bottling); however, wine produced with strain AWRI2969 was not included in the second study. Two panels of 10 judges with average ages of 48 (SD = 9.2, nine females, one male) and 50 (SD = 6.8, eight females, two males) years, respectively, were convened for each study. All panellists were part of the external AWRI trained descriptive analysis panel and had extensive experience in wine QDA. For both evaluations, assessors attended three two-hour training sessions to determine appropriate descriptors for rating in the formal sessions. For the 15-month evaluation, attributes used from the 3-month study were presented for consideration. No other information about the samples was given to the assessors at the second time point. All the wines from the study were progressively used during training sessions and appropriate attributes and definitions describing the appearance, aroma, and palate were agreed upon by judges in a consensus-based approach. Sensory standards for these descriptive attributes were presented, discussed, and recipes refined to represent attributes rated for the wines closely. These standards were available during all subsequent sessions and panellists revisited them at the beginning of each formal assessment session. The attributes rated, definitions, and standard recipes can be found in Tables S8 and S9, while the chemical composition of the wines is summarised in Tables S10 and S11. In both studies, samples were presented to panellists in 30 mL aliquots in 3-digit-coded, covered, ISO standard wine glasses at 22–24 °C, in isolated booths under daylight-type lighting, with randomised presentation order (modified Williams Latin Square). In the 3-month evaluation, wines were presented to the panel in duplicate while in the 15-month evaluation, the wines were presented in triplicate. Assessors were forced to have a 60 s rest between samples and were encouraged to rinse with water, and a minimum 10 min rest between sets of three samples. During the 10 min break, assessors were requested to leave the booths. For the 3-month evaluation, 12 samples were presented per day while for the 15-month evaluation, 15 samples were assessed per day. All samples were expectorated. Compusense Cloud sensory evaluation software (Compusense Inc., Guelph, Canada) was used on both occasions to generate presentation replicate designs and collect sensory data. The intensity of each attribute was rated using an unstructured 15 cm line scale (numericised 0 to 10), with indented anchor points of 'low' and 'high' placed at 10% and 90%, respectively. Panel performance was assessed using Compusense software and R with the SensomineR (sensominer.free.fr/) and FactomineR (factominer.free.fr/) packages.

4.8. Statistical Analysis

Minitab 19 (Minitab Inc., Sydney, NSW, Australia) was used for statistical analysis of the compositional data which were analysed by one-way analysis of variance (ANOVA). Multiple comparisons of the analyte concentration with respect to treatment were undertaken using Tukey's honestly significant difference (HSD) test (alpha = 0.05), and p values were determined by a two-tailed Student's t test. For the sensory data, ANOVA was carried out using Minitab 19. The effects of the yeast strain treatment, judge, judge by strain, ferment replicate nested into strain, judge by ferment replicate nested into strain, presentation replicate nested into strain, and ferment replicate were assessed, treating judge as a random effect. Following ANOVA, a protected HSD value was calculated using the mean square term of the judge × strain interaction at a 95% confidence level for attributes with a significant ($p < 0.05$) treatment effect. To explore the relationship between wine chemical composition and sensory attributes, PLS-R was conducted for each wine replicate, as described in [53] with some modifications. Sensory attribute responses (Ys) were included in the models if some statistical evidence ($p < 0.10$) signalled a treatment effect or a high F-ratio was found indicating potential treatment effects which may have been overshadowed by judge, fermentation, or presentation replicate variation.

5. Conclusions

This study confirmed that the higher alcohol overproduction phenotype of five variants derived from the commercial wine strain AWRI796 is maintained in pilot-scale white winemaking conditions. This overproduction was associated with meaningful changes in wine sensory profiles, especially after some period of bottle storage. The effect of these strains on wine chemical composition was not just limited to the overproduction of 2-PE but also to an increase in the concentration of the higher alcohols TyrOH and/or TOL and to the formation of VSCs. Associations between these compounds and 'sweet', 'sour' and 'bitter' tastes, and 'cooked vegetable/potato aroma', were identified. These results highlight the intricate connections between the metabolism of aromatic amino acids and the sulfur-containing amino acid methionine during fermentation, ultimately influencing wine flavour.

The various yeast strains isolated in this study provide novel tools for winemakers to adjust and preserve wine style. In particular, AWRI2965 has excellent potential as a white wine winemaking yeast, imparting accentuated and lasting rose/floral aromas to wines.

More research will be needed to understand the compositional drivers of bitterness. In particular, the role of higher alcohols derived from the metabolism of aromatic amino acids TyrOH, TOL, and its sulfonated derivative TOL-SO$_3$H, needs to be elucidated along with the physico-chemical conditions such as pH, temperature, storage time, and SO$_2$ concentration, which might influence the equilibrium between these compounds in the finished wine.

Supplementary Materials: The following are available online, Table S1. Strains used in this study; Table S2. Higher alcohols and esters produced following alcoholic fermentation of Chardonnay; Table S3. Basic wine composition at bottling of Chardonnay wines; Table S4. F-ratios, probability values, degrees of freedom, and mean square error from the analysis of variance conducted following sensory analysis of wines after 3 months in-bottle; Table S5. Mean scores and Tukey's HSD values for sensory attributes after 3 months in-bottle; Table S6. F-ratios, probability values, degrees of freedom, and mean square error from the analysis of variance conducted following sensory analysis of wines after 15 months in-bottle; Table S7. Mean scores and Tukey's HSD values for sensory attributes after 15 months in-bottle; Table S8. Sensory attributes, definitions and composition of reference standards after 3 months in-bottle; Table S9. Sensory attributes, definitions and composition of reference standards after 15 months in-bottle; Table S10. Composition of the wines after 3 months in-bottle; Table S11. Composition of the wines after 15 months in-bottle. Figure S1. Relationship between 2-PE and 2-PEA production at the end of alcoholic fermentation in the Chardonnay wines; Figure S2. Relationship between concentrations of TOL at the end of alcoholic fermentation and its maximum percentage of conversion into TOL-SO$_3$H in the Chardonnay wines; Figure S3. Relationship between the decrease in free and total SO$_2$ concentrations and TOL-SO$_3$H formation in the Chardonnay wines at different time points during ageing; Figure S4. Quantitative assay of the activation of the *ARO9* promoter by using a reporter system.

Author Contributions: Conceptualisation, A.G.C., S.A.S.; methodology, A.G.C., D.E.N., M.S., A.S.; formal analysis, M.S., A.S., D.E.N.; investigation, A.G.C., R.K.; data curation, M.S., A.S., D.E.N.; writing—original draft preparation, A.G.C., S.A.S., D.E.N.; writing—review and editing A.G.C., S.A.S., D.E.N., I.L.F.; visualisation, A.G.C. All authors have read and agreed to the published version of the manuscript.

Funding: The Australian Wine Research Institute (AWRI), a member of the Wine Innovation Cluster in Adelaide, is supported by Australia's grapegrowers and winemakers through their investment body Wine Australia, with matching funds from the Australian Government.

Institutional Review Board Statement: Not applicable.

Informed Consent Statement: Not applicable.

Data Availability Statement: All data has been made available through the manuscript itself or via supplemental materials.

Acknowledgments: The AWRI external sensory panellists and sensory analysts Eleanor Bilogrevic and Desireé Likos are thanked for their involvement in the sensory evaluations. The authors also thank John Gledhill for his valuable contribution to the winemaking.

Conflicts of Interest: The authors declare no conflict of interest. The funders had no role in the design of the study; in the collection, analyses, or interpretation of data; in the writing of the manuscript, or in the decision to publish the results.

Sample Availability: Samples are available from the authors for a limited time. Wine will not be stored indefinitely.

References

1. Vilanova, M.; Genisheva, Z.; Graña, M.; Oliveira, J.M. Determination of odorants in varietal wines from international grape cultivars (*Vitis vinifera*) grown in NW Spain. *S. Afr. J. Enol. Vitic.* **2013**, *34*, 212–222. [CrossRef]
2. Holt, S.; Miks, M.H.; de Carvalho, B.T.; Foulquié-Moreno, M.R.; Thevelein, J.M. The molecular biology of fruity and floral aromas in beer and other alcoholic beverages. *FEMS Microbiol. Rev.* **2019**, *43*, 193–222. [CrossRef] [PubMed]
3. Kitagaki, H.; Kitamoto, K. Breeding research on sake yeasts in Japan: History, recent technological advances, and future perspectives. *Annu. Rev. Food Sci. Technol.* **2013**, *4*, 215–235. [CrossRef]
4. Hazelwood, L.A.; Daran, J.M.; van Maris, A.J.; Pronk, J.T.; Dickinson, J.R. The Ehrlich pathway for fusel alcohol production: A century of research on *Saccharomyces cerevisiae* metabolism. *Appl. Environ. Microbiol.* **2008**, *74*, 2259–2266. [CrossRef] [PubMed]
5. Urrestarazu, A.; Vissers, S.; Iraqui, I.; Grenson, M. Phenylalanine- and tyrosine-auxotrophic mutants of Saccharomyces cerevisiae impaired in transamination. *Mol. Gen. Genet.* **1998**, *257*, 230–237. [CrossRef]
6. Deed, R.C.; Hou, R.; Kinzurik, M.I.; Gardner, R.C.; Fedrizzi, B. The role of yeast *ARO8*, *ARO9* and *ARO10* genes in the biosynthesis of 3-(methylthio)-1-propanol from L-methionine during fermentation in synthetic grape medium. *FEMS Yeast Res.* **2019**, *19*, foy109. [CrossRef]
7. Vuralhan, Z.; Luttik, M.A.; Tai, S.L.; Boer, V.M.; Morais, M.A.; Schipper, D.; Almering, M.J.; Kotter, P.; Dickinson, J.R.; Daran, J.M.; et al. Physiological characterization of the ARO10-dependent, broad-substrate-specificity 2-oxo acid decarboxylase activity of Saccharomyces cerevisiae. *Appl. Environ. Microbiol.* **2005**, *71*, 3276–3284. [CrossRef]
8. Perpete, P.; Duthoit, O.; De Maeyer, S.; Imray, L.; Lawton, A.I.; Stavropoulos, K.E.; Gitonga, V.W.; Hewlins, M.J.; Dickinson, J.R. Methionine catabolism in Saccharomyces cerevisiae. *FEMS Yeast Res.* **2006**, *6*, 48–56. [CrossRef]
9. Fang, Y.; Qian, M. Aroma compounds in Oregon Pinot Noir wine determined by aroma extract dilution analysis (AEDA). *Flavour Fragr. J.* **2005**, *20*, 22–29. [CrossRef]
10. de-la-Fuente-Blanco, A.; Saenz-Navajas, M.P.; Ferreira, V. On the effects of higher alcohols on red wine aroma. *Food Chem.* **2016**, *210*, 107–114. [CrossRef]
11. Ferreira, V. Volatile aroma compounds and wine sensory attributes. In *Managing Wine Quality*; Reynolds, A.G., Ed.; Woodhead Publishing: New York, NY, USA, 2010; pp. 3–28.
12. Szlavko, C. Tryptophol, tyrosol and phenylethanol—the aromatic higher alcohols in beer. *J. Inst. Brew.* **1973**, *79*, 283–288. [CrossRef]
13. Sáenz-Navajas, M.-P.; Fernández-Zurbano, P.; Ferreira, V. Contribution of Nonvolatile Composition to Wine Flavor. *Food Rev. Int.* **2012**, *28*, 389–411. [CrossRef]
14. Soejima, H.; Tsuge, K.; Yoshimura, T.; Sawada, K.; Kitagaki, H. Breeding of a high tyrosol-producing sake yeast by isolation of an ethanol-resistant mutant from a *trp3* mutant. *J. Inst. Brew.* **2012**, *118*, 264–268. [CrossRef]
15. Arapitsas, P.; Ugliano, M.; Perenzoni, D.; Angeli, A.; Pangrazzi, P.; Mattivi, F. Wine metabolomics reveals new sulfonated products in bottled white wines, promoted by small amounts of oxygen. *J. Chromatogr. A* **2016**, *1429*, 155–165. [CrossRef]
16. Arapitsas, P.; Guella, G.; Mattivi, F. The impact of SO_2 on wine flavanols and indoles in relation to wine style and age. *Sci. Rep.* **2018**, *8*, 858. [CrossRef] [PubMed]
17. Álvarez-Fernández, M.A.; Carafa, I.; Vrhovsek, U.; Arapitsas, P. Modulating wine aromatic amino acid catabolites by using *Torulaspora delbrueckii* in sequentially inoculated fermentations or *Saccharomyces cerevisiae* alone. *Microorganisms* **2020**, *8*, 1349. [CrossRef]
18. Ferreira, V.; Fernández, P.; Peña, C.; Escudero, A.; Cacho, J.F. Investigation on the role played by fermentation esters in the aroma of young Spanish wines by multivariate analysis. *J. Sci. Food Agric.* **1995**, *67*, 381–392. [CrossRef]
19. Lilly, M.; Lambrechts, M.G.; Pretorius, I.S. Effect of increased yeast alcohol acetyltransferase activity on flavor profiles of wine and distillates. *Appl. Environ. Microbiol.* **2000**, *66*, 744–753. [CrossRef]
20. Waterhouse, A.L.; Sacks, G.L.; Jeffery, D.W. *Understanding Wine Chemistry*; Wiley: Chichester, UK, 2016; p. 443.
21. Rodriguez-Bencomo, J.J.; Conde, J.E.; Rodriguez-Delgado, M.A.; Garcia-Montelongo, F.; Perez-Trujillo, J.P. Determination of esters in dry and sweet white wines by headspace solid-phase microextraction and gas chromatography. *J. Chromatogr. A* **2002**, *963*, 213–223. [CrossRef]
22. Bordiga, M.; Lorenzo, C.; Pardo, F.; Salinas, M.R.; Travaglia, F.; Arlorio, M.; Coisson, J.D.; Garde-Cerdan, T. Factors influencing the formation of histaminol, hydroxytyrosol, tyrosol, and tryptophol in wine: Temperature, alcoholic degree, and amino acids concentration. *Food Chem.* **2016**, *197 Pt B*, 1038–1045. [CrossRef]
23. Cordente, A.G.; Solomon, M.; Schulkin, A.; Leigh Francis, I.; Barker, A.; Borneman, A.R.; Curtin, C.D. Novel wine yeast with *ARO4* and *TYR1* mutations that overproduce 'floral' aroma compounds 2-phenylethanol and 2-phenylethyl acetate. *Appl. Microbiol. Biotechnol.* **2018**, *102*, 5977–5988. [CrossRef] [PubMed]

24. Dueñas-Sanchez, R.; Perez, A.G.; Codon, A.C.; Benitez, T.; Rincon, A.M. Overproduction of 2-phenylethanol by industrial yeasts to improve organoleptic properties of bakers' products. *Int. J. Food Microbiol.* **2014**, *180*, 7–12. [CrossRef]
25. Fukuda, K.; Watanabe, M.; Asano, K. Altered Regulation of Aromatic Amino Acid Biosynthesis in β-Phenylethyl-alcohol-overproducing Mutants of Sake Yeast Saccharomyces cerevisiae. *Agric. Biol. Chem.* **1990**, *54*, 3151–3156. [CrossRef]
26. Tofalo, R.; Perpetuini, G.; Battistelli, N.; Tittarelli, F.; Suzzi, G. Correlation between IRC7 gene expression and 4-mercapto-4-methylpentan-2-one production in Saccharomyces cerevisiae strains. *Yeast* **2020**, *37*, 487–495. [CrossRef] [PubMed]
27. Belda, I.; Ruiz, J.; Alastruey-Izquierdo, A.; Navascues, E.; Marquina, D.; Santos, A. Unraveling the Enzymatic Basis of Wine "Flavorome": A Phylo-Functional Study of Wine Related Yeast Species. *Front. Microbiol.* **2016**, *7*, 12. [CrossRef]
28. Cordente, A.G.; Heinrich, A.; Pretorius, I.S.; Swiegers, J.H. Isolation of sulfite reductase variants of a commercial wine yeast with significantly reduced hydrogen sulfide production. *FEMS Yeast Res.* **2009**, *9*, 446–459. [CrossRef] [PubMed]
29. Takahashi, K.; Tadenuma, M.; Kitamoto, K.; Sato, S. l-Prolyl-l-leucine Anhydride A Bitter Compound Formed in Aged Sake. *Agric. Biol. Chem.* **1974**, *38*, 927–932. [CrossRef]
30. Singleton, V.L.; Noble, A.C. Wine Flavor and Phenolic Substances. In *Phenolic, Sulfur, and Nitrogen Compounds in Food Flavors*; American Chemical Society: Washington, DC, USA, 1976; Volume 26, pp. 47–70.
31. Zhang, Q.; Jia, K.Z.; Xia, S.T.; Xu, Y.H.; Liu, R.S.; Li, H.M.; Tang, Y.J. Regulating Ehrlich and demethiolation pathways for alcohols production by the expression of ubiquitin-protein ligase gene HUWE1. *Sci. Rep.* **2016**, *6*, 20828. [CrossRef]
32. Isogai, A.; Kanda, R.; Hiraga, Y.; Nishimura, T.; Iwata, H.; Goto-Yamamoto, N. Screening and identification of precursor compounds of dimethyl trisulfide (DMTS) in Japanese sake. *J. Agric. Food Chem.* **2009**, *57*, 189–195. [CrossRef]
33. Bell, S.J.; Henschke, P.A. Implications of nitrogen nutrition for grapes, fermentation and wine. *Aust. J. Grape Wine Res.* **2008**, *11*, 242–295. [CrossRef]
34. Sapis, J.C.; Ribereau-Gayon, P. Étude dans les vins du tyrosol, du tryptophol, de l'alcool phényléthylique et de la γ-butyrolactone, produits secondaires de la fermentation alcoolique. II-Présence et signification. *Ann. Technol. Agric.* **1969**, *18*, 221–229.
35. Peña-Neira, A.; Hernández, T.; García-Vallejo, C.; Estrella, I.; Suarez, J.A. A survey of phenolic compounds in Spanish wines of different geographical origin. *Eur. Food Res. Technol.* **2000**, *210*, 445–448. [CrossRef]
36. Godden, P.; Wilkes, E.; Johnson, D. Trends in the composition of Australian wine 1984–2014. *Aust. J. Grape Wine Res.* **2015**, *21*, 741–753. [CrossRef]
37. Keast, R.S.J.; Breslin, P.A.S. An overview of binary taste–taste interactions. *Food Qual. Prefer.* **2003**, *14*, 111–124. [CrossRef]
38. Van Gemert, L.J. *Flavour Thresholds. Compilations of Flavour Threshold Values in Water and Other Media*, 2nd ed.; Oliemans Punter & Partner: Utrecht, The Netherlands, 2011.
39. Aso, K.; Nakayama, T.; Maki, M. Studies on the bitter components in alcoholic drinks (I): The tyrosol content in sake. *J. Ferment. Technol.* **1953**, *31*, 43–47.
40. Koseki, T.; Kudo, S.; Matsuda, Y.; Ishigaki, H.; Anshoku, Y.; Muraoka, Y.; Wada, Y. A high tyrosol-producing sake yeast mutant and alcohol beverage utilising the mutant. *Jap. Open Pat. Gaz.* **2004**, 215644.
41. Guth, H. Quantitation and sensory studies of character impact odorants of different white wine varieties. *J. Agric. Food Chem.* **1997**, *45*, 3027–3032. [CrossRef]
42. Escudero, A.; Hernández-Orte, P.; Cacho, J.; Ferreira, V. Clues about the role of methional as character impact odorant of some oxidized wines. *J. Agric. Food Chem.* **2000**, *48*, 4268–4272. [CrossRef]
43. Lee, K.; Hahn, J.S. Interplay of Aro80 and GATA activators in regulation of genes for catabolism of aromatic amino acids in Saccharomyces cerevisiae. *Mol. Microbiol.* **2013**, *88*, 1120–1134. [CrossRef]
44. Chen, H.; Fink, G.R. Feedback control of morphogenesis in fungi by aromatic alcohols. *Genes Dev.* **2006**, *20*, 1150–1161. [CrossRef]
45. Smith, M.E.; Bekker, M.Z.; Smith, P.A.; Wilkes, E.N. Sources of volatile sulfur compounds in wine. *Aust. J. Grape Wine Res.* **2015**, *21*, 705–712. [CrossRef]
46. Goniak, O.J.; Noble, A.C. Sensory Study of Selected Volatile Sulfur Compounds in White Wine. *Am. J. Enol. Vitic.* **1987**, *38*, 223.
47. Cordente, A.G.; Borneman, A.R.; Bartel, C.; Capone, D.; Solomon, M.; Roach, M.; Curtin, C.D. Inactivating Mutations in Irc7p Are Common in Wine Yeasts, Attenuating Carbon-Sulfur β-Lyase Activity and Volatile Sulfur Compound Production. *Appl. Environ. Microbiol.* **2019**, *85*, e02684-18. [CrossRef]
48. Siebert, T.E.; Smyth, H.E.; Capone, D.L.; Neuwohner, C.; Pardon, K.H.; Skouroumounis, G.K.; Herderich, M.J.; Sefton, M.A.; Pollnitz, A.P. Stable isotope dilution analysis of wine fermentation products by HS-SPME-GC-MS. *Anal. Bioanal. Chem.* **2005**, *381*, 937–947. [CrossRef]
49. Siebert, T.E.; Solomon, M.R.; Pollnitz, A.P.; Jeffery, D.W. Selective determination of volatile sulfur compounds in wine by gas chromatography with sulfur chemiluminescence detection. *J. Agric. Food Chem.* **2010**, *58*, 9454–9462. [CrossRef]
50. Mayr, C.M.; Capone, D.L.; Pardon, K.H.; Black, C.A.; Pomeroy, D.; Francis, I.L. Quantitative analysis by GC-MS/MS of 18 aroma compounds related to oxidative off-flavor in wines. *J. Agric. Food Chem.* **2015**, *63*, 3394–3401. [CrossRef]
51. Nissen, T.L.; Schulze, U.; Nielsen, J.; Villadsen, J. Flux distributions in anaerobic, glucose-limited continuous cultures of Saccharomyces cerevisiae. *Microbiology* **1997**, *143 Pt 1*, 203–218. [CrossRef]
52. Heymann, H.; King, E.S.; Hopfer, H. Classical descriptive analysis. In *Novel Techniques in Sensory Characterization and Consumer Profiling*; CRC Press: Boca Raton, FL, USA, 2014; pp. 9–40.
53. Bekker, M.Z.; Espinase Nandorfy, D.; Kulcsar, A.C.; Faucon, A.; Bindon, K.; Smith, P.A. Comparison of remediation strategies for decreasing 'reductive' characters in Shiraz wines. *Aust. J. Grape Wine Res.* **2021**, *27*, 52–65. [CrossRef]

Article

Air-Depleted and Solvent-Impregnated Cork Powder as a New Natural and Sustainable Fining Agent for Removal of 2,4,6-Trichloroanisole (TCA) from Red Wines

Fernanda Cosme [1], Sara Gomes [2], Alice Vilela [1], Luís Filipe-Ribeiro [2] and Fernando M. Nunes [3,*]

[1] Chemistry Research Centre-Vila Real (CQ-VR), Food and Wine Chemistry Lab, Department of Biology and Environment, School of Life Sciences and Environment, University of Trás-os-Montes and Alto Douro, 5000-801 Vila Real, Portugal; fcosme@utad.pt (F.C.); avimoura@utad.pt (A.V.)
[2] Chemistry Research Centre-Vila Real (CQ-VR), Food and Wine Chemistry Lab, School of Life Sciences and Environment, University of Trás-os-Montes and Alto Douro, 5000-801 Vila Real, Portugal; sara_aras94@hotmail.com (S.G.); fmota@utad.pt (L.F.-R.)
[3] Chemistry Research Centre-Vila Real (CQ-VR), Food and Wine Chemistry Lab, Chemistry Department, School of Life Sciences and Environment, University of Trás-os-Montes and Alto Douro, 5000-801 Vila Real, Portugal
* Correspondence: fnunes@utad.pt

Abstract: Trichloroanisole (TCA) in wine results in a sensory defect called "cork taint", a significant problem for the wine industry. Wines can become contaminated by TCA absorption from the atmosphere through contaminated wood barrels, cork stoppers, and wood pallets. Air-depleted solvent-impregnated (ADSI) cork powder (CP) was used to mitigate TCA in wines. The ADSI CP (0.25 g/L) removed 91% of TCA (6 ng/L levels), resulting in an olfactory activity value of 0.14. A Freundlich isotherm described ADSI CP TCA adsorption with irreversible adsorption and a $K_F = 33.37$. ADSI CP application had no significant impact on the phenolic profile and chromatic characteristics of red wine. Using headspace sampling with re-equilibration, an average reduction in the volatile abundance of 29 ± 15%, 31 ± 19%, and 37 ± 24% was observed for the 0.10, 0.25, and 0.50 g/L ADSI CP, respectively. The alkyl esters and acids were the most affected. The impact observed was much lower when using headspace sampling without re-equilibration. Isoamyl acetate, ethyl hexanoate, ethyl hexanoate, and ethyl decanoate abundances were not significantly different from the control wine and 0.25 g/L ADSI CP application. Thus, ADSI CP can be a new sustainable fining agent to remove this "off-flavor" from wine, with a reduced impact on the wine characteristics.

Keywords: 2,4,6-Trichloroanisole (TCA); wine; ADSI cork powder; fining agent; phenolic profile; chromatic characteristics; volatile profile

Citation: Cosme, F.; Gomes, S.; Vilela, A.; Filipe-Ribeiro, L.; Nunes, F.M. Air-Depleted and Solvent-Impregnated Cork Powder as a New Natural and Sustainable Fining Agent for Removal of 2,4,6-Trichloroanisole (TCA) from Red Wines. *Molecules* 2022, 27, 4614. https://doi.org/10.3390/molecules27144614

Academic Editor: Maurizio Ugliano

Received: 5 June 2022
Accepted: 15 July 2022
Published: 19 July 2022

Publisher's Note: MDPI stays neutral with regard to jurisdictional claims in published maps and institutional affiliations.

Copyright: © 2022 by the authors. Licensee MDPI, Basel, Switzerland. This article is an open access article distributed under the terms and conditions of the Creative Commons Attribution (CC BY) license (https://creativecommons.org/licenses/by/4.0/).

1. Introduction

2,4,6-Trichloroanisole (TCA) is a fungal metabolite with an unpleasant moldy odor that can contaminate wine, producing the so-called "cork taint" or "corked taste". The "corked taste" is usually a musty, moldy, mildew, or earthy smell and is sometimes described as burnt rubber, smoke, or even camphor [1]. Other chloroanisoles, such as 2,4-dichloroanisole, 2,6-dichloroanisole, 2,3,4,6-tetrachloroanisole (TeCA), and pentachloroanisole (PCA), may also contribute to the "cork taste" but do not play a dominant role in this sensory defect. 2,4,6-Tribromoanisole (TBA) may also play a significant role in wine's musty/mold odor [2]. TCA can be produced by different metabolic pathways. However, the formation of TCA from 2,4,6-trichlorophenol (TCP) by biomethylation reactions is the only scientifically proven origin. This biomethylation reaction is carried out through the enzyme chlorophenol O-methyltransferase, which is present in filamentous fungi of different families (*Streptomyces* spp., *Aspergillus* spp., *Trichoderma* spp., *Penicillium* spp. and *Cephalouscus* spp., among others). These fungi grow on different materials, such as cork and wood. Under high humidity and limited ventilation conditions, fungi can transform odorless chlorophenols with a high threshold of perception into chloroanisols with a low perception threshold. This enzyme catalyzes the reaction that converts halophenols into haloanisols [3]. Wine can be contaminated by TCA or other haloanisols even before it is bottled if it comes into contact with contaminated materials (such as wood barrels) and/or cellars that have a contaminated atmosphere [2]. According to Sefton and Simpson [4], the proportion of affected bottles is estimated to be between 1 and 5% and occasionally up to 30%. Contamination with chlorophenols caused by fungicides or insecticides can involve woody materials used for building cellars, wooden pallets for bottles, paints, boxes, and other materials such as polluted bottles, corks, and wines. TCA has also been identified as a contaminant of oak barrels [5].

TCA has an extremely low detection threshold of nanograms per liter (ng/L), which indicates that it will be easily detectable by the consumer, even at low concentrations. According to several authors, the sensory threshold of TCA in wine ranges from 1.4 ng/L to 4 ng/L [6–11], with values found in the literature that differ from author to author; for example, Vestner et al. [12] report that the sensory limit of TCA is around 4 ng/L (in wine). In contrast, Juanola et al. [13] refer to a 5 ng/L sensory threshold. Sefton and Simpson [4] mention that the detection limit can be between 1.4 and 4.6 ng/L and the recognition limit between 4.2 and 10 ng/L. On the other hand, Fontana et al. [14] state that the threshold of perception of TCA is greater than 0.03 ng/L. However, the threshold value in wine strongly depends on the type of wine, the wine's style, and the taster's experience [15].

Due to the sensory impact of TCA on wine, and the fact that TCA does not only originate from cork stoppers, it is necessary to find an effective technological solution that can eliminate or minimize the TCA in the wine with a minimal impact on its characteristics. A patent describes using an aqueous suspension of activated carbon obtained from coconut to remove the "cork taste" [16]. Another patent proposes contacting wine with synthetic aliphatic polymers (ultra-high molecular weight polyethylene) to reduce the TCA concentration. According to the data described in the patent, the TCA concentration of the treated wine is reduced from about 10 (ng/L) to preferably less than 5 (ng/L) or less, with the taste and smell of TCA in wine being undetectable below these values [17]. Vuchot et al. [18] used highly absorbent yeast cell extracts. Yeast cells were able to remove TCA (27%), TeCA (55%), and PCA (73%) without analytical or sensory modification of the wines. Doubling the dose yielded better results, allowing for a reduction by 45%, 73%, and 83%, respectively [18]. Molecularly imprinted polymers are synthetic materials with artificially generated recognition sites capable of specifically rebinding a target molecule. Molecularly imprinted polymers and non-molecularly imprinted polymers have been used with good results in wines for TCA removal with about 90% TCA removal [19].

The latest European Union legislation (EU Regulation 2019/934) allows for a filter plate treatment that contains Y-faujasite zeolites solely to adsorb haloanisols and is applied during filtration to reduce the concentration of the haloanisols responsible for flavor in wines below the threshold of perception. This treatment must be carried out on clarified wines, and the filter plates must be cleaned and disinfected before passing the wine through them and applying Y-faujasite zeolites [20].

Paraffin wax can absorb chloroanisols from wine, and absorption by polyethylene film can be even more effective, but TeCA was removed more efficiently than TCA. Polyethylene film offers an inexpensive and effective means of reducing trichloroanisole in wines, with only a slight impact on their characteristics. However, a loss of floral/fruity aroma was observed [21].

A plastic film composed of a mixture of synthetic polymers and certified for food use (where there is no migration of plastic molecules to the wine) was added to the wine at a dose of 20 m^2 film/hL to study its efficiency in the removal of TCA from wine [22]. The removal of TCA from wine became more noticeable as the film–wine contact time increased. In barrels with contamination of 3 ng/L, the TCA concentration decreased by 47% after 8 h of treatment with the film. A more extended treatment of 24 h and 48 h led to a 73% and

83% reduction in TCA concentration, respectively. Furthermore, according to the results of this study, it can be observed that, globally, the use of plastic film to eliminate/reduce the content of haloanisols in wines did not impact the content of phenolic compounds (proanthocyanidins and anthocyanins) for up to 24 h of treatment with the film [22]. Valdés et al. [23] also studied the possibility of applying two polyaniline-based materials (100 to 500 mg/L) to remove TCA and TBA in methanol at a concentration of 20 ng/L. The results of these authors showed that the removal percentages of TCA and TBA were 68–72% and 84–85%, respectively, for the two materials tested in methanol, and their effectiveness varied with the interaction time and with the amount of polymer used.

Cork residues and cork powders have been used as bio-adsorbents to remove pesticides and other pollutants from wastewater with promising results [24]. This by-product obtained from the cork industry is an abundant, natural, and cheap material recently exploited in its raw form and after optimizing its adsorption properties by simple physicochemical treatments, such as air removal and simultaneous impregnation with ethanol. This treatment makes the cell wall components more accessible, demonstrating an increase of at least 4 times its adsorption capacity after treatment, which could be a new sustainable fining agent for wines [25,26]. This by-product was a good solution for the removal of volatile phenols without affecting the wine quality and sensory profile [25,26]. The use of cork dust waste produced in the cork stopper industry can increase its economic value and thus reduce the entry of new materials into the wine production chain. Due to its improved adsorption properties, air-depleted solvent-impregnated (ADSI) cork powder has a similar potential to other wine fining agents. The different adsorption mechanisms driven by hydrophobicity represent an alternative solution to be employed [25,26].

Therefore, this work aimed to study the efficiency of ADSI cork powder in the removal of TCA from red wines and the impact of its application on red wine characteristics, namely the chromatic characteristics, phenolic composition, and volatile profile.

2. Results and Discussion

2.1. Performance of Air-Depleted and Solvent-Impregnated Cork Powder in the Removal of Trichloroanisole (TCA)

The hydrophobic cork extractives were first removed by sequential treatment with dichloromethane and ethanol to increase the performance of natural cork powder in terms of its ability to remove TCA from the wine, as described by Filipe Ribeiro et al. [25]. As raw cork material contains significant amounts of trapped air, and water has a very low diffusion coefficient in cork, the air from the extracted cork powder was removed and impregnated with ethanol under vacuum by repeated degassing cycles (11 times) immersed in ethanol [25]. It was then sieved to obtain a particle size below 75 µm.

Wines were contaminated with two levels of TCA (3 and 6 ng/L). The treatment of the contaminated wines with air-depleted and solvent-impregnated cork powder at different doses (0.1, 0.25, and 0.5 g/L) decreased the wine's TCA concentration significantly (Table 1). It was also observed that the higher the amount of ADSI cork powder applied to the wine, the greater its effectiveness in reducing the wine's TCA concentration. Table 1 shows the percentage of TCA removal after the application of ADSI cork powder. There was observed an increase in the percentage of TCA removal with the increase in the applied dose of ADSI powder, and, as expected, the higher the concentration of TCA in wines the higher the removal percentage. Additionally, shown in Table 1 is the odor activity value (OAV) of TCA in the wines treated with ADSI cork powder. The OAV is a measure of the importance of a specific compound to the odor of the sample [27]. The odor detection threshold of TCA in wines varies widely in the literature, ranging from 1.4 to 22 ng/L depending on the study and also on the wine matrix. A more recent study using different white and red wine matrixes established a detection threshold of 4 and 5 ng/L of TCA both for aroma and flavor, respectively, while for 3 ng/L it was not considered significant; therefore, a detection threshold of 4 ng/L was used for calculating the OAV [15]. For all the application doses of ADSI cork powder for both TCA contamination levels, the OAV was well below

1 (Table 1); therefore, the impact of TCA on the aroma of wines treated with ADSI cork powder is expected to be negligible.

Table 1. TCA remaining in wine contaminated with 3 ng/L of TCA and 6 ng/L of TCA after applying different doses of ADSI cork powder (0.10, 0.25, and 0.50 g/L) and the corresponding TCA odor activity values (OAVs) in the final wines.

Wine	TCA Remaining (ng/L)	OAV
Wine with 3 ng/L of TCA		
0.10 g/L	2.25 ± 0.35 [a]	0.56
0.25 g/L	1.95 ± 0.25 [a]	0.49
0.50 g/L	1.35 ± 0.25 [a]	0.34
Wine with 6 ng/L of TCA		
0.10 g/L	3.30 ± 0.40 [a]	0.83
0.25 g/L	0.55 ± 1.05 [a]	0.14
0.50 g/L	1.40 ± 0.30 [a]	0.35

Values in the same column for each contamination level (3 ng/L TCA or 6 ng/L TCA) followed by the same letter are not significantly different (Tukey's HSD, $p \leq 0.05$).

When compared with other research works that studied the removal of TCA from wine using different materials, it can be concluded that ADSI cork powder is one of the most effective materials for TCA removal. For example, with the application of highly absorbent yeast cell extract (0.4 g/L) added to wine containing 6 ng/L TCA, the removal was 27% of TCA, and doubling the application dose of yeast cell extract achieved better removal results (45%) [18]. For the use of molecularly imprinted polymers and non-molecularly imprinted polymers, good results were obtained (a TCA removal percentage of about 90%) [19]. Some plastics quickly absorb chloroanisols, and the absorption efficiency increases with the increase in the number of chlorine atoms in the molecule. Chloroanisols are hydrophobic substances and are therefore particularly soluble in non-polar media. Absorption of chloroanisoles from wine contaminated by non-polar substances such as food-grade paraffin wax or food-grade polyethylene film could be a viable way to reduce or even remove the odor of trichloroanisole from wine. Thus, the use of polyethylene film described by Capone et al. [21] showed that, after 4 days, it removed 90% of the TCA and 97% of the TeCA from white wine artificially contaminated with 100 ng/L of TCA and 100 ng/L of TeCA, respectively.

Valdes et al. [23] also showed that the application of two polyaniline-based materials (0.1 to 0.5 g/L) to wine contaminated with TCA and TBA (20 ng/L) had TCA and TBA removal percentages of 68–72%, and 84–85%, respectively. A recent study of the application of plastic film to wines stored in wooden barrels with 3 ng/L and 9 ng/L of TCA contamination showed that immersion of plastic film in wine for 8 h reduced the TCA concentration by 47% to 57%, and that after 24 h the TCA reductions were 73% and 75%, respectively. After 48 h of treatment, TCA concentration reductions of 83% and 81% were observed [22].

The results obtained in the present work using ADSI cork powder show that it was possible to remove 91% of the TCA with 0.25 g of ADSI cork powder/L of wine with an initial contamination of 6 ng/L of TCA (Table 1), which indicates that, compared with the other materials described in the literature, it is one of the most effective treatments in the removal of TCA from contaminated wines.

2.2. TCA Adsorption Isotherms of Air-Depleted Solvent-Impregnated Cork Powder in Model Wine

The adsorption isotherm of TCA to the ADSI cork powder was determined in a model wine solution at 25 °C for a 0.25 g/L application dose. As shown in Figure 1, the ADSI cork powder adsorption capacity increased in the entire concentration range assayed (2.5–50 ng/L of TCA in the model wine solution). To analyze the equilibrium data obtained experimentally, three isothermal models were used to characterize the adsorption system: the Langmuir, Freundlich, and Langmuir–Freundlich isotherm models [28]. The Langmuir isotherm is usually used for ideal monolayer adsorption on a homogeneous surface [29]. The Freundlich isotherm is generally suitable for nonideal adsorption on heterogeneous surfaces. It assumes that there are large numbers and many different types of available sites acting simultaneously, each with a different free energy of sorption [30]. Only the Freundlich model yielded high correlation coefficients (>0.999). The type of Freundlich isotherm is indicated by the value of n, in which both the K_F and n parameters are dependent on temperature. The $1/n$ value is the intensity of the adsorption or surface heterogeneity and indicates the energy distribution and the adsorbate sites' heterogeneity. When $1/n$ is greater than zero ($0 < 1/n < 1$), the adsorption is favorable; when $1/n$ is greater than 1, the adsorption process is unfavorable, and it is irreversible when $1/n = 1$ [31–33]. Therefore, the adsorption of TCA on ADSI cork powder seems to be irreversible, showing a K_F of 33.37.

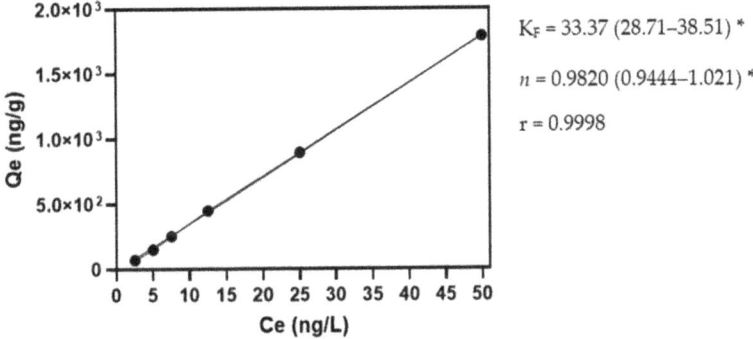

Figure 1. Freundlich adsorption isotherm of ADSI cork powder for TCA in a model wine solution. Qe is the amount of TCA adsorbed at equilibrium; Ce is the equilibrium concentration; * denotes the 95% confidence interval.

2.3. Impact of ADSI Cork Powder on Wine Quality

To obtain a deeper insight into the impact of ADSI cork powder on the wine's chemical composition, besides its TCA removal efficiency, the effects on the phenolic composition, chromatic characteristics, and volatile profile of the wine after application of increasing doses of ADSI cork powder were determined.

2.3.1. Impact of ADSI Cork Powder on the Chromatic Characteristics and Phenolic Composition of the Wine

Table 2 shows the total phenolic compounds, color intensity, hue, and chromatic characteristics of red wines after the application of increasing doses of ADSI cork powder (0.10, 0.25, and 0.5 g/L). It can be observed that there are no significant differences in total polyphenols after the application of the ADSI cork powder compared with the control wine.

Table 2. Total phenolic compounds, color intensity, hue, and chromatic characteristics of red wines after the application of different doses of ADSI cork powder (0.10, 0.25, and 0.50 g/L).

Wine	Total Phenolic Compounds (mg/L)	Color Intensity a.u.	Hue	L*	a*	b*	C*	h°	ΔE*
Control	1544 ± 187 [a]	15.02 ± 0.24 [a]	0.71 ± 0.00 [a]	70.1 ± 0.5 [a]	35.05 ± 0.96 [a]	7.06 ± 0.15 [a]	35.75 ± 0.96 [a]	0.20 ± 0.01 [a]	-
0.10 g/L	1694 ± 263 [a]	15.14 ± 0.53 [a]	0.71 ± 0.02 [a]	69.6 ± 0.7 [a]	35.21 ± 1.95 [a]	7.08 ± 0.49 [a]	35.92 ± 2.00 [a]	0.20 ± 0.01 [a]	1.79 ± 1.11 [a]
0.25 g/L	1425 ± 199 [a]	14.79 ± 0.07 [a]	0.71 ± 0.00 [a]	70.2 ± 0.7 [a]	34.21 ± 0.31 [a]	7.06 ± 0.13 [a]	35.04 ± 0.32 [a]	0.20 ± 0.00 [a]	1.45 ± 0.33 [a]
0.50 g/L	1513 ± 224 [a]	14.83 ± 0.29 [a]	0.71 ± 0.01 [a]	70.1 ± 0.7 [a]	34.21 ± 0.31 [a]	7.11 ± 0.35 [a]	34.94 ± 0.36 [a]	0.20 ± 0.01 [a]	1.11 ± 0.64 [a]

L* (lightness), a* (redness), b* (yellowness) coordinates, C* (chroma), h° (hue-angle), ΔE* (total color difference in relation to control wine). Values in the same column followed by the same letter are not significantly different ($n = 8$) (Tukey's HSD, $p \leq 0.05$). a.u. (Absorbance unit).

Gonzàlez-Centeno et al. [22], using plastic film to remove TCA, found that this material had little impact on the total phenolic compounds of the wine, with only a slight decrease (4.4%) in the total phenolic compounds concerning the untreated wine after 48 h of contact with the plastic film. In addition, the application of yeast cell extract at a dose of 400 mg/L for TCA removal did not significantly decrease the red wine's color intensity [18].

The application of the ADSI cork powder did not significantly alter the color intensity and hue of the red wine (Table 2). These results agree with Filipe-Ribeiro et al. [25], who applied ADSI cork powder in red wine to remove volatile phenols and observed that the ADSI cork powder did not change the color intensity of red wines significantly. In line with the results obtained for the color intensity and hue, there were no significant changes in the chromatic characteristics of the wine (Table 2). These results agree with those obtained by Filipe-Ribeiro et al. [25], who applied ADSI cork powder to remove volatile phenols from red wines and also did not observe significant changes in the chromatic characteristics compared with the control wine.

In the wine treated with plastic film for the removal of TCA as described by Gonzàlez-Centeno et al. [22], the chromatic characteristics were not altered after the treatment of the wine in contact with the plastic film. Although there were significant differences between untreated and plastic-film-treated wines and even between plastic-film-treated wines with different contact times, these differences were not visually perceived by any taster during the sensory analysis.

Table 3 shows the total pigments, polymeric pigments, small polymeric pigments (SPPs), large polymeric pigments (LPPs), monomeric anthocyanins, and tannins of the red wine treated with ADSI cork powder for TCA removal. The data clearly show no significant impact on these wine parameters after application of the ADSI cork powder compared with the control wine.

After applying plastic film to remove TCA from red wine, Gonzàlez-Centeno et al. [22] observed that the total proanthocyanidin values remained constant regardless of the film–wine contact time. The results of anthocyanins in this study show that wines treated with plastic film exhibited a small but significant increase in the total anthocyanin concentration, both after 48 h and after 24 h of contact with the plastic film. This increase suggests that the plastic wrap can absorb certain compounds in wine that anthocyanins combine with. Additionally, Gonzàlez-Centeno et al. [22] showed that using plastic film to eliminate/reduce the TCA content in wines did not significantly affect their levels of proanthocyanidins and anthocyanins for up to 24 h of treatment with film or plastic film.

These results indicate that ADSI cork powder has a low impact on the phenolic profile of red wine. The content of individual phenolic acids and catechin did not show significant differences after applying the different doses of ADSI cork powder, except for the ethyl ether of coumaric acid, which showed a significant decrease (Table 4). These data agree with those obtained by Filipe-Ribeiro et al. [25], who also observed few significant changes in phenolic acids and catechin compared with untreated wine.

The data on monomeric anthocyanin levels are shown in Table 5. Generally, no significant differences were observed, except for malvidin-3-O-glucoside. However, the total monomeric anthocyanins did not show significant differences from untreated wine. These data also agree with those obtained by Filipe-Ribeiro et al. [25], who, when applying ADSI cork powder to red wine, observed few significant differences in the monomeric anthocyanin profiles of wines treated with ADSI cork powder compared with the untreated wine.

Table 3. Total pigments, polymeric pigments, small polymeric pigments (SPPs), large polymeric pigments (LPPs), monomeric anthocyanins, and tannins of red wines after applying different doses of ADSI cork powder (0.10, 0.25, and 0.50 g/L).

Wine	Total Pigments a.u.	Polymeric Pigments (SPPs + LPPs) a.u.	SPPs a.u.	LPPs a.u.	Monomeric Anthocyanins a.u.	Tannins a.u.
Control	5.27 ± 0.09 [a]	2.54 ± 0.05 [a]	1.53 ± 0.05 [a]	1.00 ± 0.08 [a]	2.74 ± 0.04 [a]	0.81 ± 0.47 [a]
0.10 g/L	5.29 ± 0.09 [a]	2.54 ± 0.06 [a]	1.67 ± 0.30 [a]	0.87 ± 0.31 [a]	2.75 ± 0.04 [a]	0.87 ± 0.03 [a]
0.25 g/L	5.26 ± 0.15 [a]	2.52 ± 0.08 [a]	1.51 ± 0.07 [a]	1.01 ± 0.10 [a]	2.74 ± 0.07 [a]	0.80 ± 0.24 [a]
0.50 g/L	5.38 ± 0.23 [a]	2.59 ± 0.10 [a]	1.51 ± 0.14 [a]	1.08 ± 0.16 [a]	2.79 ± 0.14 [a]	1.15 ± 0.75 [a]

Values in the same column followed by the same letter are not significantly different ($n = 8$) (Tukey's HSD, $p \leq 0.05$). a.u. (Absorbance unit).

Table 4. Phenolic acid profile and flavonoids in mg/L of red wines after applying different doses of ADSI cork powder (0.10, 0.25, and 0.50 g/L).

Wine	Catechin	Gallic Acid	trans-Caftaric Acid	Coutaric Acid Isomer	Coutaric Acid	Caffeic Acid	p-Coumaric Acid	Ferulic Acid	Ethyl Ester of Caffeic Acid	Ethyl Ester of Coumaric Acid
Control	7.36 ± 1.43 [a]	20.34 ± 0.49 [a]	5.57 ± 0.20 [a]	6.56 ± 0.14 [a]	2.20 ± 0.28 [a]	3.16 ± 0.15 [a]	1.53 ± 0.08 [a]	0.76 ± 0.12 [a]	0.37 ± 0.01 [a]	1.66 ± 0.06 [b]
0.10 g/L	7.73 ± 1.14 [a]	20.63 ± 0.24 [a]	5.53 ± 0.38 [a]	6.46 ± 0.73 [a]	1.90 ± 0.43 [a]	3.18 ± 0.16 [a]	1.56 ± 0.24 [a]	0.96 ± 0.43 [a]	0.36 ± 0.03 [a]	1.48 ± 0.11 [a]
0.25 g/L	8.29 ± 0.93 [a]	20.31 ± 0.35 [a]	5.76 ± 0.05 [a]	6.64 ± 0.24 [a]	2.20 ± 0.34 [a]	3.10 ± 0.24 [a]	1.68 ± 0.14 [a]	0.82 ± 0.14 [a]	0.37 ± 0.03 [a]	1.49 ± 0.19 [b]
0.50 g/L	8.02 ± 0.65 [a]	20.65 ± 0.21 [a]	5.48 ± 0.39 [a]	6.45 ± 0.63 [a]	2.22 ± 0.37 [a]	3.02 ± 0.20 [a]	1.47 ± 0.17 [a]	0.70 ± 0.04 [a]	0.36 ± 0.01 [a]	1.52 ± 0.10 [ab]

Values in the same column followed by the same letter are not significantly different ($n = 4$) (Tukey's HSD, $p \leq 0.05$).

Table 5. Monomeric anthocyanin profile in mg/L of red wines after applying different doses of ADSI cork powder (0.10, 0.25, and 0.50 g/L).

Wine	D-3-G	C-3-G	Pet-3-G	Peo-3-G	M-3-G	D-3-A	Pet-3-A	Peo-3-A	M-3-A	C-3-C	M-3-C	Total Monomeric Anthocyanins
Control	1.11 ± 0.06 [a]	4.65 ± 0.35 [a]	6.88 ± 0.34 [a]	5.56 ± 0.11 [a]	32.22 ± 0.43 [a]	0.28 ± 0.03 [a]	0.46 ± 0.04 [a]	0.08 ± 0.16 [a]	4.27 ± 0.56 [a]	0.38 ± 0.03 [a]	5.15 ± 0.37 [a]	61.05 ± 0.72 [a]
0.10 g/L	1.04 ± 0.17 [a]	4.46 ± 0.17 [a]	6.96 ± 0.18	5.55 ± 0.58 [a]	32.10 ± 1.75 [a]	0.24 ± 0.16 [a]	0.51 ± 0.21 [a]	0.08 ± 0.16 [a]	4.18 ± 0.48 [a]	0.36 ± 0.06 [a]	4.79 ± 0.96 [a]	60.27 ± 1.20 [a]
0.25 g/L	1.01 ± 0.14 [a]	4.65 ± 0.16 [a]	6.78 ± 0.31 [a]	5.59 ± 0.34 [a]	31.98 ± 1.08 [a]	0.21 ± 0.15 [a]	0.51 ± 0.11 [a]	0.18 ± 0.20 [a]	4.26 ± 0.34 [a]	0.35 ± 0.07 [a]	5.17 ± 0.20 [a]	60.69 ± 1.40 [a]
0.50 g/L	0.98 ± 0.12 [a]	4.32 ± 0.25 [a]	6.70 ± 0.14 [a]	5.30 ± 0.60 [a]	31.92 ± 1.29 [a]	0.22 ± 0.17 [a]	0.59 ± 0.14 [a]	0.23 ± 0.17 [a]	4.55 ± 0.18 [a]	0.42 ± 0.12 [a]	5.24 ± 0.28 [a]	60.46 ± 1.89 [a]

Delphinidin-3-O-glucoside (D-3-G), Cyanidin-3-O-glucoside (C-3-G), Petunidin-3-O-glucoside (Pet-3-G), Peonidin-3-O-glucoside (Peo-3-G), Malvidin-3-O-glucoside (M-3-G), Delphinidin-3-O-acetylglucoside (D-3-A), Petunidin-3-O-acetylglucoside (Pet-3-A), Peonidin-3-O-acetylglucoside (Peo-3-A), Malvidin-3-O-acetylglucoside (M-3-A), Cyanidin-3-O-coumaroylglucoside (C-3-C), Malvidin-3-O-coumaroylglucoside (M-3-C). Values in the same column followed by the same letter are not significantly different ($n = 4$) (Tukey's HSD, $p \leq 0.05$).

In the use of plastic film for the removal of TCA described by Gonzàlez-Centeno et al. [22], the duration of the plastic film treatment did not lead to significant differences between the plastic-film-treated wines regarding monomeric anthocyanins. However, compared with untreated wine (the control), plastic-film-treated wines had slightly higher concentrations of some monomeric anthocyanins after 8 h of contact with the plastic film (2–14%), with malvidin-3-*O*-glucoside and delphinidin-3-*O*-glucoside the main compounds responsible for these increases. These observations agree with what was previously described for total anthocyanins. They could be explained by the potential absorption by the plastic film of certain carbonyl compounds that tend to combine with anthocyanins. This absorption of anthocyanins by the ADSI cork powder was not verified in the present study.

2.3.2. Impact of ADSI Cork Powder on Wine Volatile Composition

We used two methods of SPME headspace sampling to study the impact of the application of ADSI cork powder on the volatile profile of red wine. A standard lengthy steady-state extraction method, in which the extraction time allows for the re-equilibration of volatiles between the liquid matrix, headspace volatile, and SPME fiber, was used to extract the maximum amount of analyte. A fast snapshot method, whose reduced extraction time avoids/diminishes the re-equilibration of the headspace volatile composition above the wine, was also used without agitation and heating. Roberts and coworkers [34] found that HS-SPME with a short sampling time can determine the "true headspace" concentration at equilibrium between the headspace and water, which can minimize the disruption caused by the fiber/headspace partition. The "true headspace" discussed by Roberts et al. [34] reflects the volatile compounds in the air space at equilibrium between the headspace and the sample solution. Figure 2a and Table 6 show the volatile profile of red wines after applying three different doses of ADSI cork powder (0.1, 0.25, and 0.5 g/L), analyzed by two methods: headspace extraction with and without re-equilibration. When re-equilibration was allowed, the volatile abundance decreased with the increase in the ADSI cork powder dose applied. Even for the lowest dose of ADSI cork powder, there was observed a decrease in the abundance of almost all compounds analyzed, except for isoamyl alcohol, 3-methylbutanoic acid, diethylbutanoate, benzyl alcohol, phenylethanol, and decanoic acid (Table 6). For the 0.1 g/L ADSI cork powder application dose, an average reduction of 29 ± 15% was observed. The decline increased to 31 ± 19% and 37 ± 24% for the 0.25 g/L and 0.50 g/L ADSI cork powder application doses, respectively. The alkyl esters and acids were the most affected, resulting in average reductions of 48 ± 20% for the highest application dose. These results agree with those described by Filipe-Ribeiro et al. [25], who used ADSI cork powder for the removal of volatile phenols and observed a decrease in the total abundance of volatile compounds in the headspace with an increasing application dose of ADSI cork powder.

The use of the fast extraction method without re-equilibration, as expected, decreased the total abundance of the compounds extracted to only 4.46% (Table 6) but also changed the relative abundance of the extracted volatile compounds (Figure 2b and Table 6). When using this headspace sampling method, with few exceptions, significant reductions in the headspace volatile abundance were only significant for the 0.50 g/L ADSI cork powder application dose. For *p*-cymene, 3-methylbutanoic acid, 1,1,6-trimethyl-1,2-dihydronaphatalene, phenylethylacetate, phenylethanol, β-caryophyllene oxide, ethyl hexanoate, and decanoic acid, we observed a reduction in the headspace abundance with the application dose. A decrease in the abundance below the method detection limit for the less-abundant volatiles, such as phenylethylacetate, β-caryophyllene oxide, ethyl hexadecanoate, and decanoic acid, was also observed. Interestingly, for the low-molecular-weight alkyl esters, such as isoamyl acetate, ethyl hexanoate, ethyl hexanoate, and ethyl decanoate, the abundance observed for the 0.25 g/L ADSI cork powder application dose was not significantly different from that of the control wine. Therefore, although a substantial impact was observed on the abundance of the volatile compounds when headspace sampling with re-equilibration was employed, the apparent impact of ADSI cork powder application on the "true headspace" composition seems to be lower.

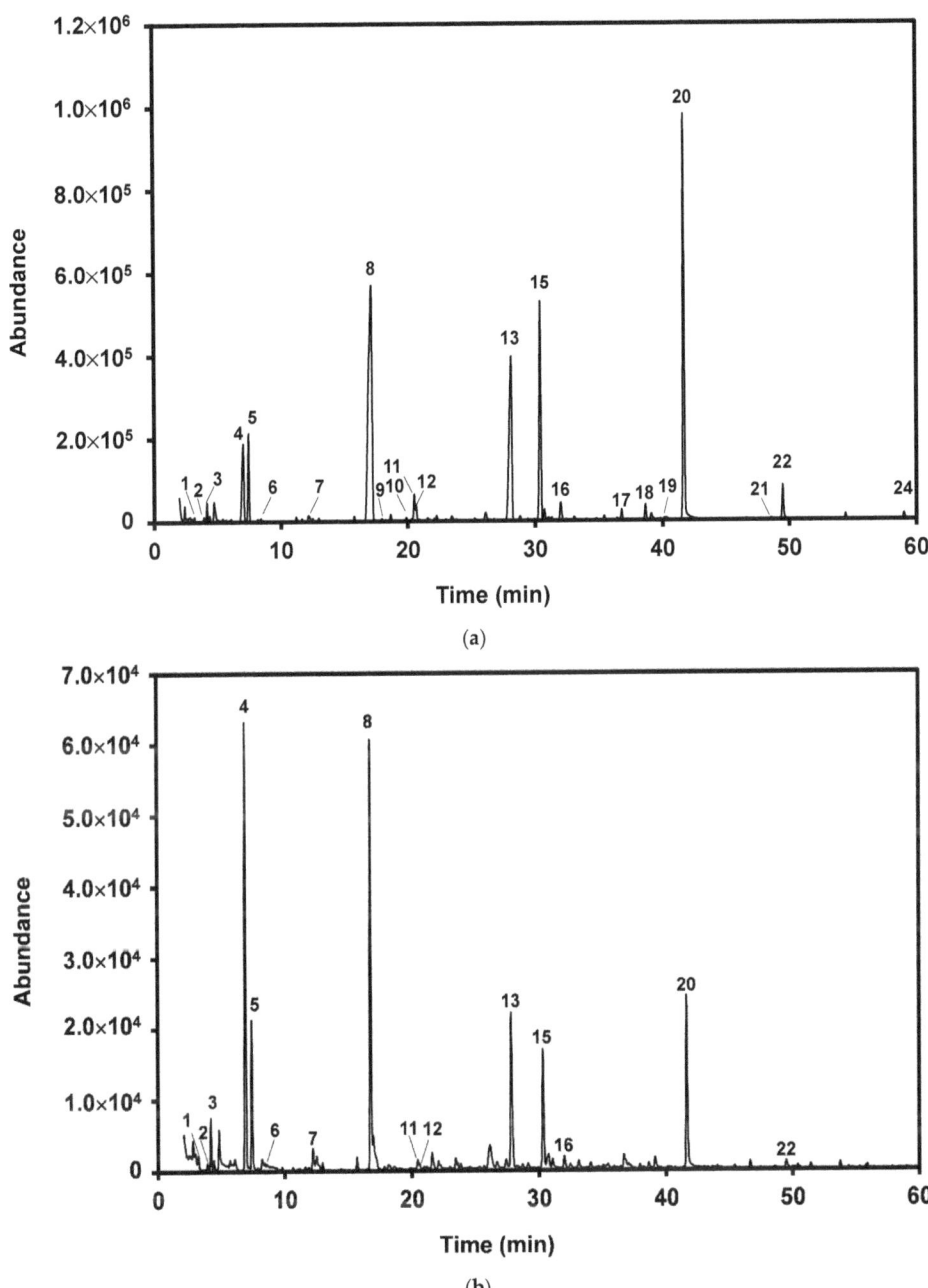

Figure 2. Typical chromatograms of red wines without ADSI cork powder addition using SPME headspace sampling with re-equilibration (**a**) and without re-equilibration (**b**). Only the major peaks are highlighted. For peak identification, refer to Table 6.

Table 6. Volatile components identified in the red wine headspace by SPME extraction with re-equilibration and without re-equilibration and the effect of ADSI cork powder application dose (0.10, 0.25, and 0.50 g/L) on the volatile abundance (area × 10^5).

Peak	Compound	RI	Aroma Descriptors	Control	With Re-Equilibration 0.10 g/L	With Re-Equilibration 0.25 g/L	With Re-Equilibration 0.50 g/L	Control	Without Re-Equilibration 0.10 g/L	Without Re-Equilibration 0.25 g/L	Without Re-Equilibration 0.50 g/L
1	Hexanal **	1083	Green, woody, vegetative, apple, grassy, citrus, and orange	0.19 ± 0.01 [a]	0.13 ± 0.01 [b]	0.14 ± 0.01 [b]	0.13 ± 0.01 [b]	0.038 ± 0.002 [a]	0.034 ± 0.003 [a,b]	0.029 ± 0.001 [a,b]	0.027 ± 0.004 [b]
2	Acetaldehyde ethyl amyl acetal **	1098		0.31 ± 0.02 [a]	0.26 ± 0.00 [a,b]	0.26 ± 0.00 [a,b]	0.24 ± 0.01 [b]	0.023 ± 0.000 [a]	0.022 ± 0.001 [a]	0.024 ± 0.000 [a]	0.021 ± 0.003 [a]
3	Isoamyl acetate *	1144	Banana	5.09 ± 0.47 [a]	3.67 ± 0.42 [b]	3.37 ± 0.12 [b]	3.16 ± 0.24 [b]	1.02 ± 0.07 [a]	1.16 ± 0.04 [a]	1.01 ± 0.04 [a]	0.71 ± 0.02 [b]
4	Isoamyl alcohol *	1194	Alcohol, floral, cheese	40.6 ± 4.8 [a]	34.1 ± 1.1 [a]	33.9 ± 0.3 [a]	32.0 ± 1.7 [a]	7.89 ± 0.25 [a]	8.19 ± 0.07 [a]	8.26 ± 0.13 [a]	8.06 ± 0.02 [a]
5	Ethyl hexanoate *	1203	Fruity, strawberry, green apple, anise	29.7 ± 3.4 [a]	23.5 ± 0.4 [a,b]	22.2 ± 0.4 [b]	20.2 ± 0.6 [b]	2.90 ± 0.07 [a]	3.05 ± 0.06 [a]	3.07 ± 0.09 [a]	2.86 ± 0.03 [a]
6	*p*-Cymene **	1223	Fruity, sweet	0.20 ± 0.01 [a]	0.17 ± 0.00 [c]	0.15 ± 0.00 [c]	0.12 ± 0.01 [d]	0.030 ± 0.001 [a]	0.024 ± 0.001 [b]	0.025 ± 0.000 [b]	0.019 ± 0.001 [c]
7	Hexanol *	1340	Green grass	0.21 ± 0.0 [a]	0.18 ± 0.01 [a,b]	0.17 ± 0.00 [b]	0.17 ± 0.00 [b]	0.031 ± 0.001 [a]	0.031 ± 0.001 [a]	0.029 ± 0.001 [a]	0.029 ± 0.001 [a]
8	Ethyl octanoate *	1418	Sweet, fruit, fresh, pineapple, pear, floral	320 ± 29 [a]	248 ± 3 [b]	220 ± 2 [b]	190 ± 2 [b]	10.4 ± 0.1 [a]	10.3 ± 0.4 [a]	9.88 ± 0.47 [a,b]	8.76 ± 0.23 [b]
9	Isopentyl hexnaoate **	1420	Fruity, banana, apple, pineapple, green	0.04 ± 0.00 [a]	0.02 ± 0.01 [a,b]	0.02 ± 0.00 [a,b]	0.02 ± 0.00 [b]	0.001 ± 0.000 [a]	0.001 ± 0.000 [a]	0.001 ± 0.000 [a]	0.001 ± 0.000 [a]
10	Terpinen-4-ol acetate **	1462	Peppery, woody, earthy, musty, sweet	0.21 ± 0.00 [a]	0.19 ± 0.00 [b]	0.17 ± 0.00 [b]	0.15 ± 0.01 [c]	0.003 ± 0.000 [a]	0.003 ± 0.000 [a]	0.002 ± 0.000 [a,b]	0.002 ± 0.000 [b]
11	Vitispirane A **	1475	Fruity, floral, earthy, woody, camphor, eucalyptus, spice	1.50 ± 0.10 [a]	1.25 ± 0.00 [b]	1.10 ± 0.00 [b,c]	0.95 ± 0.02 [c]	0.025 ± 0.000 [a]	0.026 ± 0.001 [a]	0.023 ± 0.000 [b]	n.d. [c]
12	Vitispirane B **	1487	Floral, camphor, eucalyptus, spice, wood	0.51 ± 0.03 [a]	0.45 ± 0.00 [a,b]	0.41 ± 0.01 [b,c]	0.34 ± 0.01 [d]	0.009 ± 0.000 [a]	0.010 ± 0.000 [a]	0.009 ± 0.001 [a]	n.d. [b]
13	Ethyl decanoate *	1625	Grape, pleasant, soap	153 ± 16 [a]	105 ± 3 [b]	80.9 ± 1.8 [c,d]	57.0 ± 1.6 [d]	6.65 ± 0.14 [a]	6.47 ± 0.45 [a]	5.35 ± 0.47 [a,b]	4.48 ± 0.36 [b]
14	3-methylbutanoic acid **	1672	Cheese, fatty, rancid	0.07 ± 0.00 [a]	0.07 ± 0.00 [a]	0.07 ± 0.00 [a]	0.07 ± 0.01 [a]	0.006 ± 0.000 [a]	0.004 ± 0.000 [b]	0.004 ± 0.000 [b,c]	0.003 ± 0.000 [c]
15	Diethyl succinate *	1683	Fruity, apple, cooked apple, ylang	104 ± 4 [a]	95.6 ± 0.6 [a]	96.1 ± 5.2 [a]	91.1 ± 9.1 [a]	3.76 ± 0.21 [a]	4.27 ± 0.06 [b]	3.23 ± 0.02 [c]	3.18 ± 0.04 [c]
16	1,1,6-trimethyl-1,2-dihydronaphtalene (TDN) **	1716	Floral, fruit, pleasant,	8.59 ± 0.75 [a]	5.02 ± 0.26 [b]	3.66 ± 0.03 [c,d]	2.70 ± 0.16 [d]	0.59 ± 0.03 [a]	0.50 ± 0.02 [b]	0.46 ± 0.01 [b,c]	0.42 ± 0.01 [c]
17	Phenylethyl acetate *	1815	Roses, flowery	1.49 ± 0.00 [a]	1.50 ± 0.23 [a]	1.17 ± 0.10 [a]	1.06 ± 0.04 [a]	0.012 ± 0.001 [a]	n.d. [b]	n.d. [b]	n.d. [b]
18	Ethyl dodecanoate *	1819	Flowery, fruity	7.31 ± 0.44 [a]	3.76 ± 0.00 [b]	2.21 ± 0.40 [c]	1.17 ± 0.15 [d]	0.39 ± 0.04 [a]	0.32 ± 0.04 [a,b]	0.19 ± 0.03 [b]	0.19 ± 0.03 [b]
19	Benzyl alcohol *	1885	Floral, citrusy, sweet	0.18 ± 0.01 [a]	0.19 ± 0.00 [a]	0.18 ± 0.00 [a]	0.18 ± 0.03 [a]	0.007 ± 0.000 [a]	0.006 ± 0.000 [a]	0.006 ± 0.000 [a]	0.006 ± 0.001 [a]
20	Phenylethanol *	1919	Roses, sweet	163 ± 3 [a]	145 ± 9 [a]	145 ± 4 [a]	158 ± 24 [a]	3.81 ± 0.28 [a]	2.75 ± 0.16 [b]	2.68 ± 0.28 [b]	2.40 ± 0.02 [b]
21	β-Caryophyllene oxide **	2005	Sweet, fresh, dry, woody, spicy	1.70 ± 0.05 [a]	1.21 ± 0.10 [b]	0.88 ± 0.01 [c]	0.68 ± 0.03 [c]	0.042 ± 0.001 [a]	n.d. [b]	n.d. [b]	n.d. [b]
22	Octanoic acid *	2061	Fatty acid, rancid	10.4 ± 0.7 [a]	6.44 ± 0.18 [b]	5.97 ± 0.37 [b]	5.71 ± 0.63 [b]	0.25 ± 0.02 [a]	0.21 ± 0.01 [a]	0.12 ± 0.01 [b]	0.13 ± 0.02 [b]
23	Ethyl hexadecanoate **	2255	Fatty, rancid, fruity, sweet	0.06 ± 0.00 [a]	0.03 ± 0.01 [b]	0.03 ± 0.00 [b]	0.02 ± 0.00 [b]	0.006 ± 0.001 [a]	n.d. [b]	n.d. [b]	n.d. [b]
24	Decanoic acid *	2281	Fatty, rancid, soap	3.35 ± 0.15 [a]	2.82 ± 0.98 [a]	1.89 ± 0.13 [a]	1.54 ± 0.22 [a]	0.10 ± 0.01 [a]	n.d. [b]	n.d. [b]	n.d. [b]
25	Ethyl hydrogen succinate **	2378	Sweet, sour, fruity	2.24 ± 0.09 [a]	1.99 ± 0.50 [a,b]	1.10 ± 0.01 [a]	2.10 ± 0.16 [a,b]	0.096 ± 0.007 [a]	0.071 ± 0.003 [b]	0.056 ± 0.005 [b]	n.d. [c]
26	Dodecanoic acid *	2464	Fatty, acidic, soapy, waxy	0.03 ± 0.00 [a]	0.02 ± 0.01 [a,b]	0.01 ± 0.01 [a,b]	0.01 ± 0.00 [b]	0.003 ± 0.001 [a]	0.003 ± 0.000 [a]	0.002 ± 0.000 [a,b]	n.d. [b]

Values are presented as the mean ± standard deviation (n = 2). Values in the same column for each headspace sampling method used followed by the same letter are not significantly different (One-way ANOVA, Tukey's HSD post hoc test, $p \leq 0.05$). RI, Kovats retention index. Odor descriptor from [35–42]. n.d, not detected. The reliability of the identification or structural proposal is indicated by the following: (*) mass spectrum and retention time consistent with those of an authentic standard; (**) structural proposals are given based on mass spectral data (Wiley 275) or are consistent with spectra found in the literature.

Compared with plastic film for TCA removal, for the longest contact time, Gonzàlez-Centeno et al. [22] observed an 82% reduction for ethyl octanoate, ethyl decanoate, and ethyl dodecanoate.

3. Materials and Methods

3.1. Cork Powder Sample Preparation

Cork powder with an average granulometry of 372 µm was obtained from a local cork stopper producer free of TCA and supplied by SAI. Lda. (Paredes, Portugal). To extract the extractives, the natural cork powder was subjected to a dichloromethane extraction by soxhlet for 24 h, followed by a second extraction with ethanol by soxhlet for 24 h. To obtain extractive-free cork powder with a particle size of less than 75 µm, the cork powder was sieved through a sieve. To remove the air contained in the cork powder and simultaneously impregnate the material with ethanol, proportions of 0.01 g, 0.025 g, and 0.05 g of cork powder were immersed in 5 mL of ethanol, and the suspension was vacuum-degassed (0.00131 atm) by repeated cycles (11 times). The number of degassing cycles was chosen by observing the sedimentation of the cork powder at the bottom of the container. After impregnation, the cork powder was left in contact with ethanol (96% v/v) for 12 h. After this period, the ethanol was removed by centrifugation for 10 min at 10.956 g and 20 °C. The ADSI cork powder was used for the wine fining experiments [25].

3.2. Wine Contamination with TCA

A red wine from the Douro region (vintage 2019) was used, with an alcohol content of 13.0 (% v/v), a total acidity of 5.4 g/L of tartaric acid, a volatile acidity of 0.38 g/L of acetic acid, and a pH of 3.70. Six liters of wine were divided into three parts (2 L each), in which one part was artificially contaminated with 3 ng/L of TCA, another part with 6 ng/L of TCA, and a third part was not contaminated with TCA, which was used as a control wine. These contamination levels were chosen by taking into account the "consumer rejection threshold" of 3.1 ng/L of TCA as described by Prescott et al. [8]. The free sulfur dioxide in the wine was adjusted to 50 mg/L.

3.3. Fining Experiment

To study the performance of the cork powder in removing TCA, red wine samples were spiked with 3 ng/L and 6 ng/L of TCA. Different doses of cork powder (0 g, 0.10 g, 0.25 g, and 0.50 g) were added to 1 L of contaminated wine. The wine was left in contact with the cork powder for 6 days at room temperature, without stirring. After 6 days, the wine was centrifuged for 10 min at 10.956 g and 20 °C for analysis. All experiments were performed in duplicate.

3.4. Determination of 2,4,6-Trichloroanisole Extractable by Solid-Phase Microextraction (SPME) Using Gas Chromatography Coupled to Mass Spectrometry (GC-MS)

To determine 2,4,6-trichloroanisole, we used a 10 mL wine sample containing 3 g of NaCl and 100 µL of internal standard solution. D5-TCA (2 µg/L) was placed in 20 mL SPME vials, which were immediately sealed. Samples were analyzed using a GC-MS instrument equipped with an autosampler configured in SPME mode. The flasks were incubated for 2 min and extracted for 8 min, under agitation (250 rpm) at 50 °C, using a 100 µm PDMS fiber. The fiber was desorbed in the injector at 270 °C for 4 min in splitless mode. Compounds were separated on a 5 MS capillary column (30 m × 0.25 mm × 0.25 µm). The detection and quantification limits of this method are 0.2 ng/L and 0.5 ng/L, respectively. This analysis was carried out in cooperation with the company Souto & Castro. All analyses were performed in duplicate.

3.5. Quantification of Total Phenolic Compounds

The wine's total phenolic compounds were determined using the absorbance at 280 nm according to Ribéreau-Gayon et al. [43]. The results are expressed as gallic acid equivalents through calibration curves with standard gallic acid. All analyses were performed in duplicate.

3.6. Color Intensity, Hue, and Chromatic Characteristics

The red wine's color intensity and hue were quantified as described in the OIV methods [44]. For the chromatic characteristics of red wine, the absorption spectra of wine samples were scanned from 380 to 780 nm using a 1 cm path length quartz cell, and the wine's chromatic characteristics (L* (lightness), a* (redness), and b* (yellowness) coordinates) were calculated using the International Commission on Illumination (CIE) method using the L*, a*, and b* coordinates according to the OIV [44]. The chroma ($C^* = [(a^*)^2 + (b^*)^2]^{1/2}$) and hue-angle ($h° = \tan^{-1}(b^*/a^*)$) values were also determined. To distinguish the color more accurately, the color difference was calculated using the following equation: $\Delta E^* = [(\Delta L^*)^2 + (\Delta a^*)^2 + (\Delta b^*)^2]^{1/2}$. This parameter allows for the reliable quantification of the overall color difference in a sample compared to a control sample (untreated wine). Analyses were performed in duplicate.

3.7. High-Performance Liquid Chromatography (HPLC) Analysis of Anthocyanins, Catechin, and Phenolic Acids

Analyses were carried out with an Ultimate 3000 Dionex HPLC system equipped with a PDA-100 photodiode array detector (Dionex. Sunnyvale, CA, USA) and an Ultimate 3000 Dionex pump. The separation was performed on a C18 column (250 mm × 4.6 mm, 5 μm particle size, ACE, Aberdeen, Scotland) with a 1 mL/min flow rate at 35 °C. The injection volume was 50 μL, and the detection was performed in the wavelength range of 200 to 650 nm. The analysis was carried out using 5% aqueous formic acid (A) and methanol (B), and the gradient was as follows: 5% B from zero to 5 min, followed by a linear gradient up to 65% B until 65 min and from 65 to 67 min down to 5% B [45]. Quantification was performed with calibration curves with caffeic acid, coumaric acid, ferulic acid, gallic acid, and catechin as standards. *trans*-Caftaric acid, 2-*S*-glutathionylcaftaric acid (GRP), and caffeic acid ethyl ester are expressed as caffeic acid equivalents, and coutaric acid and coumaric acid ethyl ester are expressed as coumaric acid equivalents. A calibration curve of malvidin-3-glucoside, peonidin-3-glucoside, and cyanidin-3-glucoside was used to quantify these anthocyanins. Using the coefficient of molar absorptivity (ε) and extrapolation, it was possible to obtain the slopes for delphinidin-3-glucoside, petunidin-3-glucoside, and malvidin-3-coumaroylglucoside to perform the quantification. The results on delphinidin-3-acetylglucoside, petunidin-3-acetylglucoside, peonidin-3-acetylglucoside, cyanidin-3-acetylglucoside, and cyanidin-3-coumaroylglucoside are expressed as the respective glucoside equivalent [46,47].

3.8. Total Pigments, Polymeric Pigments, Small Polymeric Pigments (SPPs), Large Polymeric Pigments (LPPs), Anthocyanins, and Tannins

For profiling, the phenolic fractions responsible for the red wine color, the method described by Adams et al. [48] was used. This method combines the protein precipitation (BSA) assay and the bisulfite bleaching assay to distinguish monomeric anthocyanins from polymeric pigments, and two classes of polymeric pigments in wines can also be measured: small polymeric pigments (SPPs) that do not precipitate with proteins and large polymeric pigments (LPPs) that precipitate with proteins. The combination of SPPs and LPPs is equivalent to the sulfur-dioxide-resistant pigments in wine. In the first tube, 500 μL of wine was mixed with 1 mL of acetic acid–NaCl buffer (200 mM acetic acid and 170 mM NaCl, adjusted to pH 4.9 with sodium hydroxide). The absorbance at 520 nm (in a 1 mm path length cuvette) of the mixture was measured (A value). Then, 80 μL of a 0.36 M potassium metabisulfite solution was added. After 10 min of incubation, the absorbance at 520 nm was measured again (B value). The absorbance due to monomeric pigments can be calculated as (A-B), where the B value represents the total amount of polymeric pigment (SPPs + LPPs). In a second tube, 500 μL of wine was mixed with 1 mL of acetic acid–NaCl buffer containing bovine serum albumin (BSA) (1 mg/mL). The mixture was allowed to stand at room temperature for 15 min with slow stirring, and then the tube was centrifuged for 5 min at 13.500 g to sediment the tannin–protein precipitate. One milliliter of the supernatant was mixed with 80 µL of a 0.36 M potassium metabisulfite solution. After 10 min of incubation, the absorbance at 520 nm was measured (C value). This absorbance (the C value) corresponds to the polymeric pigment that did not precipitate with the tannin and the protein. The absorbance is due to small polymeric pigments (SPPs), and this C value was used to calculate the amount of polymeric pigment that precipitated with the tannin and the protein (B-C) absorbance due to large polymeric pigments (LPPs). Total polymeric pigments (PPs) are the sum of the small polymeric pigments and the large polymeric pigments. The supernatant from the second experiment described above was discarded, and the remaining pellet was washed with 250 µL of acetic acid–NaCl buffer to remove residual monomeric anthocyanins. The tube was centrifuged for 1 min at 13.500 g, and the supernatant was discarded. Then, the pellet was dissolved in 875 µL of buffer containing 5% (v/v) triethanolamine (TEA) and 5% (w/v) sodium dodecyl sulfate (SDS). The buffer dissolves the precipitate containing tannins, proteins, and any polymeric pigments that precipitated with the tannin and the protein. After incubation, the tube was vortexed to dissolve any remaining precipitate. The absorbance at 510 nm (in a 10 mm path length cuvette) was measured after allowing the solution to stand at room temperature for 10 min (value D). To calculate the tannin absorbance, 125 µL of a ferric chloride solution was added (10 mM ferric chloride and 10 mM hydrochloric acid in water). The absorbance at 510 was reread after 10 min (value E). All analyses were performed in duplicate.

3.9. Wine Volatile Composition Determined by SPME-GC-MS

Two methods were used to analyze the volatile profile of wines, namely headspace extraction with and without re-equilibration [30].

To determine the headspace volatile composition of red wines with re-equilibration, a validated method was confirmed in our laboratory [49]. Briefly, the Divinylbenzene/Carboxen/Polydimethylsiloxane (DVB/CAR/PDMS) 50/30 µm fiber was conditioned before use by insertion into the GC injector at 270 °C for 60 min. To a 20 mL headspace vial, we added 10 mL of wine and 2.5 g of NaCl. The vial was sealed with a Teflon septum. The fiber was inserted through the vial septum previously conditioned at 35 °C and exposed for 60 min with agitation to perform the extraction by an automatic CombiPal system. The fiber was inserted into the injection port of the GC for 3 min at 270 °C. All analyses were performed in duplicate.

To determine the headspace volatile composition of red wines without re-equilibration, the extraction time was initially evaluated by measuring the headspace abundance and profile after extraction during 1, 2, and 3 min. The abundance of the obtained chromatograms increased as the extraction time increased. As the relative abundance of the peaks did not change significantly between 1 and 3 min, the extraction time of 3 min was used. The Divinylbenzene/Carboxen/Polydimethylsiloxane (DVB/CAR/PDMS) 50/30 µm fiber was conditioned before use by insertion into the GC injector at 270 °C for 60 min. To a 20 mL headspace vial, 10 mL of wine was added. The vial was sealed with a Teflon septum. The fiber was inserted through the vial septum previously conditioned at 25 °C (room temperature) and exposed for 3 min without agitation to perform the extraction by an automatic CombiPal system. The fiber was inserted into the injection port of the GC for 3 min at 270 °C. All analyses were performed in duplicate.

Analyses were performed by gas chromatography using a Trace GC Ultra system with a Polaris Q mass spectrometer. Separation was performed using a DB-FFAP column (30 m × 0.25 mm, and 0.25 µm film thickness) with a 1 mL/min helium flow. The oven temperature program was: 40 °C for 5 min, increased to 155 °C at 5 °C/min, then increased to 300 °C at 20 °C/min, and held at that temperature for 1 min. All analyses were performed in duplicate.

3.10. Modeling of the Adsorption Isotherms

After determining the amount of cork powder that best removed TCA (0.025 g), a model wine solution was prepared (ethanol at 12.0% v/v with 3.5 g/L of tartaric acid; the pH of the solution was adjusted to 3.60 with NaOH). A total of 0.025 g of ADSI cork powder was placed per 100 mL of model wine solution, and an increasing concentration of TCA (2.5 ng/L, 5 ng/L, 7.5 ng/L, 12.5 ng/L, 25 ng/L, and 50 ng/L) was used. Three isothermal models were used to characterize the adsorption systems, namely the Langmuir, Freundlich, and Langmuir–Freundlich isotherm models, to analyze the equilibrium data obtained experimentally. The Langmuir model is the simplest and the most frequently used in adsorption studies. This model assumes that adsorption occurs on a homogeneous surface with identical active sites and uniform energies [28]. In the Langmuir model, the Langmuir isotherm expression is represented by the following equation [28]:

$$Q_e = (Q_{max} \times K_L \times C_e)/(1 + K_L \times C_e) \tag{1}$$

where K_L is the Langmuir constant related to the affinity of the active sites, Q_{max} is the theoretical maximum monolayer capacity, C_e is the equilibrium concentration, and Q_e is the amount of TCA adsorbed at equilibrium.

The Freundlich model assumes that adsorption occurs on a heterogeneous surface with an exponential distribution of active sites and energies [28], and it is expressed by the equation:

$$Q_e = K_F \times C_e^{1/n} \tag{2}$$

where K_F is the Freundlich constant, and C_e and Q_e are defined as above and related to the adsorption favorability and adsorption capacity, respectively.

The Freundlich constant (K_F) is related to the adsorption capacity, and the constant n is related to the adsorption intensity. Values of n in the range $1 < n < 10$ indicate favorable adsorption.

The Langmuir–Freundlich isotherm—also known as the Sips equation—is capable of modeling homogeneous and heterogeneous bonding surfaces and is expressed by [50]:

$$Q_e = (Q_m \times K_s \times C_e^n)/(1 + (K_s \times C_e^n)) \tag{3}$$

where Q_e and C_e are described as above, Q_m is the total number of binding sites, and n represents the system's heterogeneity index, which can vary from 0 to 1. If $n = 1$, the system is homogeneous and can be equated to the Langmuir model, and $n < 1$ represents a heterogeneous material. K_s is a parameter related to the median binding affinity (K_0) via $K_0 = a1/n$, where n is the heterogeneity index, which ranges from 0 to 1.

The Langmuir–Freundlich isotherm is composed of the Langmuir isotherm and the Freundlich isotherm and can be reduced to either one in its limits. When $n = 1$, the Langmuir–Freundlich isotherm reduces to the Langmuir isotherm, which corresponds directly to the binding affinity (K_L). Alternatively, as Ce or a approaches 0, the Langmuir–Freundlich isotherm reduces to the Freundlich isotherm. Furthermore, the Langmuir–Freundlich isotherm reduces to the Freundlich isotherm for all systems at low concentrations.

3.11. Statistical Treatment

Statistically significant differences between means were determined by analysis of variance (ANOVA, one-way) followed by Tukey's honestly significant difference (HSD, 5% level) post-hoc test for the physicochemical data. All analyses were performed using Statistica 10 software (StatSoft, Tulsa, OK, USA).

4. Conclusions

The application of air-depleted solvent-impregnated cork powder in a 0.25 g/L dose to red wine contaminated with TCA (6 ng/L) resulted in a significant decrease in TCA levels (a 91% reduction). Applying ADSI cork powder up to 0.50 g/L did not result in a significant change in the red wine's phenolic composition and chromatic characteristics. On the other hand, the application of ADSI cork powder resulted in a significant decrease in the red wine's volatile composition when determined by exhaustive headspace extraction. However, the impact on the "true headspace" concentration was much lower. This natural material may represent a new and efficient technological solution with a low environmental impact, contributing to a more sustainable wine industry.

Author Contributions: Conceptualization, F.C., L.F.-R. and F.M.N.; methodology, F.C., L.F.-R. and F.M.N.; validation, F.C., L.F.-R. and F.M.N.; formal analysis, F.C., S.G., L.F.-R. and F.M.N.; investigation, S.G.; resources. F.C., L.F.-R. and F.M.N.; data curation, F.C., S.G., L.F.-R. and F.M.N.; writing—original draft preparation, F.C., L.F.-R. and F.M.N.; writing—review and editing, F.C., S.G., A.V., L.F.-R. and F.M.N.; supervision, F.C. and A.V.; project administration, F.C., L.F.-R. and F.M.N.; funding acquisition, F.C., L.F.-R. and F.M.N. All authors have read and agreed to the published version of the manuscript.

Funding: This research was funded by CQ-VR—Chemistry Research Center—Vila Real (UIDB/00616/2020 and UIDP/00616/2020) by FCT—Portugal and COMPETE. The financial support of the project Agri-Food XXI (NORTE-01-0145-FEDER-000041), co-financed by the European Regional Development Fund through NORTE 2020 (Programa Operacional Regional do Norte 2014/2020), is also acknowledged.

Institutional Review Board Statement: Not applicable.

Informed Consent Statement: Not applicable.

Data Availability Statement: Not applicable.

Acknowledgments: We thank Souto & Castro for the determination of 2,4,6-trichloroanisole (TCA) and SAI for supplying the cork powder.

Conflicts of Interest: The authors declare no conflict of interest.

Sample Availability: Not applicable.

References

1. Simpson, R.F.; Sefton, M.A. Origin and fate of 2,4,6-trichloroanisole in cork bark and wine corks. *Aust. J. Grape Wine Res.* **2007**, *13*, 106–116. [CrossRef]
2. Chatonnet, P.; Bonnet, S.; Boutou, S.; Labadie, M.D. Identification and responsibility of 2,4,6-tribromoanisole in musty. corked odors in wine. *J. Agric. Food Chem.* **2004**, *52*, 1255–1262. [CrossRef] [PubMed]
3. INBIOTEC; Coque, J.J.R.; Perez, E.R.; dos Santos Marques, S.; Ferreira, J.R.; APCOR. *Contaminação do Vinho por Haloanisóis: Desenvolvimento de Estratégias Biotecnológicas para Prevenir a Contaminação de Rolhas de Cortiça por Cloroanisóis*; APCOR: Santa Maria de Lamas, Portugal, 2006.
4. Sefton, M.A.; Simpson, R.F. Compounds causing cork taint and the factors affecting they transfer from natural cork closures to wine—A review. *Aust. J. Grape Wine Res.* **2005**, *11*, 226–240. [CrossRef]
5. Chatonnet, P.; Fleury, A.; Boutou, S. Identification of a new source of contamination of *Quercus* sp. oak wood by 2,4,6-trichloroanisole and its impact on the contamination of barrel-aged wines. *J. Agric. Food Chem.* **2010**, *58*, 10528–10538. [CrossRef] [PubMed]
6. Duerr, P. Wine quality evaluation. In Proceedings of the International Symposium on Cool Climate Viticulture and Enology, Corvallis, OR, USA, 25–28 June 1985; pp. 257–266.
7. Hervé, E.; Price, S.; Burns, G.; Weber, P. Chemical Analysis of TCA as a Quality Control Tool for Natural Cork. 2004. Available online: https://docplayer.net/146328650-Chemical-analysis-of-tca-as-a-quality-control-tool-for-natural-corks.html (accessed on 4 June 2022).
8. Prescott, J.; Norris, L.; Kunst, M.; Kim, S. Estimating a 'consumer rejection threshold' for cork-taint in white wine. *Food Qual. Prefer.* **2005**, *16*, 345–349. [CrossRef]
9. Liacopoulos, D.; Barker, D.; Howland, P.R.; Alcorso, D.C.; Pollnitz, A.P.; Skouroumounis, G.K.; Pardon, K.H.; McLean, H.J.; Gawel, R.; Sefton, M.A. Chloroanisole taint in wines. In Proceedings of the Tenth Australian Wine Industry Technical Conference, Sydney, NSW, Australia, 2–5 August 1998; Australian Wine Industry Technical Conference Inc.: Adelaide, SA, Australia, 1999; pp. 224–226.
10. Amon, J.M.; Vandeepeer, J.M.; Simpson, R.F. Compounds Responsible for Cork Taint. *Aust. N. Z. Wine Ind.* **1989**, *4*, 62–69.
11. APCOR. Associação Portuguesa de Cortiça—Manual Técnico APCOR. Cork Information Bureau. 2010. Available online: https://silo.tips/download/cork-information-bureau-2010-cork-sector-in-numbers (accessed on 4 June 2022).
12. Vestner, J.; Fritsch, S.; Rauhut, D. Development of a microwave assisted extraction method for the analysis of 2,4,6-trichloroanisole in cork stoppers by SIDA–SBSE–GC–MS. *Anal. Chim. Acta* **2010**, *660*, 76–80. [CrossRef]

13. Juanola, R.; Guerrero, L.; Subirà, D.; Salvadó, V.; Insa, S.; Garcia Regueiro, J.A.; Anticó, E. Relationship between sensory and instrumental analysis of 2,4,6-trichloroanisole in wine and cork stoppers. *Anal. Chim. Acta* **2004**, *513*, 291–297. [CrossRef]
14. Fontana, A.R.; Patil, S.H.; Banerjee, K.; Altamirano, J.C. Ultrasound- Assisted Emulsification Microextraction for Determination of 2.4.6- Trichloroanisole in Wine Samples by Gas Chromatography Tandem Mass Spectrometry. *J. Agric. Food Chem.* **2010**, *58*, 4576–4581. [CrossRef]
15. Mazzoleni, V.; Maggi, L. Effect of wine style on the perception of 2,4,6-trichloroanisole, a compound related to cork taint in wine. *Food Res. Int.* **2007**, *40*, 694–699. [CrossRef]
16. Cioni, G.A.; Cadinu, T. Physical Chemical Method to Remove the Cork Taste and in General Anomalous Smells of Cork Materials. WIPO Patent 2001 WO/2001/041989A2, 14 June 2001.
17. Swan, J.S. Process for Removing Off-Flavors and Odors from Foods and Beverages. U.S. Patent US6610342B22003, 26 August 2003.
18. Vuchot, P.; Puech, C.; Fernandez, O.; Fauveau, C.; Pellerin, P.; Vidal, S. Elimination des goûts de bouchon/moisi et de l'OTA à l'aide d'écorces de levures hautement adsorbantes. *Rev. Internet Vitic. Œnol.* **2007**, *2*, 62–72.
19. Garde-Cerdán, T.; Zalacain, A.; Lorenzo, C.; Alonso, J.L.; Salinas, M.R. Molecularly imprinted polymer-assisted simple clean-up of 2.4.6-TCA and ethylphenols from aged red wines. *Am. J. Enol. Vitic.* **2008**, *59*, 396–400.
20. Cravero, M.C. Musty and Moldy Taint in Wines: A Review. *Beverages* **2020**, *6*, 41. [CrossRef]
21. Capone, D.L.; Skouroumounis, G.K.; Barker, D.A.; Mclean, H.J.; Pollnitz, A.P.; Sefton, M.A. Absorption of chloroanisoles from wine by corks and by other materials. *Aust. J. Grape Wine Res.* **1999**, *5*, 91–98. [CrossRef]
22. Gonzàlez-Centeno, M.R.; Tempère, S.; Teissedre, P.; Chira, K. Use of alimentary film for selective sorption of haloanisoles from contaminated red wine. *Food Chem.* **2021**, *350*, 128364. [CrossRef] [PubMed]
23. Valdés, O.; Marican, A.; Avila-Salas, F.; Ignacio Castro, R.; Amalraj, J.; Felipe Laurie, V.; Santos, L.S. Polyaniline Based Materials as a Method to Eliminate Haloanisoles in Spirits Beverage. *Ind. Eng. Chem.* **2018**, *57*, 8308–8831. [CrossRef]
24. Pintor, A.M.A.; Ferreira, C.I.A.; Pereira, J.C.; Silva, P.C.; Vilar, S.P.; Botelho, V.J.P.; Boaventura, C.M.S. Use of cork powder and granules for the adsorption of pollutants: A review. *Water Res.* **2012**, *1*, 1–5. [CrossRef]
25. Filipe-Ribeiro, L.; Cosme, F.; Nunes, M.F. Air Depleted and Solvent Impregnated Cork Powder as a New Natural and Sustainable Wine Fining Agent. In *Advances in Grape and Wine Biotechnology*; IntechOpen: London, UK, 2019. [CrossRef]
26. Filipe-Ribeiro, L.; Cosme, F.; Nunes, F.M. A Simple Method to Improve Cork Powder Waste Adsorption Properties: Valorization as a New Sustainable Wine Fining Agent. *ACS Sustain. Chem. Eng.* **2019**, *7*, 1105–1111. [CrossRef]
27. Guth, H. Quantitation and Sensory Studies of Character Impact Odorants of Different White Wine Varieties. *J. Agric. Food Chem.* **1997**, *45*, 3027–3032. [CrossRef]
28. García-Calzon, J.A.; Díaz-García, M.E. Characterization of binding sites in molecularly imprinted polymers. *Sens. Actuators B Chem.* **2007**, *123*, 1180–1194. [CrossRef]
29. Saeed, A.; Iqbal, M.; Höll, W.H. Kinetics, equilibrium and mechanism of Cd^{2+} removal from aqueous solution by mungbean husk. *J. Hazard. Mater.* **2009**, *168*, 1467–1475. [CrossRef] [PubMed]
30. García-Zubiri, I.X.; González-Gaitano, G.; Isasi, J.R. Sorption models in cyclodextrin polymers: Langmuir, Freundlich, and a dual-mode approach. *J. Colloid Interface Sci.* **2009**, *337*, 11–18. [CrossRef] [PubMed]
31. Ayawei, N.; Ebelegi, A.N.; Wankasi, D. Modelling and interpretation of adsorption isotherms. *J. Chem.* **2017**, *2017*, 3039817. [CrossRef]
32. Chen, X. Modeling of experimental adsorption isotherm data. *Information* **2015**, *6*, 14–22. [CrossRef]
33. Do, D. Adsorption analysis: Equilibria and kinetics. *Ser. Chem. Eng.* **1998**, *2*, 1–916.
34. Roberts, D.D.; Pollien, P.; Milo, C. Solid-Phase Microextraction Method Development for Headspace Analysis of Volatile Flavor Compounds. *J. Agric. Food Chem.* **2000**, *48*, 2430–2437. [CrossRef] [PubMed]
35. Perestrelo, R.; Fernandes, A.; Albuquerque, F.F.; Marques, J.C.; Câmara, J.S. Analytical characterization of the aroma of Tinta Negra Mole red wine: Identification of the main odorants compounds. *Anal. Chim. Acta* **2006**, *563*, 154–164. [CrossRef]
36. Dragone, G.; Mussato, S.I.; Oliveira, J.M.; Teixeira, J.A. Characterization of volatile compounds in an alcoholic beverage produced by whey fermentation. *Food Chem.* **2009**, *112*, 929–935. [CrossRef]
37. Jiang, B.; Zhang, Z. Volatile compounds of young wines from Cabernet Sauvignon, Cabernet Gernischet and Chardonnay varieties grown in the Loess Plateau Region of China. *Molecules* **2010**, *15*, 9184–9196. [CrossRef]
38. Vararu, F.; Moreno-García, J.; Cotea, V.V.; Moreno, J. Grape musts differentiation based on selected aroma compounds using SBSE-GC-MS and statistical analysis. *Vitis* **2015**, *54*, 97–105.
39. López de Lerma, N.; Peinado, R.A.; Puig-Pujol, A.; Mauricio, J.C.; Moreno, J.; García-Martínez, T. Influence of two yeast strains in free, bioimmobilized or immobilized with alginate forms on the aromatic profile of long aged sparkling wines. *Food Chem.* **2018**, *250*, 22–29. [CrossRef]
40. Perestrelo, R.; Silva, C.; Câmara, J.S. Madeira Wine Volatile Profile. A Platform to Establish Madeira Wine Aroma Descriptors. *Molecules* **2019**, *24*, 3028. [CrossRef]
41. Li, H.; Sheng Tao, Y.; Wang, H.; Zhang, L. Impact odorants of Chardonnay dry white wine from Changli County (China). *Eur. Food Res. Technol.* **2008**, *227*, 287–292. [CrossRef]
42. Pereira, V.; Cacho, J.; Marques, J.C. Volatile profile of Madeira wines submitted to traditional accelerated ageing. *Food Chem* **2014**, *162*, 122–134. [CrossRef]

43. Ribéreau-Gayon, P.; Glories, Y.; Maujean, A.; Dubourdieu, D. Clarification and stabilization treatments: Fining wine. In *Handbook of Enology: The Chemistry of Wine Stabilization and Treatments*, 2nd ed.; Wiley: Hoboken, NJ, USA, 2006; Volume 2.
44. International Organization of Vine and Wine. *Compendium of International Methods of Wine and Must Analysis*; Edition 2020; International Organization of Vine and Wine: Paris, France, 2020; Volume 1.
45. Guise, R.; Filipe-Ribeiro, L.; Nascimento, D.; Bessa, O.; Nunes, F.M.; Cosme, F. Comparison between different types of carboxylmethylcellulose and other oenological additives used for white wine tartaric stabilization. *Food Chem.* **2014**, *156*, 250–257. [CrossRef]
46. Filipe-Ribeiro, L.; Milheiro, J.; Matos, C.C.; Cosme, F.; Nunes, F.M. Data on changes in red wine phenolic compounds. headspace aroma compounds and sensory profile after treatment of red wines with activated carbons with different physicochemical characteristics. *Data Brief* **2017**, *12*, 188–202. [CrossRef]
47. Filipe-Ribeiro, L.; Milheiro, J.; Matos, C.C.; Cosme, F.; Nunes, F.M. Reduction of 4-ethylphenol and 4-ethylguaiacol in red wine by activated carbons with different physicochemical characteristics: Impact on wine quality. *Food Chem.* **2017**, *229*, 242–251. [CrossRef]
48. Adams, D.O.; Harbertson, J.F.; Picciotto, E.A.; Avenue, O.S. Fractionation of Red Wine Polymeric Pigments by Protein Precipitation and Bisulfite Bleaching. *Red Wine Color* **2004**, *17*, 275–288. [CrossRef]
49. Vás, G.; Gál, L.; Harangi, J.; Dobó, A.; Vékey, K. Determination of volatile compounds of Blaufrankisch wines extracted by solid-phase microextraction. *J. Chromatogr. Sci.* **1998**, *36*, 505–510. [CrossRef]
50. Umpleby, R.J., II; Baxter, S.C.; Chen, Y.; Shah, R.N.; Shimizu, K.D. Characterization of Molecularly Imprinted Polymers with the Langmuir-Freundlich Isotherm. *Anal. Chem.* **2001**, *73*, 4584–4591. [CrossRef]

MDPI
St. Alban-Anlage 66
4052 Basel
Switzerland
Tel. +41 61 683 77 34
Fax +41 61 302 89 18
www.mdpi.com

Molecules Editorial Office
E-mail: molecules@mdpi.com
www.mdpi.com/journal/molecules

www.ingramcontent.com/pod-product-compliance
Lightning Source LLC
LaVergne TN
LVHW070701100526
838202LV00013B/1009